WENNER-GREN INTERNATIONAL SERIES

Volume 65

Active Hearing

Wenner-Gren International Series

Active Hearing

Edited by

Å. FLOCK
D. OTTOSON
and
M. ULFENDAHL

PERGAMON

U.K.   Elsevier Science Ltd, The Boulevard, Langford Lane,
          Kidlington, Oxford, OX5 1GB, U.K.

U.S.A.   Elsevier Science Inc., 660 White Plains Road,
          Tarrytown, New York 10591-5153, U.S.A.

JAPAN   Elsevier Science Japan, Tsunashima Building Annex,
          3-20-12 Yushima, Bunkyo-ku, Tokyo 113, Japan

First edition 1995

British Library Cataloguing in Publication Data

A catalogue record for this book is available from the British Library

Library of Congress Cataloging in Publication Data

Active hearing/edited by Å. Flock, D. Ottoson, M. Ulfendahl.
  p. cm. — (Wenner-Gren international series; v. 65)
"Proceedings of a Wenner-Gren international symposium held in
Stockholm in May, 1994"—Pref.
Includes index.
1. Auditory pathways—Congresses. 2. Hearing—Physiological
aspects—Congresses. 3. Ear—Physiology—Congresses. I. Flock, Å.
II. Ottoson, David, 1918– . III. Ulfendahl, M. IV. Series:
Wenner-Gren Center international symposium series; v. 65.
QP461.A25 1995
612.8′5—dc20   94-41161

ISBN 0 08 042514 3

Printed in Great Britain by Galliard (Printers) Ltd, Great Yarmouth

Contents

Preface

In recent years a number of observations have made important contributions to our knowledge of the processes involved in the transmission of signals in the auditory system. These events include the mechanical motion pattern of the cochlear partition, the motion of hair cell cilia, gating of receptor currents, hair cell motility, synaptic transmission and signal processing in central nuclei and in the auditory cortex. Hearing may therefore be considered to arise from a series of active processes which modify the signals sent to the brain.

This book is the edited proceedings of a Wenner-Gren International Symposium held in Stockholm in May, 1994. Focusing on the dynamic aspects of the hearing process, hence *Active Hearing*, it contains chapters by authors representing major research laboratories in several fields of auditory research. The chapters are organized to proceed from the brain "backwards", level by level, out towards the periphery. This was done to stimulate active thinking about how things are interrelated.

Åke Flock
David Ottoson
Mats Ulfendahl

Stockholm, September, 1994

Acknowledgements

The symposium was made possible by generous support from the Wenner-Gren Center Foundation and by grants from the Swedish Medical Research Council K94-045-11035-01 and the Foundation Tysta Skolan. We would also like to acknowledge the assistance of our secretaries Katarina Florin and Gun Lennerstrand, without whom the symposium would have been far less successful.

List of Contributors

Richard A. ALTSCHULER
Kresge Hearing Research Institute, University of Michigan, 1301 East Ann Street, Ann Arbor, MI 48109-0506, USA

J. F. ASHMORE
Department of Physiology, School of Medical Sciences, University of Bristol, Bristol BS8 1TD, UK

W. E. BROWNELL
Center for Hearing Sciences, The Johns Hopkins University School of Medicine, Baltimore, MD 21205, USA. Present address: Department of Otorhinolaryngology and Communication Sciences, Baylor College of Medicine, Houston, TX 77030, USA

John A. BUTMAN
Department of Biology, Washington University, One Brookings Drive, St Louis, MO 63130-4899, USA

B. J. CONNOR
Department of Physiology, School of Medicine, University of Auckland, Private Bag 92019, Auckland, New Zealand

N. P. COOPER
Department of Neurophysiology, University of Wisconsin, Madison, WI 53706, USA

David P. COREY
Howard Hughes Medical Institute and Massachusetts General Hospital, WEL414, Boston, MA 02114, USA

Christine Gervais D'ALDIN
INSERM U. 254, Neurobiologie de l'Audition – Plasticité Synaptique, CHU Hôpital St Charles, 34295 Montpellier Cedex 5, France

Bruce H. DERFLER
Howard Hughes Medical Institute amd Massachusetts General Hospital, WEL414, Boston, MA 02114, USA

D. F. DOLAN
Kresge Hearing Research Institute, University of Michigan, 1301 East Ann Street, Ann Arbor, MI 48109-0506, USA

Didier DULON
Laboratoire d'Audiologie Expérimentale, INSERM et Université de Bordeaux II, Hôpital Pellegrin, 33076 Bordeaux, France

Jerome DUPONT
Kresge Hearing Research Institute, University of Michigan, 1301 East Ann Street, Ann Arbor, MI 48109-0506, USA

Geoffrey M. DUYK
Millennium Pharmaceuticals, Inc., 640 Memorial Drive, Cambridge, MA 02139, USA

Michel EYBALIN
INSERM U. 254, Neurobiologie de l'Audition – Plasticité Synaptique, CHU Hôpital St Charles, 34295 Montpellier Cedex 5, France

James D. FESSENDEN
Kresge Hearing Research Institute, University of Michigan, Ann Arbor, MI 48109-0506, USA

Åke FLOCK
Department of Physiology and Pharmacology, Division of Physiology II, Karolinska Institutet, S-171 77 Stockholm, Sweden

David N. FURNESS
Department of Communication and Neuroscience, University of Keele, Keele, Staffordshire ST5 5BG, UK

J. E. GALE
Department of Physiology, School of Medical Sciences, University of Bristol, Bristol BS8 1TD, UK

S. GHOSHAL
Division of Otolaryngology, Surgical Research Center, Center for Neurological Sciences, University of Connecticut Health Center, Farmington, CT 06030-1110, USA

Anthony W. GUMMER
Section of Physiological Acoustics and Communication, Department of
Otolaryngology, University of Tübingen, Silcherstraße 5, 72076
Tübingen, Germany

Carole M. HACKNEY
Department of Communication and Neuroscience, University of Keele,
Keele, Staffordshire ST5 5BG, UK

Richard HALLWORTH
Department of Otolaryngology – Head and Neck Surgery, University of
Texas Health Science Center, San Antonio, TX 78284, USA

R. HARI
Low Temperature Laboratory, Helsinki University of Technology, 02150
Espoo, Finland

Werner HEMMERT
Section of Physiological Acoustics and Communication, Department of
Otolaryngology, University of Tübingen, Silcherstraße 5, 72076
Tübingen, Germany

Conor HENEGHAN
Department of Electrical Engineering, Columbia University, New York,
NY 10032, USA

G. D. HOUSLEY
Department of Physiology, School of Medicine, University of Auckland,
Private Bag 92019, Auckland, New Zealand

Fernán JARAMILLO
Department of Physiology, Emory University School of Medicine,
Atlanta, GA 30322, USA

Bechara KACHAR
Section on Structural Cell Biology, Laboratory of Cellular Biology,
NIDCD, NIH, MD, USA

Federico KALINEC
Section on Structural Cell Biology, Laboratory of Cellular Biology,
NIDCD, NIH, MD, USA

Shyam M. KHANNA
Department of Otolaryngology, College of Physicians and Surgeons,
Columbia University, New York, NY 10032, USA

D. O. KIM
Division of Otolaryngology, Surgical Research Center, Center for Neurological Sciences, University of Connecticut Health Center, Farmington, CT 06030-1110, USA

W. J. KONG
Kresge Hearing Research Institute, University of Michigan, 1301 East Ann Street, Ann Arbor, MI 48109-0506, USA

M. KÖSSL
Zoologisches Institut, Universität München, 80333 München, Germany

C. J. KROS
School of Biological Sciences, The University of Sussex, Falmer, Brighton BN1 9QG, UK

G. W. T. LENNAN
School of Biological Sciences, The University of Sussex, Falmer, Brighton BN1 9QG, UK

S. LEVÄNEN
Low Temperature Laboratory, Helsinki University of Technology, 02150 Espoo, Finland

Hyun Ho LIM
Kresge Hearing Research Institute, University of Michigan, 1301 East Ann Street, Ann Arbor, MI 48109-0506, USA

Brenda L. LONSBURY-MARTIN
Department of Otolaryngology, University of Miami Ear Institute, PO Box 016960, Miami, FL 33101, USA

J. P. MÄKELÄ
Low Temperature Laboratory, Helsinki University of Technology, 02150 Espoo, Finland

F. MAMMANO
Department of Physiology, School of Medical Sciences, University of Bristol, Bristol BS8 1TD, UK

Glen K. MARTIN
Department of Otolaryngology, University of Miami Ear Institute, PO Box 016960, Miami, FL 33101, USA

Josef M. MILLER
Kresge Hearing Research Institute, University of Michigan, 1301 East Ann Street, Ann Arbor, MI 48109-0506, USA

Peter M. NARINS
Department of Physiological Science, UCLA, 405 Hilgard Avenue, Los Angeles, CA 90095, USA

A. L. NUTTALL
Kresge Hearing Research Institute, University of Michigan, 1301 East Ann Street, Ann Arbor, MI 48109-0506, USA

Harunori OHMORI
Department of Physiology, Faculty of Medicine, Kyoto University, Kyoto, 606-01 Japan

John F. OLSEN
National Institute of Health Animal Center, Emer. School Rd, Bldg 220, Box 289, Pollesville, MD 20837, USA

K. PARHAM
Division of Otolaryngology, Surgical Research Center, Center for Neurological Sciences, University of Connecticut Health Center, Farmington, CT 06030-1110, USA

Robert PATUZZI
The Auditory Laboratory, Department of Physiology, University of Western Australia, Nedlands, Australia 6009

A. S. POPEL
Department of Biomedical Engineering, The Johns Hopkins University School of Medicine, Baltimore, MD 21205, USA

Jean-Luc PUEL
INSERM U. 254, Neurobiologie de l'Audition – Plasticité Synaptique, CHU Hôpital St Charles, 34295 Montpellier Cedex 5, France

Rémy PUJOL
INSERM U. 254, Neurobiologie de l'Audition – Plasticité Synaptique, CHU Hôpital St Charles, 34295 Montpellier Cedex 5, France

Yehoash RAPHAEL
Kresge Hearing Research Institute, University of Michigan, 1301 East Ann Street, Ann Arbor, MI 48109-0506, USA

J. T. RATNANATHER
Center for Hearing Sciences, The Johns Hopkins University School of Medicine, Baltimore, MD 21205, USA

N. P. RAYBOULD
Department of Physiology, School of Medicine, University of Auckland, Private Bag 92019, Auckland, New Zealand

T. Y. REN
Kresge Hearing Research Institute, University of Michigan, 1301 East
Ann Street, Ann Arbor, MI 48109-0506, USA

W. S. RHODE
Department of Neurophysiology, University of Wisconsin, Madison, WI
53706, USA

G. P. RICHARDSON
School of Biological Sciences, The University of Sussex, Falmer,
Brighton BN1 9QG, UK

Mario A. RUGGERO
Department of Communication Sciences and Disorders, Northwestern
University, 2299 North Campus Drive, Evanston, IL 60208-3550, USA

I. J. RUSSELL
School of Biology, University of Sussex, Falmer, Brighton BN1 9QG, UK

Saaid SAFIEDDINE
Laboratory of Neurochemistry, National Institute on Deafness and
Other Communication Disorders, National Institute of Health,
Bethesda, MD 20892, USA

M. SAMS
Department of Psychology, University of Tampere, Tampere, Finland

Kazuo SATO
Kresge Hearing Research Institute, University of Michigan, 1301 East
Ann Street, Ann Arbor, MI 48109-0506, USA

Jochen SCHACHT
Kresge Hearing Research Institute, University of Michigan, Ann Arbor,
MI 48109-0506, USA

P. S. SIT
Department of Biomedical Engineering, The Johns Hopkins University
School of Medicine, Baltimore, MD 21205, USA

Norma B. SLEPECKY
Department of Bioengineering and Neuroscience, Institute for Sensory
Research, Syracuse University, Syracuse, NY 13244-5290, USA

Charles F. SOLC
Howard Hughes Medical Institute and Massachusetts General
Hospital, WEL414, Boston, MA 02114, USA. Present address: 2140 Ash
Street, Palo Alto, CA 94306, USA

Nobuo SUGA
Department of Biology, Washington University, One Brookings Drive, St Louis, MO 63130-4899, USA

Haibing TENG
Department of Biology, Washington University, One Brookings Drive, St Louis, MO 63130-4899, USA

M. J. TUNSTALL
Department of Neurobiology and Anatomy, University of Texas Medical School, 6431 Fannin, Houston, TX 77030, USA

Mats ULFENDAHL
Department of Physiology and Pharmacology, Division of Physiology II, Karolinska Institutet, S-171 77 Stockholm, Sweden

J.-P. VASAMA
Low Temperature Laboratory, Helsinki University of Technology, 02150 Espoo, Finland

Martin L. WHITEHEAD
Department of Otolaryngology, University of Miami Ear Institute, PO Box 016960, Miami, FL 33101, USA

Jun YAN
Department of Biology, Washington University, One Brookings Drive, St Louis, MO 63130-4899, USA

Gary ZAJIC
Kresge Hearing Research Institute, University of Michigan, Ann Arbor, MI 48109-0506, USA

Hans-Peter ZENNER
Section of Physiological Acoustics and Communication, Department of Otolaryngology, University of Tübingen, Silcherstraße 5, 72076 Tübingen, Germany

H. ZHAO
Division of Otolaryngology, Surgical Research Center, Center for Neurological Sciences, University of Connecticut Health Center, Farmington, CT 06030-1110, USA

M. ZHI
Center for Hearing Sciences, The Johns Hopkins University School of Medicine, Baltimore, MD 21205, USA

Exploration of Human Auditory Cortical Functions: Neuromagnetic Approach

R. HARI, S. LEVÄNEN, J. P. MÄKELÄ, M. SAMS AND
J.-P. VASAMA

*Low Temperature Laboratory, Helsinki University of Technology,
02150 Espoo, Finland*

Much of the present understanding of sensory processing by the human brain has been derived from studies of patients with sensory impairments. These investigations have led to a far better knowledge of the cortical mechanisms responsible for information processing in the visual and somatosensory modalities than in the auditory. One reason for this is that in the past auditory studies have focused on the mechanisms of peripheral hearing disorders. This has led to an overuse of stimuli that are good for testing the transfer functions of simple acoustic features, but which are not suitable for understanding the perception of complex auditory stimuli. Another reason is that the bilateral cortical representation of both ears permits a lesion in the auditory cortex of one hemisphere to remain unnoticed by the patient. Moreover, processing of sounds and sound sequences requires temporal buffers which are inherently more complicated to investigate than stationary neural firing patterns.

During the last decade, magnetoencephalography (MEG) has become available to monitor cortical activity in awake humans. MEG recordings are based on the detection of weak magnetic fields produced by cerebral electric currents, and they allow differentiation between signals from various cortical regions with good temporal and spatial resolution. Due to its sensitivity to tangential current sources, MEG is well suited for non-invasive investigation of the auditory cortical regions embedded within the Sylvian fissures. In humans, these regions include the primary koniocortex and several surrounding belt areas.

Moreover, signals from either hemisphere can be readily resolved in the MEG records.

Here we first present a brief review of the principles of MEG recordings and data interpretation, and then provide examples of neuromagnetic exploration of the human auditory cortex. Further details about the MEG method in general, and its application to auditory studies in particular, can be found in previous reviews (Hari and Lounasmaa, 1989; Grandori *et al.*, 1990; Sato, 1990; Hämäläinen *et al.*, 1993).

Measurement and Interpretation of Magnetic Signals

Electric currents flowing in the brain generate weak magnetic signals, typically only one part in 10^8 or 10^9 of the Earth's geomagnetic field. The measurements are therefore typically carried out in a magnetically shielded room. The signals are first detected by an array of superconducting flux transformers, and then sensed by SQUIDs (Superconducting QUantum Interference Devices), which are sensitive to extremely small magnetic fields. The electronics connected to each SQUID converts the strength of the sensed magnetic field to an electrical signal.

To locate the underlying neural activity, the magnetic field pattern over the scalp must be sampled. Nowadays this can be done simultaneously with devices containing tens of sensors. Recently, the first helmet-type magnetometers, covering the whole scalp, have been introduced. Figure 1 illustrates the Finnish 122-channel device (Neuromag Ltd, Espoo, Finland). Planar gradiometers incorporated in this instrument detect the strongest signals just above a locally restricted area of cortical activity.

Figure 2 shows an example of auditory evoked magnetic fields (AEFs) recorded with the 122-channel neuromagnetometer. Prominent responses occur over the right and left temporal lobes, reflecting activity of the auditory cortices. The signals are usually interpreted with the help of source models; equivalent current dipoles (ECDs) within a sphere are commonly applied. The orientation, strength, and the three-dimensional location of the ECD, best accounting for the measured signal distribution, are found by means of a least-squares fit to the experimental data. In Fig. 2, two ECDs have been superimposed on a magnetic resonance image (MRI) to combine the functional and structural information. Such an integration localizes sources of the AEFs onto the lower surface of each Sylvian fissure. These locations agree with the generator sites of intracranially recorded long-latency auditory evoked potentials (Liegeois-Chauvel *et al.*, 1990). In physiological interpretation of the MEG results, even a good account of the data by an ECD does not necessarily imply simple dipolar source currents in the cortex. More adequate methods to characterize multiple simultaneously active sources are under development.

Activity in the supratemporal cortex, deduced from the peak latencies of AEF deflections, starts 19 msec after a sound onset and continues for a few hundred milliseconds, with source areas changing slightly as a function of time (for a review, see Hari, 1990). The most prominent AEF deflection, N100m, peaks about 100 msec after the stimulus onset (cf. Fig. 2). Comparisons of cortical current densities and observed dipole strengths suggest activation of 1–2.5 cm^2 of cortex during N100m. An infarction extending to deep parts of the temporal lobe abolishes N100m, suggesting that normally this cortical area either generates or modifies this response (Mäkelä *et al.*, 1991). N100m has a larger amplitude and peaks 4–10 msec earlier for contralateral than ipsilateral stimuli.

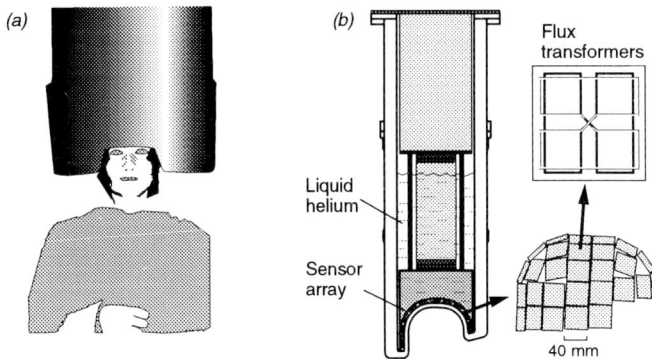

Fig. 1. (a) Measurement of the cerebral magnetic fields with a 122-channel SQUID gradiometer (Ahonen *et al.*, 1993). The dewar, filled with liquid helium, is brought close to the subject's head without direct contact. (b) Schematic presentation of the device. At each of the 61 recording locations of the helmet-shaped sensor array (insert at lower right), two orthogonal derivatives of the magnetic field component normal to the sensor surface are measured simultaneously. The planar flux transformers of this device are in a figure-of-eight configuration; two orthogonal flux transformers (black and white wires) are shown on one chip in the upper insert.

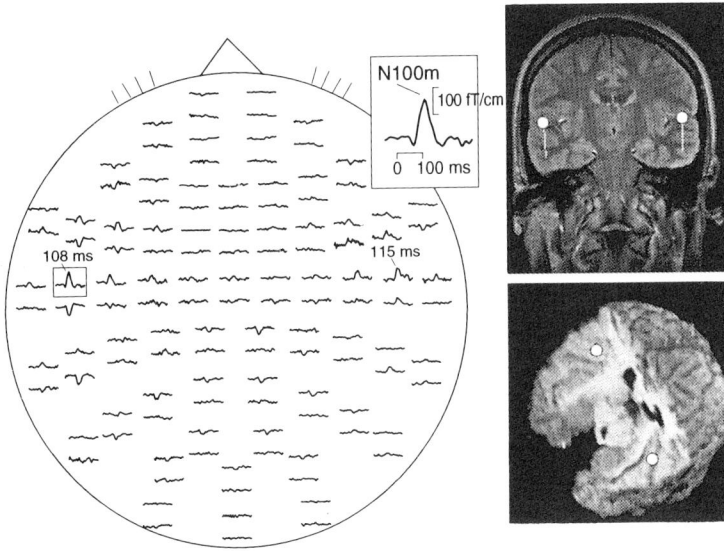

Fig. 2. (Left) Auditory evoked magnetic fields recorded with the 122-channel device when 50 msec tones were presented to the subject's right ear once every 4 sec. The latitudinal and longitudinal derivatives are shown over each other. The head is viewed from above and the helmet has been 'flattened' to show responses from the whole head simultaneously; the nose points upwards. The main response, N100m, peaks at 93 msec over the contralateral, left hemisphere and at 100 msec over the right hemisphere. The passband is 0.05–40 Hz. In the amplitude scale, fT refers to femtotesla (= 10^{-15} Tesla) and the unit fT/cm indicates that the field derivatives are measured as a function of distance. (Right above) ECDs for N100m, superimposed on a coronal MRI slice of the same subject. The dots illustrate the locations and the tails the orientations of the dipoles. (Right below) The same sources projected on an MRI surface rendering, viewed from above and with the frontal lobes removed.

Correlates of Sensory Memory

The timing and history of the stimulus sequence influence the characteristics of the evoked responses. The amplitude of N100m increases with increasing interstimulus intervals (ISIs) and reaches a plateau with ISIs of 4–8 sec (for a review, see Hari, 1990). This ISI dependence has been reported to parallel the duration of the psychophysically determined pitch memory (Lu et al., 1992). We recently observed that the ISI dependence of the N100m amplitude is similar in both hemispheres, suggesting similar durations for sensory memory traces in the left and right auditory cortices (Mäkelä et al., 1993).

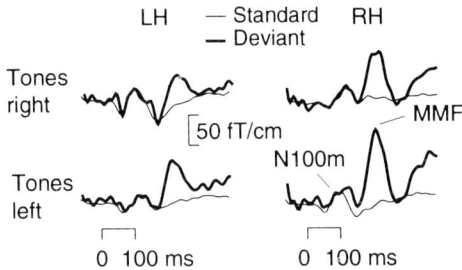

Fig. 3. Results of an oddball task when the subject ignored the stimuli. Responses to standard stimuli (1 kHz 100 msec tones, probability 0.85; thin lines) and to deviant stimuli differing in duration (50 msec; thick lines). The interstimulus interval was 600 msec. Responses from the left and right hemispheres (LH, RH) are expanded below for both right-ear and left-ear stimuli. Adapted from Levänen *et al.* (1995).

In addition to recovery cycle studies, the few-second storage of auditory information can be also demonstrated by presenting infrequent deviant stimuli among a series of monotonously repeated standards. Various such deviations evoke magnetic responses, called mismatch fields (MMFs; for reviews, see Näätänen and Picton, 1987; Hari, 1990). MMF seems not to be related to the stimulus features *per se* but rather to their changes. Studies of MMFs can, therefore, reveal memory processes needed for comparing the similarity of successive stimuli. Both the MMF and behavioral data suggest that the duration of the auditory sensory ('echoic') memory is about 10 sec (Sams *et al.*, 1993), in good agreement with the estimates based on the ISI dependence of N100m.

Figure 3 illustrates AEFs to standard tones and to infrequent changes in tone duration (Levänen *et al.*, 1994). MMFs elicited by the deviant tones are stronger over the right hemisphere for both left-ear and right-ear tones. Satisfactory explanation of the response distributions necessitated, in many subjects, two sources in the right hemisphere, whereas one source was sufficient in the left. These data suggest stronger involvement of the right than the left hemisphere in change detection. Accordingly, amplitude and frequency modulations in the middle of a 620 msec sound elicited stronger responses in the right than in the left auditory cortex (Pardo *et al.*, 1994).

Recent MEG recordings imply that infarction of anteromedial thalamus may affect the formation of auditory sensory memory traces: MMFs were either abolished or dampened in five patients (Mäkelä *et al.*, 1994). Thalamic lesions also disturb the cortical spontaneous rhythms in widespread areas.

Cortical Mechanisms of Directional Hearing

In audition, information on stimulus location does not exist on the receptor surface, in contrast to vision and somesthesis; sound source locations have to be calculated centrally from the interaural time and intensity differences. There has been a long-standing debate whether the auditory cortex contains an orderly representation of sound source locations, similar to that observed in the owl midbrain (Konishi *et al.*, 1988). We recently studied auditory cortical mechanisms of sound lateralization by recording AEFs to click trains containing interaural time differences (ITDs).

In one study (McEvoy *et al.*, 1993) , the sounds were 168 msec trains of 1 msec clicks presented at 70 Hz. ITDs varied between +0.7 and –0.7 msec (+ referring to left-ear leading sounds and – to right-ear leading sounds) in different trains and the subjects perceived the stimuli at different locations, as illustrated in Fig. 4(a). In an oddball experiment (Fig. 4b), the stimuli giving rise to the leftmost perception were presented as standards with infrequent deviants perceived at other locations. The MMF amplitude was smallest when the deviants were

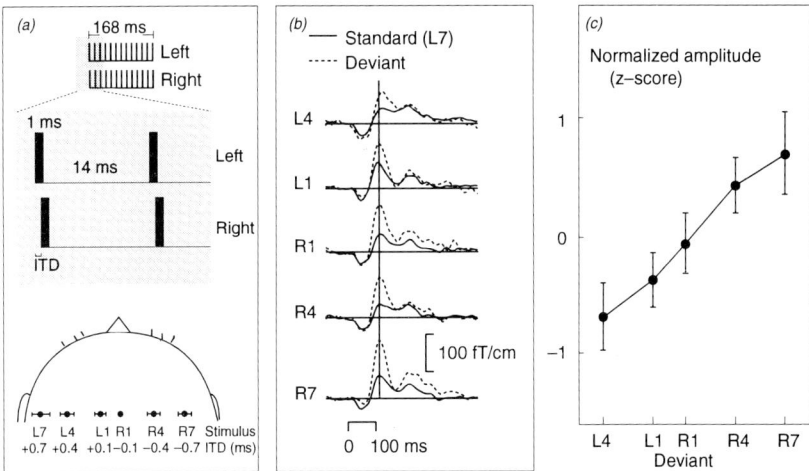

Fig. 4. Directional hearing experiment. (a) Composition of the binaural stimulus trains; the ITD was constant throughout the train. The mean (± SEM; six subjects) perceived locations with trains of different ITDs are shown below. (b) Responses to leftmost standard stimuli (L7) and different deviants in an oddball experiment with 0.15 probability for the deviants. (c) Relative amplitudes (mean ± SEM; five subjects) of responses to deviants with different ITDs. Adapted from McEvoy *et al.* (1993).

close in perceived location to the standard stimulus and increased as the spatial separation between the sounds increased (Fig. 4c). These results can be interpreted to reflect a functional separation between neuronal groups reacting to sounds with different ITDs; evidently neuronal populations activated by spatially close stimuli overlap and the overlap decreases with increasing spatial separation. However, source locations in the auditory cortex did not show any signs of a large-scale spatial mapping of the sound location. These data fit with recent results showing that, in the auditory cortex of an anesthetized cat, the sound locations are coded in the time patterns of the firing neurons rather than in a spatial map (Middlebrooks *et al.*, 1994). More recent MEG studies (McEvoy *et al.*, 1994) imply segregation of neuronal populations sensitive and nonsensitive to ITDs already during the middle-latency responses.

Monaural stimuli, although commonly used in studies of auditory physiology, are of course very unphysiological and do not occur in ecologically valid situations if the auditory system of the organism is intact.

Processing of Incongruent Speech Stimuli

Visual cues are known to modify perceived localization of sound sources, as happens in motion pictures and in listening to ventriloquists. The interaction of visual and auditory stimulation in speech perception has been studied with MEG (Sams *et al.*, 1991) utilizing 'McGurk illusion', first reported in behavioral experiments (McGurk and MacDonald, 1976). The subject viewed a video tape showing a female face articulating either a frequent (84%) syllable /pa/ or an infrequent syllable (16%) /ka/. Auditory /pa/ syllables were dubbed to all visual articulation movements. The perception of the sound changed during the incongruent stimuli (visual /ka/, auditory /pa/) and most subjects reported that they heard /ta/.

Both congruent and incongruent stimuli evoked responses in the auditory cortex, but the responses differed from 150 msec onwards. The ECD locations suggested that the difference was due to a change in the activity of the supratemporal auditory cortex. Recent whole-head MEG recordings from our laboratory (Sams and Levänen, unpublished) have demonstrated the same phenomenon for equiprobable congruent and incongruent stimuli, thereby confirming that the observed difference between the responses was not caused by an MMF to the infrequent stimuli. The results thus suggest that seeing the articulatory movements may affect the activity of the auditory cortex, thereby modulating the

Active hearing

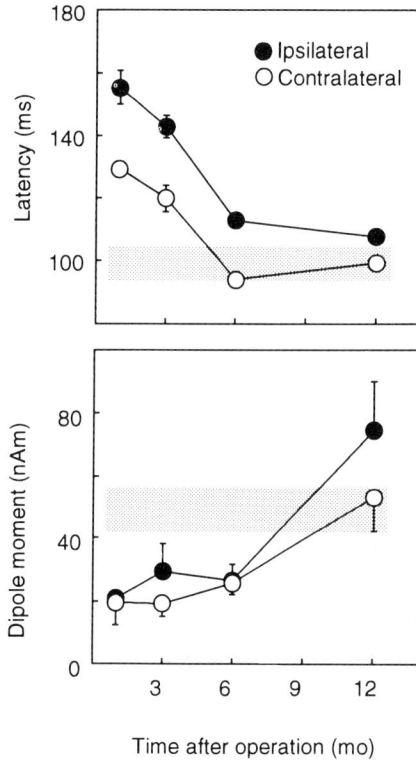

Fig. 5. Mean (± SEM) peak latencies (over interstimulus intervals of 1–16 sec) and amplitudes of one patient as a function of time for right-ear stimulation, measured from the ipsi- and contralateral hemispheres. The left ear of the patient became deaf during the operation of an acoustic neuroma. The shaded area represents the mean (± SEM) variability of the control group. Adapted from Vasama *et al.* (1995).

perception of speech sounds. This type of intermodal combination of information facilitates speech perception in noisy environments.

Cortical Reorganization

Unilateral peripheral auditory damage is known to produce central effects in experimental animals, with histological changes at several levels of the auditory pathway and deficits in binaural interaction. In humans such information is sparse. Our first MEG observation about modification of central auditory pathways took place in a study of a deaf patient with cochlear prosthesis (Pelizzone *et al.*, 1986). The 57-year-old patient had been deaf in the right ear since early childhood (she did not remember having ever heard with that ear) and had lost the hearing from the left ear around the age of 7 years. With a cochlear implant in the right ear she started to understand simple sounds. MEG responses peaking at 65 msec were obtained to electric stimulation of the implant, but only on the right, ipsilateral hemisphere, which is in strong contrast to normal-hearing subjects who typically have stronger responses over the contralateral hemisphere. This finding suggested that due to the early deafening of the right ear, the central auditory pathways had been modified, with the left-ear input receiving a prominent role. Interestingly, when the patient was implanted a couple of years later to the other ear, her speech comprehension improved considerably, supporting the view that the left-ear pathway really was the most functional one.

We have recently studied small patient groups with unilateral deafness or hearing impairment to unravel evidence for cortical reorganization. In patients with congenital unilateral hearing loss, hemispheric differences and latencies of cortical responses remained normal up to hearing losses of 60–70 dB (Vasama *et al.*, 1994), implying that the input imbalance has to be considerable for macroscopic changes to occur in the cortical organization.

In patients who became unilaterally deaf after acoustic neuroma removal, N100m to stimulation of the healthy ear was clearly delayed (by up to 30–70 msec) and dampened immediately after the operation but recovered to normal level within 1 year (Vasama *et al.*, 1995). The results imply that functions of the central auditory pathways may be modified for a considerably long time after changes in the input (Fig. 5).

Conclusions

MEG provides a non-invasive method to study macroscopic functional organization of the human auditory cortex, thereby complementing information obtained from other brain imaging methods and from scalp electroencephalograms. MEG recordings have indicated that the human auditory cortex is extremely sensitive to many kinds of events in the acoustic environment, such as onsets, offsets and changes within the sounds, including frequency and amplitude modulations. The auditory cortex also seems to be important for sensory memory and for directional hearing. All these are examples of bottom-up studies which tempt to characterize feature extraction and processing of stimulus attributes in functionally specialized auditory cortical areas.

Top-down approaches, with identical stimuli but with varying tasks, have now become feasible with the whole-head instruments. Previous MEG recordings have already shown that identical sounds produce different activation in the auditory cortex depending on the attentional state of the subject (Hari *et al.*, 1989; Rif *et al.*, 1991).

Whole-head MEG recordings allow comparisons of hemispheric differences in the processing of complex stimuli, such as speech sounds. The ability to monitor simultaneously activity of several cortical regions facilitates research into the neural basis of human cognitive activity, and it is also possible to investigate subjects displaying special abilities or disorders of auditory processing. In addition to evoked responses, important information can be obtained from recordings of ongoing spontaneous activity.

The focus of auditory studies is changing from traditional cochleo-centric thinking to more integrative approaches. For example, Gestalt principles are now frequently discussed in the assessment of central auditory processes (Bregman, 1990) although the underlying neural mechanisms are still unknown. We hope that methods that characterize behavior of large neuronal populations, as MEG does, will open new insights into some key issues of auditory perception, such as extraction of auditory objects from the acoustic inflow.

Acknowledgements

This work was supported by the Academy of Finland and by the Sigrid Jusélius Foundation. We thank Dr Norman Loveless for comments on the manuscript. The MR images were recorded at the Department of Radiology, University of Helsinki.

References

Ahonen, A., Hämäläinen, M. S., Kajola, M. J., Knuutila, J. E. T., Laine, P. P., Lounasmaa, O. V., Parkkonen, L. T., Simola, J. T. and Tesche, C. D. (1993). 122-channel SQUID instrument for investigating the magnetic signals from the human brain. *Phys. Scripta* **T49**: 198–205.

Bregman, A. (1990). *Auditory Scene Analysis*. MIT Press, Cambridge, MA.

Grandori, F., Hoke, M. and Romani, G.-L. (eds) (1990). *Auditory Evoked Magnetic Fields and Potentials. Advances in Audiology*, Vol. 6. Karger, Basel.

Hari, R. (1990). The neuromagnetic method in the study of the human auditory cortex. In *Auditory Evoked Magnetic Fields and Potentials. Advances in Audiology*, Vol. 6 (F. Grandori, M. Hoke and Romani G. eds), pp. 222–282. Karger, Basel.

Hari, R. and Lounasmaa, O. V. (1989). Recording and interpretation of cerebral magnetic fields. *Science* **244**: 432–436.

Hari, R., Hämäläinen, M., Kaukoranta, E., Mäkelä, J. P., Joutsiniemi, S. L. and Tiihonen J. (1989). Selective listening modifies activity of the human auditory cortex. *Exp. Brain Res.* **74**: 463–470.

Hämäläinen, M., Hari, R., Ilmoniemi, R., Knuutila, J. and Lounasmaa, O. V. (1993). Magnetoencephalography – theory, instrumentation, and applications to noninvasive studies of the working human brain. *Rev. Mod. Phys.* **41**: 413–497.

Konishi, M., Takahashi, T. T., Wagner, H., Sullivan, W. E. and Carr C. E. (1988). Neurophysiological and anatomical substrates of sound localization in the owl. In *Auditory Function: Neurobiological Bases of Hearing* (G. M. Edelman, W. E. Gall and W. M. Cowan eds), pp. 721–745. Wiley, New York.

Levänen, S., Ahonen, A., Hari, R., McEvoy, L. and Sams, M. (1995). Deviant auditory stimuli activate human left and right auditory cortex differently. *Cerebral Cortex* (in press).

Liegeois-Chauvel, C., Musolino, A. and Chauvel, P. (1990). Générateurs des potentiels évoqués auditifs corticaux chez l'homme. *Coll. Phys.* **51**(C2): 135–138.

Lu, Z.-L., Williamson, S. and Kaufman L. (1992). Behavioral lifetime of human auditory sensory memory predicted by physiological measures. *Science* **258**: 1668–1670.

McEvoy, L., Hari, R., Imada, T. and Sams, M. (1993). Human auditory cortical mechanisms of sound lateralization: II. Interaural time differences at sound onset. *Hear Res.* **67**: 98–109.

McEvoy, L., Mäkelä, J. P., Hämäläinen, M. and Hari, R. (1994). Effect of interaural time differences on middle-latency and late auditory evoked magnetic fields. *Hearing Res.* **78**: 249–257.

Middlebrooks, J., Clock, A., Xu, L. and Greeen, D. (1994). A panoramic code for sound location by cortical neurons. *Science* **264**: 842–844.

Mäkelä, J. P., Hari, R., Valanne, L. and Ahonen, A. (1991). Auditory evoked magnetic fields after ischemic brain lesions. *Ann. Neurol.* **30**: 76–82.

Mäkelä, J. P., Ahonen, A., Hämäläinen, M., Hari, R., Ilmoniemi, R., Kajola, M., Knuutila, J., Lounasmaa, O. V., McEvoy, L., Salmelin, R., Sams, M., Simola, J., Tesche, C. and Vasama, J.-P. (1993). Functional differences between auditory cortices of the two hemispheres revealed by whole-head neuromagnetic recordings. *Hum. Brain Mapping* **1**: 48–56.

Mäkelä, J. P., Salmelin, R., Kotila, M. and Hari, R. (1994). Neuromagnetic correlates of memory disturbance caused by infarction in anterior thalamus. *Soc. Neurosci. Abstr.* **20**(1): 810.

Näätänen, R. and Picton, T. (1987). The N1 wave of the human electric and magnetic response to sound: a review and analysis of the component structure. *Psychophysiology* **24**: 375–425.

Pardo, P., Mäkelä, J., Sams, M. and Hari, R. (1994). Neuromagnetic measurements of hemispheric differences in auditory processing of frequency and amplitude modulations. *Soc. Neurosci. Abstr.* **20**(1): 325.

Pelizzone, M., Hari, R., Mäkelä, J. P., Kaukoranta, E. and Montandon, P. (1986). Activation of the auditory cortex by cochlear stimulation in a deaf patient. *Neurosci. Lett.* **68**: 192–196.

Rif, J., Hari, R., Hämäläinen, M. and Sams, M. (1991). Auditory attention affects two different areas in the human auditory cortex. *Electroenceph. Clin. Neurophysiol.* **79**: 464–472.

Sams, M., Aulanko, R., Hämäläinen, M., Hari, R., Lounasmaa, O. V., Lu, S.-T. and Simola, J. (1991). Seeing speech: visual information from lip movements modifies activity in the human auditory cortex. *Neurosci. Lett.* **127**: 141–145.

Sams, M., Hari, R., Rif, J. and Knuutila, J. (1993). The human auditory sensory memory trace persists about 10 s: neuromagnetic evidence. *J. Cogn. Neurosci.* **5**: 363–370.

Sato, S. (1990). *Magnetoencephalography. Advances in Neurology*, Vol. 54. Raven Press, New York.

Vasama, J.-P., Mäkelä, J. P., Parkkonen, L. and Hari, R. (1994). Auditory cortical responses in humans with congential unilateral conductive hearing loss. *Hearing Res.* **78**: 249–257.

Vasama, J.-P., Mäkelä, J. P., Pyykkö, I. and Hari R. (1995). Modification of central auditory pathways after unilateral permanent deafness due to removal of acoustic neuroma. *Hearing Res.* (in press).

Neural Processing of Target-Distance Information in the Mustached Bat

NOBUO SUGA, JOHN A. BUTMAN, HAIBING TENG, JUN YAN
AND JOHN F. OLSEN

Department of Biology, Washington University, One Brookings Drive, St Louis, MO 63130-4899, USA

Introduction

For echolocation (biosonar), bats emit orientation sounds (biosonar pulses) and listen to echoes. Target-distance information is carried by the delay of an echo from an emitted biosonar pulse. How does the auditory system extract distance information from a pair of a biosonar pulse and its echo?

A neural model for processing target-distance information carried by echo delay was first proposed with the data obtained from the little brown bat, Myotis lucifugus (Suga and Schlegel, 1973). However, the data obtained from the mustached bat, Pteronotus parnellii, are different from those of the little brown bat (Suga *et al.*, 1978; O'Neill and Suga, 1979; Suga and O'Neill, 1979; O'Neill, 1985; Olsen, 1986; Olsen and Suga, 1991). Accordingly, a neural model which is different from that for the little brown bat has been proposed (Suga, 1990). Both models, however, consist of three basic elements: phasic constant latency neurons, delay lines (or recovery periods) and coincidence detectors. The data obtained from the mustached bat were comprehensive and we considered that we basically understood where and how neurons tuned to specific echo delays were created for the processing of target-distance information (e.g. Suga's review article, 1990). However, the recent data obtained by Wenstrup's and Suga's laboratories clearly indicate that where and how response properties of delay-tuned neurons are created are different from the model proposed in 1990, although the above three elements are necessary as proposed in the 1990s model. These recent data and the old data cited above were confusing at first, but now are found to fit each other like pieces of a jigsaw puzzle. The aim of the present article is to propose a revised neural model for the processing of target-distance information. The model reconciles the new anatomical and physiological data with the previous findings and clearly indicates the experiments to be performed for further study. In particular, the model clearly defines a functional role for the corticothalamic projection. Our hope is that the model will stimulate the research on the descending auditory system.

Extraction of Distance Information by FM–FM Combination-Sensitive Neurons

The delay of an echo from the pulse is a primary cue for measuring target distance. A 1.0 msec echo delay corresponds to a target distance of 17.3 cm at 25 deg C. At the periphery, neurons respond to both an emitted pulse and its echo, because they are tonic on-responders with a very short recovery period (e.g. Suga, 1964). The time interval between the two grouped discharges evoked by either the pulse or the echo is directly related to the echo delay (e.g. Grinnell, 1963). In the central

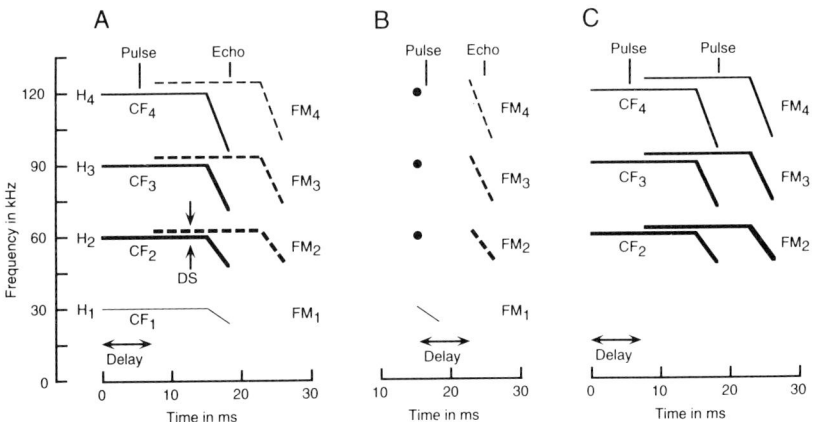

Fig. 1. Schematized sonagrams of a biosonar pulse of the mustached bat and its echo (A), essential signal elements in the pulse–echo pair to evoke a facilitative response of FM–FM neurons (B) and biosonar pulses emitted by conspecifics (C). Four harmonics, constant-frequency (CF) components and frequency-modulated (FM) components are indicated by H_{1-4}, CF_{1-4} and FM_{1-4}, respectively. DS: Doppler shift.

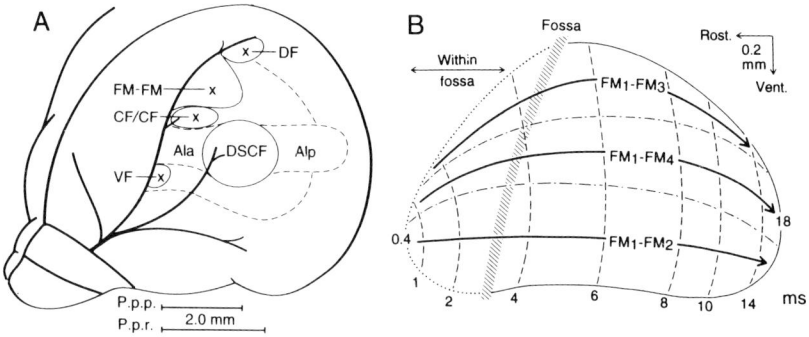

Fig. 2. Several areas in the auditory cortex of the mustached bat (A) and echo-delay axes in the FM_1–FM_2, FM_1–FM_3 and FM_1–FM_4 subdivisions of the FM–FM area of the auditory cortex (B). In (A), AIa, DSCF and AIp are the three subdivisions of the tonotopically organized primary auditory cortex. The FM–FM, DF and VF areas consist of delay-tuned neurons and have a distance (echo delay) axis. The CF/CF area consists of velocity-tuned neurons and has a velocity axis. In (B), vertically oriented dashed lines indicate iso-best delay lines which are labeled with best delays from 0.4 to 18 msec.

auditory system, however, echo delays shorter than 20 msec are represented by neurons tuned to specific echo delays (in Eptesicus, Feng *et al.*, 1978; in Pteronotus, Suga *et al.*, 1978, 1983; Suga and O'Neill, 1979; O'Neill and Suga, 1979, 1982; in Myotis, Sullivan, 1982a,b; Suga and Horikawa, 1986; Berkowitz and Suga, 1989; Edamatsu *et al.*, 1989; Wong *et al.*, 1992; Dear *et al.*, 1993; Edamatsu and Suga, 1993).

Neural mechanisms to create *delay-tuned neurons* differ among different species of bats, reflecting the physical properties of their species-specific biosonar pulses. In the mustached bat, pulses are complex, consisting of eight major components (CF_{1-4} and FM_{1-4} in Fig. 1A). In the FM–FM, DF and VF areas of the auditory cortex of this species (Fig. 2A), neurons are tuned to specific delays of the $FM_{2, 3\ or\ 4}$ of an echo ($EFM_{2\ or\ 4}$) from the FM_1 of the emitted pulse (PFM_1) (Fig. 1B). Therefore, there are three major types of *delay-tuned FM–FM (combination-sensitive) neurons*: FM_1–FM_2, FM_1–FM_3 and FM_1–FM_4 (Suga *et al.*, 1978, 1983; O'Neill and Suga, 1979, 1982). In each of the FM-FM, DF and VF areas, these three types of FM–FM neurons are separately clustered, forming three subdivisions. In each subdivision, echo delay is systematically represented. In other words, each subdivision has an echo-delay axis or map (Fig. 2B). Since echo delay is the primary cue for target-distance measurement, we concluded that the FM–FM, DF and VF areas are specialized for processing target-distance information. In the FM–FM area, echo delays from 0.4 to 18 msec, i.e. target distances from 7 to 310 cm are systematically represented along its rostrocaudal axis (Fig. 2B; Suga and O'Neill, 1979). Along this distance axis, distance information changes at a rate of 20 mm per cortical minicolumn, which is about 20 μm wide.

In the emitted pulse, the second harmonic is most intense. The third and fourth harmonics are weaker than the second in this order. The first harmonic (CF_1 and FM_1) is the weakest component. It has less than 1% of the total energy of the emitted pulse. However, it is used as the reference to measure target distances (Fig. 1B). The higher three harmonic FM components ($FM_{2, 3 \text{ or } 4}$) are hereafter called FM_n to simplify their description.

Why does the auditory system create FM_1–FM_n combination-sensitive neurons to extract distance information? As explained below, this 'heteroharmonic' sensitivity is one of the mechanisms used to protect the neural processing of biosonar information from masking by the sounds produced by conspecifics (Suga and O'Neill, 1979). Let us consider the situation in which hundreds of bats are flying in a cave and the first harmonic of their pulses is completely suppressed by the anti-resonance of the vocal tract. Then in the air, there are many FM_n produced by conspecifics (Fig. 1C). The combinations of these harmonics in their pulses cannot excite FM_1–FM_n neurons, because FM_1 is absent. When the animal itself emits pulses, however, FM_1 is present at the larynx and stimulates both the ears by bone conduction (Kawasaki *et al.*, 1988). The FM_1–FM_n neurons are then activated by FM_1 to process target-distance information. In this way, the processing of distance information is protected from the masking which would otherwise be caused by the sounds emitted by many conspecifics flying nearby.

The Basic Principles for Creating an Array of Delay-Tuned FM–FM Neurons

A phasic on-response to PFM_1 (top of Fig. 3A) is delayed different amounts by delay lines created in prethalamic auditory nuclei (Fig. 3B). Individual coincidence detectors (circles in Fig. 3B) have two inputs: one input carries activity evoked by PFM_1 through a delay line and the other input carries activity evoked by EFM_n without delay lines (bottom of Fig. 3A). An echo always follows the emitted pulse. When an echo delay is 4.0 msec, for example, the impulses evoked by PFM_1 and EFM_n arrive at the same time only at the coincidence detector that is associated with a 4.0 msec delay line (asterisked circle in Fig. 3B). Then, this coincidence detector shows a facilitative response and conveys it to

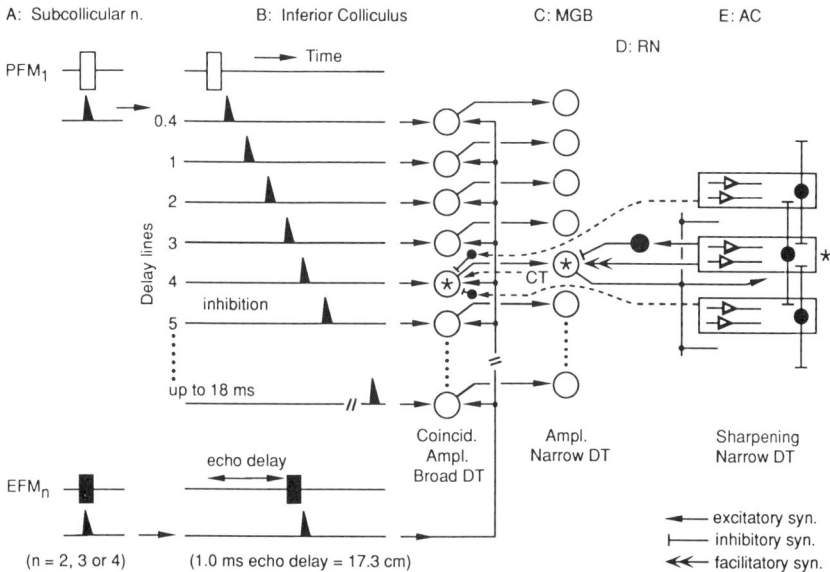

Fig. 3. A neural model to explain the basic neural mechanisms for processing target-distance information in the mustached bat. AC: auditory cortex. DT: delay tuning. EFM_n: $FM_{2, 3 \text{ or } 4}$ of an echo. MGB: medial geniculate body. PFM_1: FM_1 of an emitted pulse. RN: reticular nucleus. This figure is complementary to Fig. 4. For simplicity, neural nets for creating amplitude and FM selectivities are not included in the model.

higher-order neurons (circles in Fig. 3C). The best delay to excite this detector is 4.0 msec. The neural net to create delay lines and an array of coincidence detectors consequently produces a map or axis for the systematic representation of echo delays, i.e. target distances. Delay-dependent coincidence detectors of the mustached bat have been called *FM-FM (combination-sensitive) neurons* or *delay-tuned neurons.* Their response properties are more complex than those described above, indicating that these are created with neural elements more than the above three. The updated theory (neural model) for ranging is explained below.

Subthalamic Mechanisms for the Processing of Distance Information

The FM_1 and FM_n of a biosonar pulse emitted by the bat (open squares in Fig. 4, AS and CM) excite peripheral auditory neurons tuned to frequencies within either PFM_1 or PFM_n (hereafter, *FM_1 or FM_n neurons or channels*). An echo contains almost no FM_1, but intense FM_n

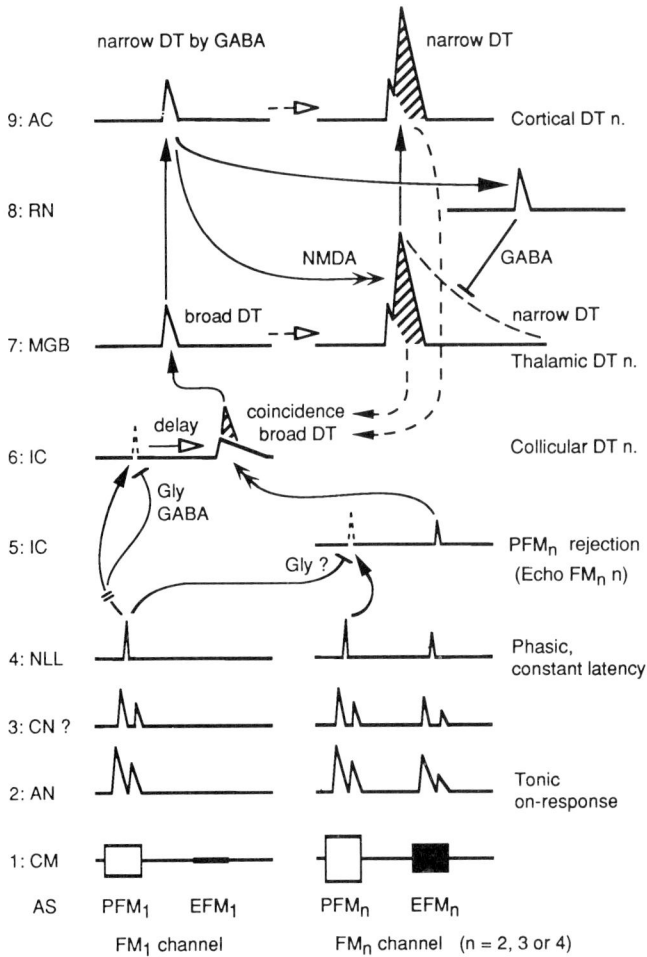

Fig. 4. A neural model to explain the basic neural mechanisms for processing target-distance information in the mustached bat. Pulse (PFM$_1$ and PFM$_n$) and echo (EFM$_1$ and EFM$_n$) stimuli evoke cochlear microphonic responses (CM). Then, auditory nerve fibers (AN) send impulses into the brain. These auditory signals are sent up to the auditory cortex (AC) through several nuclei: cochlear nucleus (CN), nucleus of lateral lemniscus (NLL), inferior colliculus (IC) and medial geniculate body (MGB). Neural responses to pulse–echo stimuli occurring in these nuclei are schematically displayed by peri-stimulus-time (PST) histograms. Filled single and double arrowheads indicate excitatory and facilitative synapses, respectively. 'T' heads indicate inhibitory synapses. Open arrowheads indicate either delayed excitatory response (6: IC) or change in response pattern (MGB and AC). AS: acoustic stimuli. DT: delay tuning or delay tuned. n: neuron. RN: reticular nucleus. Neurons excited by FM$_1$ or FM$_n$ are called the FM$_1$ or FM$_n$ neurons or channels, respectively (subscript n = 2, 3 or 4). Note that a part of these two channels merge in the IC. This figure is complementary to Fig. 3.

(filled square in Fig. 4, AS and CM), so that the FM_1 channel is not excited by the echo, but the FM_n channel is. Since the bat is active (e.g. flying) during echolocation, the auditory system is excited by external and internal noises. Action potentials of peripheral neurons evoked by these noises will be an origin of jitters in response latency of peripheral neurons to the biosonar signals. AN of Fig. 4 schematically shows the shapes of the PST histograms displaying responses of peripheral neurons to FM_1 and FM_n.

In the central auditory system, inhibition increases *frequency selectivity* of some neurons and adds *amplitude selectivity* to many neurons at higher levels (based on Suga *et al.*, 1983; Olsen and Suga, 1991). Interestingly, *FM-selectivity* is also added to some neurons through disinhibition or facilitation. Some of these 'FM-specialized' neurons selectively respond to particular FM sounds (in Myotis, Suga, 1965a,b,c, 1968, 1969; Suga *et al.*, 1983; in Pteronotus, O'Neill, 1985). *Phasic constant-latency neurons* which may be called *onset detectors* are also created in the ascending auditory system (NLL of Fig. 4). For example, the nucleus of the lateral lemniscus contains such neurons (Suga and Schlegel, 1973; Covey and Casseday, 1991; O'Neill *et al.*, 1992).

In the central nucleus of the inferior colliculus (ICc), some neurons have an inhibitory frequency-tuning curve for FM_1 and an excitatory tuning curve for FM_n. The duration of inhibition is very short (O'Neill, 1985; Mittmann and Wenstrup, 1994). These neurons are called 'echo FM_n' neurons ('5: IC' of Fig. 4), because of the following reason. When the bat emits a biosonar pulse, its ears are stimulated by the emitted pulse containing FM_1 and FM_n, but this PFM_n does not excite the echo FM_n neurons, because they are inhibited by PFM_1, presumably by inhibiting input from phasic constant-latency neurons responding to PFM_1 (Fig. 4). In an echo, FM_n is much stronger than FM_1 because the radiation of PFM_1 from the bat's mouth is so limited. Therefore, the response of the echo FM_n neurons to EFM_n is not inhibited by the neurons which do not respond to a weak EFM_1.

In the ICc, the latency of excitatory response to an acoustic stimulus systematically lengthens along the ventrodorsal axis of each iso-best frequency slab. This means that *delay lines* that spread neural responses in the time domain are formed orthogonal to the frequency axis (Hattori and Suga, 1989). A part of the delay lines may be based upon subcollicular neurons with different response latencies, but others are created within the ICc. Olsen and Suga (1991) speculated that long delay lines are mostly created by inhibition that shifts (delays) excitatory responses in the time domain. As a matter of fact, injections of inhibitory transmitter blockers (strychnine and/or bicucculline) into the ICc shortens delay tuning of FM–FM neurons (Saitoh and Suga, 1992). Park

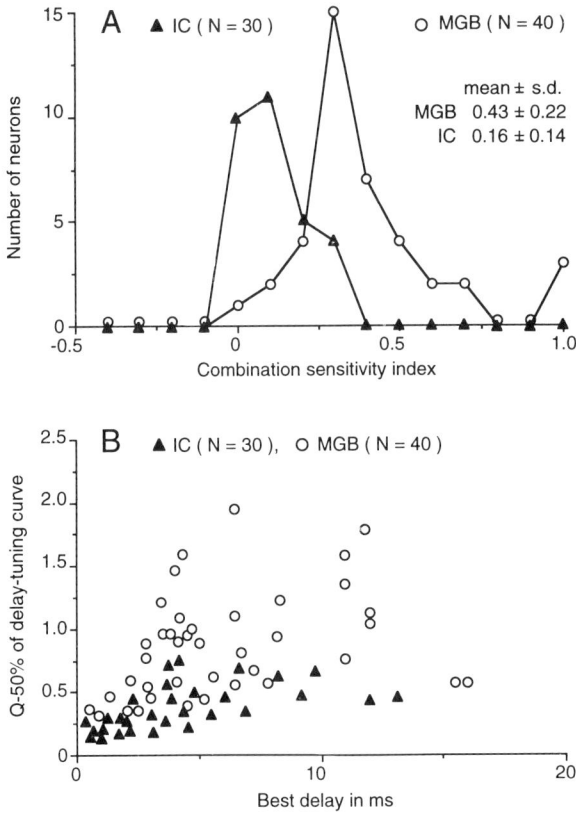

Fig. 5. Differences in magnitude of facilitation (A) and sharpness of delay tuning (B) between collicular and thalamic FM–FM neurons. Magnitude of facilitation is expressed by 'combination-sensitivity index (CSI)' which is defined as $(R_c - Rs_1 - Rs_2)/(R_c + Rs_1 + Rs_2)$. R_c, Rs_1 and Rs_2 are respectively the numbers of impulses per paired (c) and single (s_1 or s_2) stimuli. CSI is larger than zero for facilitation, smaller than zero for inhibition and zero for linear summation. Sharpness of delay tuning is expressed by Q–50% which is a best delay divided by the width of a delay-tuning curve at 50% of maximum response. In (A), the average CSI of thalamic FM–FM neurons is 2.7 times higher than that of collicular FM–FM neurons. In (B), the average Q–50% value of thalamic FM–FM neurons is 2.2 times larger than that of collicular neurons.

and Pollack (1993) studied response latencies in the CF_2 responding portion of the ICc and found that there was the latency axis and that bicucculine shortened response latencies of neurons.

Contrary to the conclusion of O'Neill (1985), it has been found that FM_1–FM_n neurons exist in the ICc (Mittmann and Wenstrup, 1994) and that the FM_1 channel in the ICc does not project to the region where thalamic FM–FM neurons are located (Wenstrup and Grose, 1993). Therefore, we have to revise the 1990s model as follows.

In the ICc, echo FM_n neurons that are not associated with delay lines project to delay-tuned (FM–FM) neurons that are associated with delay lines created in the FM_1 channel (Fig. 3B and '6: IC' of Fig. 4). When an EFM_n delay is too short to cause a coincidence of an EFM_n response with the delayed excitation to PFM_1, the EFM_n response is inhibited by inhibition which delays the excitation to PFM_1. When an echo (EFM_n) delay is equal to the amount of neural delay, however, the EFM_n response coincides with the delayed excitation to PFM_1 and evokes a facilitative response ('6: IC' of Fig. 4). Therefore, neurons which show the response illustrated in '6: IC' have been called FM–FM neurons or delay-tuned neurons.

In a neural model explained in Fig. 3, delay lines and then co-incidence detectors are created. In other words, neurons acting as coincidence detectors are neurons which are of a higher-order than those showing long response latencies. However, this may not be the case. As explained above, collicular FM–FM neurons are most likely to be the neurons where inhibition takes place to delay their excitatory response to PFM1 ('6: IC' of Fig. 4).

As shown in '6: IC' of Fig. 4, collicular and thalamic FM–FM neurons with a long best delay show a very phasic EFM_n response, but a PFM_1 response which is longer in duration than the EFM_n response (Yan and Suga, 1995). Accordingly, their delay tuning is broad, unless a mechanism for sharpening of delay tuning exists.

When collicular FM–FM neurons are created, amplitude and FM selectivities created by subcollicular and collicular nuclei are utilized by them, so that many FM–FM neurons are amplitude selective and some are FM selective (Suga *et al.*, 1983).

Thalamocortical Mechanisms for the Processing of Distance Information

Collicular FM–FM neurons project to thalamic FM–FM neurons in the medial geniculate body (MGB). This projection may contain connections mediating lateral inhibition, because delay tuning of the thalamic neurons is sharper than that of the collicular neurons (Fig. 5B; Yan and Suga, 1995). However, thus far there is no sign to support this prediction, because an application of bicuculline to thalamic FM–FM neurons does not alter the width of their delay tuning curves (Fig. 8A; Butman, 1992).

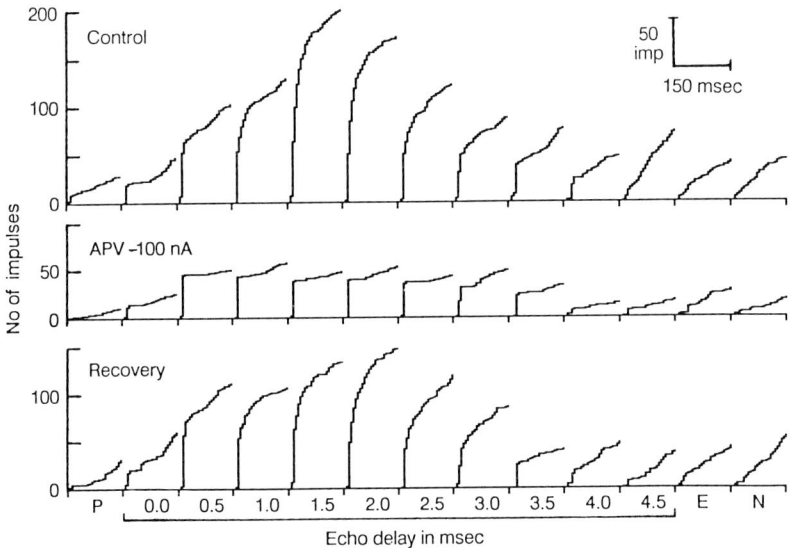

Fig. 6. An antagonist (D-2-amino-5-phosphovalerate: APV) to N-methyl-D-aspartate (NMDA) receptors specifically eliminates a late component (a burst of discharges) of facilitative responses of a thalamic FM–FM neuron to a paired pulse-echo stimuli. The facilitative responses in 'control' and 'recovery' consist of an initial spike and a burst of discharges, while those in 'APV' consist of only the initial spike. Each peri-stimulus-time-cumulative (PSTC) histogram is 150 msec long. The best delay of this neuron was 1.8 msec. P: pulse stimulus. E: echo stimulus. N: no stimulus. 0.0–4.5 ms: paired pulse–echo stimuli with different echo delays between 0.0 and 4.5 msec (Suga *et al.*, 1990).

Thalamic FM–FM neurons usually show a strong facilitative response with two components to an FM_1–FM_n stimulus: a fast component (initial spike) and a slow component (burst of dishcarges). The slow component is short in duration in many neurons, so that it does not look like a burst of impulses unless GABA-mediated inhibition is eliminated by bicucculline (Fig. 8A; MGB right of Fig. 4). Delay tuning of thalamic FM–FM neurons is sharp. *Aminophosphovalerate (APV)* is an antagonist specific to NMDA receptors. An APV application to thalamic FM–FM neurons eliminates the slow component and flattens their delay tuning curves (Fig. 6; Suga *et al.*, 1990; Butman, 1992). Therefore, it is expected that thalamic FM–FM neurons show stronger facilitative response than collicular FM–FM neurons. As a matter of fact, our recent experiments indicate that this is the case (Fig. 5A; Yan and Suga, 1995).

The strong facilitative response and sharp delay tuning of thalamic FM–FM neurons had been considered to be created by interactions occurring among neurons only in the ascending auditory system. But they are probably not, because an application of local anesthetic,

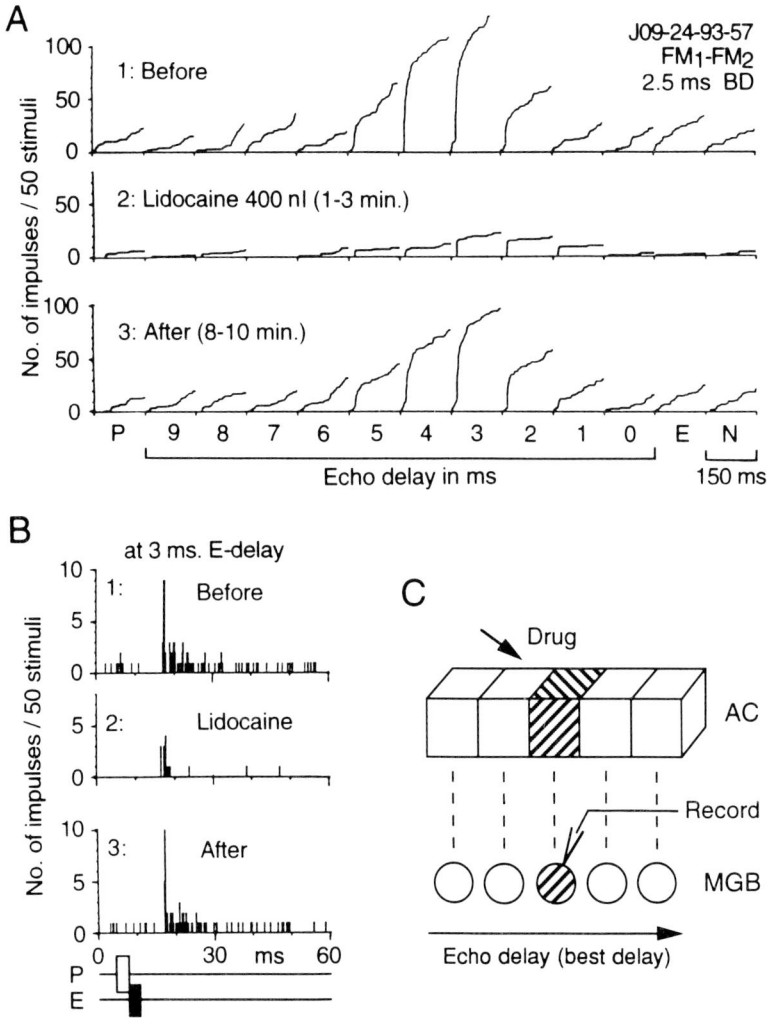

Fig. 7. Corticothalamic influence on facilitative responses of a thalamic FM–FM neuron to pulse-echo stimuli. (A) Lidocaine applied to the FM–FM area of the auditory cortex abolishes the NMDA-mediated facilitative response of the thalamic neuron. (B) PSTC histograms in (A) at a 3 msec echo delay are shown as PST histograms. (C) Recording of single unit activity was made from the medial geniculate body (MGB) and a drug was applied to the auditory cortex (AC). See the legend to Fig. 5.

lidocaine, to cortical FM–FM neurons reduces the facilitative response and the sharp portion of the delay tuning curve of thalamic FM–FM neurons (Fig. 7). The response properties of thalamic FM–FM neurons

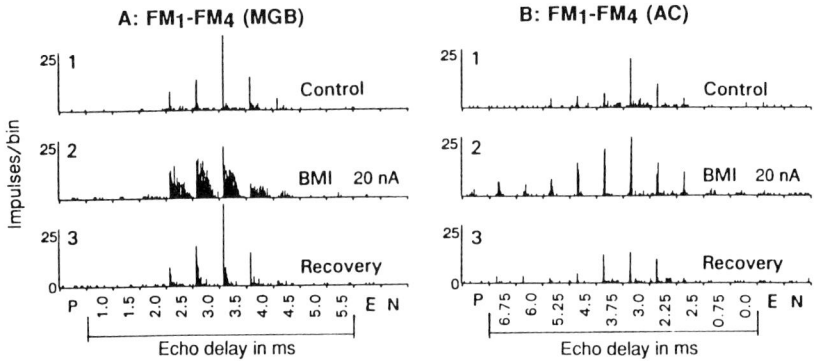

Fig. 8. Effect of an antagonist (bicuculline methiodide: BMI) to $GABA_A$ receptors on the responses of a thalamic (A) or a cortical FM–FM neuron (B) to pulse-echo stimuli. BMI increases the burst of discharges of the thalamic FM–FM neuron evoked by the stimuli, but shows no effect on its delay tuning. On the other hand, BMI shows a little effect on the discharges of the cortical FM–FM neuron evoked by the stimuli, but broadens its delay tuning. See the legend to Fig. 5. (Based on Butman and Suga, 1995.)

apparently result from the interaction between the auditory thalamus and the auditory cortex (Fig. 4). That is, the *thalamocortical complex* sculptures the response properties of thalamic FM–FM neurons, as described below in detail.

Assuming that the anatomy of the corticothalamic projection in cats (Wong, 1981; Olsen, 1986) is applicable to the mustached bat, then 1990s neural model can be revised as follows. Thalamic FM–FM neurons directly excite cortical FM–FM neurons in the fourth layer and indirectly excite neurons in the other layers ('9: AC' left of Fig. 4). FM–FM neurons in the sixth and/or deep fifth layers send descending fibers (corticothalamic fibers) directly back to thalamic FM–FM neurons which are isotopic in the echo delay domain. Then, the EPSP evoked by collicular FM–FM neurons and the EPSP evoked by the corticothalamic fibers evoke NMDA-mediated facilitation of the thalamic neurons ('7: MGB' right, shaded and dashed portions of Fig. 4). FM–FM neurons in the sixth and/or deep fifth layers of the cortex also project to the reticular nucleus, RN ('8: RN' right of Fig. 4) which consists of GABAergic inhibitory neurons. RN neurons project to thalamic FM–FM neurons and inhibit the late portion of the NMDA-mediated burst of discharges, as described below, so that the duration of the burst becomes very short ('7: MGB' right, dashed line of Fig. 4). Neurons in the fifth layer of the cortex project to the IC. Neurons in the MGB also project to the IC. We have not yet studied the corticocollicular and

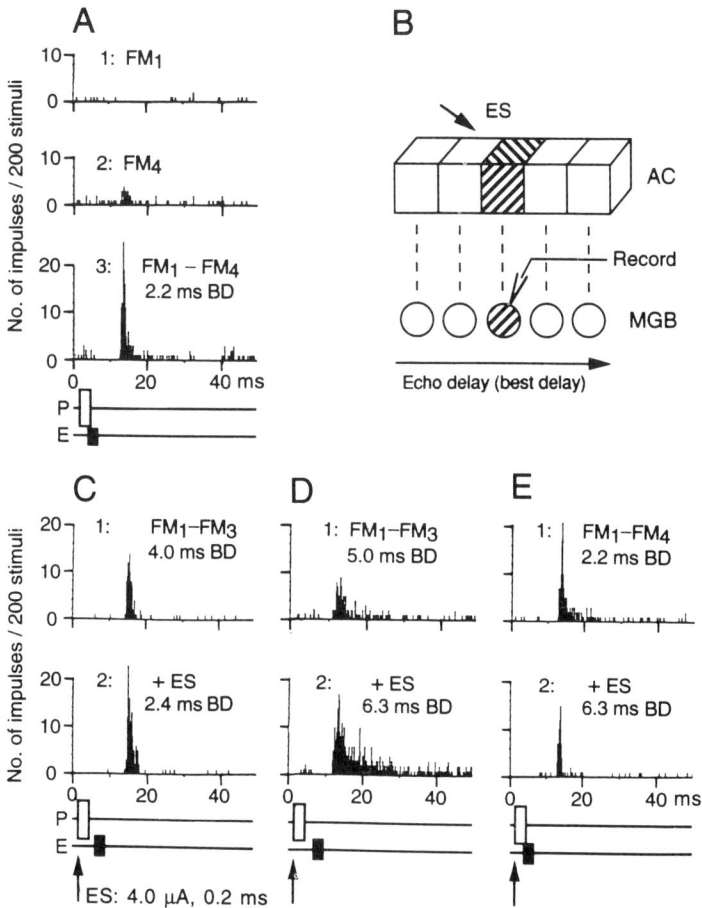

Fig. 9. Corticothalamic influence on facilitative responses of three thalamic FM–FM neurons (C, D and E) to pulse-echo stimuli. An electrical stimulus (0.2 msec long, 0.4 μA single electric pulse) is delivered to the FM–FM area of the auditory cortex (B). (A) An FM–FM neuron does not respond to a pulse FM_1 (1), but weakly responds to an echo FM_4 (2). The neuron shows a strong facilitative response when the pulse and echo stimuli are paired with a 2.2 msec echo delay (3). An electrical stimulation of the FM–FM area evokes either facilitation (C and D) or inhibition (E) of the facilitative responses of the thalamic neurons to pulse–echo stimuli. Compare the responses to the pulse–echo stimuli without (1) and with (2) the single electrical stimulus (ES) which is indicated by the arrow at the bottom. The best delays of the thalamic neurons are shown in '1', and those of cortical neurons to which ES was delivered are shown in '2'. BD: best delay.

geniculocollicular projections, but it is expected that these descending fibers evoke facilitation or inhibition of collicular neurons, as found in another species of bat, *Rhinolophus rouxi* (Sun *et al.*, 1989).

As described above, corticothalamic fibers evoke the NMDA-mediated burst of discharges of thalamic FM–FM neurons (see Figs 6 and 7). These thalamic neurons would in turn send a burst of discharges to the cortex. Therefore, if this positive feedback was not associated with negative feedback, thalamic and cortical neurons would oscillate. An application of *bicuculline* (an antagonist specific to inhibitory GABA$_A$ receptors) to thalamic FM–FM neurons dramatically increases the duration of the burst of discharges (Fig. 8A; '7: MGB' right, dashed line Fig. 4; Butman, 1992). These discharges are eliminated by an APV application to the neurons (Butman, 1992). RN neurons are a likely source of this negative feedback. RN neurons may be excited by an colateral of ascending fibers of thalamic neurons (Jones, 1975). The MGB of the mustached bat has very few, if any, intrinsic GABAergic neurons (Winer *et al.*, 1992). The remaining source of inhibition is the IC. Inhibition occurring in the MGB remains to be further studied.

Why is delay tuning of thalamic FM–FM neurons based upon NMDA-mediated facilitation sharper than that based upon non-NMDA-mediated facilitation? Why are the cortical FM–FM neurons so similar to those of thalamic FM–FM neurons? A bicucculline application to thalamic FM–FM neurons drastically changes their response pattern as described above, but does not change their delay tuning at all (Fig. 8A). However, a bicucculline application to cortical FM–FM neurons does not change their response pattern noticeably, but broadens their delay tuning in 50% of neurons studied (Fig. 8B; Butman, 1992; Butman and Suga, 1995). This indicates that sharpening of delay tuning takes place by lateral inhibition in at least 50% of the neurons within the FM–FM area of the auditory cortex, not through the reticular nucleus. Since the response properties of thalamic and cortical FM–FM neurons are sculptured by the thalamocortical complex, a delay-tuning curve is similar or the same between thalamic and cortical delay-tuned neurons (Olsen and Suga, 1991).

A focal weak electrical stimulation of the FM–FM area in the auditory cortex evokes either facilitation (Fig. 9C, D) or inhibition (Fig. 9E) of the facilitative responses of thalamic FM–FM neurons to FM$_1$–FM$_n$ stimuli or to pulse-echo stimuli. When the electrical stimulus is strengthened, its effect changes from facilitation to inhibition, or inhibition to facilitation, or inhibition to facilitation then inhibition (Teng and Suga, 1994). These data and the data obtained from bicucculline applications to the cortex can be interpreted as follows. A neuron in a cortical column of the FM–FM area, which receives an input from a thalamic FM–FM neuron, projects back to that thalamic neuron and evokes NMDA-mediated facilitation. However, the cortical column inhibits neighboring cortical columns that have best echo delays different from its own best echo delay. When an electrical stimulus is delivered to a cortical column which is isotopic in best echo delay to a thalamic FM–FM neuron, the response of the thalamic neuron to a pulse-echo pair is facilitated. However, electrical stimulation of a cortical column that is anisotopic to the thalamic neuron inhibits the cortical column which is isotopic to that thalamic neuron (Fig. 3E). Consequently, it reduces the NMDA-mediated facilitative response of the thalamic neuron. Therefore, this electrical stimulation appears to evoke 'inhibition' of the facilitative response of the thalamic neuron to the pulse-echo pair. The amount of facilitation of the thalamic neuron evoked by electrical stimulation of its isotopic cortical column is larger than the amount of 'disfacilitation' evoked by electrical stimulation of anisotopic cortical columns on one side of the isotopic cortical column, but it is smaller than the amount of indirect inhibition evoked by strong electrical stimulation that stimulates anisotopic cortical columns on both sides of the isotopic cortical column.

An electrical stimulation of the auditory cortex evokes inhibition or facilitation of collicular neurons (Sun *et al.*, 1989). Since this is also true in the system that processes target-distance information, it is expected that inhibition or facilitation that occurs in the IC changes both the fast and slow components of the response of thalamic FM–FM neurons. Several critical experiments remain to be performed to test the validity of the neural net model proposed in this article.

Acknowledgements

Our research on the bat's auditory system has been supported by the research grants from ONR (N00014–90-J-1068), NIDCD (DC 00175) and the McKnight Foundation. Suga is particularly grateful to J. J. Wenstrup (Northeastern Ohio University) for discussions with him.

References

Berkowitz, A. and Suga, N. (1989). Neural mechanisms of ranging are different in two species of bats. *Hearing Res.* **41**: 255–264.

Butman, J. A. (1992). Synaptic mechanisms for target ranging in the mustached bat. Washington University Ph.D. thesis (unpublished).

Butman, J. A. and Suga, N. (1995). Inhibitory mechanisms for delay tuning of FM–FM combination-sensitive neurons in the auditory cortex and thalamus of the mustached bat. *J. Neurophysiol.* (in press).

Covey, E. and Casseday, J. H. (1991). The monaural nuclei of the lateral lemniscus in an echolocating bat: parallel pathways for analyzing temporal features of sound. *J. Neurosci.* **11:** 3456–3470.

Dear, S. P., Simmons, J. A. and Fritz, J. (1993). A possible neural basis for representation of acoustic scenes in auditory cortex of the big brown bat. *Nature* **364:** 620–623.

Edamatsu, H. and Suga, N. (1993). Differences in response properties of neurons between two delay-tuned areas in the auditory cortex of the mustached bat. *J. Neurophysiol.* **69:** 1700–1712.

Edamatsu, H., Kawasaki, M. and Suga, N. (1989). Distribution of combination-sensitive neurons in the ventral fringe area of the auditory cortex of the mustached bat. *J. Neurophysiol.* **61:** 202–207.

Feng, A. S., Simmons, J. A. and Kick, S. A. (1978). Echo detection and target-ranging neurons in the auditory system of the bat *Eptesicus fuscus*. *Science* **202:** 645–648.

Grinnell, A. D. (1963). The neurophysiology of audition in bats: temporal parameters. *J. Physiol.* **167:** 67–96.

Hattori, T. and Suga, N. (1989). Latency map in the inferior colliculus of the mustached bat. *Abstr. Assoc. Res. Otolaryngol.* Vol. 94, p. 1989.

Kawasaki, M., Margoliash, D. and Suga, N. (1988). Delay-tuned combination-sensitive neurons in the auditory cortex of the vocalizing mustached bat. *J. Neurophysiol.* **59:** 623–635.

Mittmann, D. H. and Wenstrup, J. J. (1994). Combination-sensitive neurons in the inferior colliculus of the mustached bat. *Abstr. Assoc. Res. Otolaryngol.* Vol. 371, p. 93.

Olsen, J. F. (1986). Functional organization of the medial geniculate body of the mustached bat. Ph.D. thesis, Washington University (unpublished).

Olsen, J. F. and Suga, N. (1991). Combination-sensitive neurons in the medial geniuiate body of the mustached bat: encoding of target range information. *J. Neurophysiol.* **65:** 1275–1296.

O'Neill, W. E. (1985). Responses to pure tones and linear FM components of the CF–FM biosonar signal by single units in the inferior colliculus of the mustached bat. *J. Comp. Physiol.* **157:** 797–815.

O'Neill, W. E. and Suga, N. (1979). Target range-sensitive neurons in the auditory cortex of the mustache bat. *Science* **203:** 69–73.

O'Neill, W. E. and Suga, N. (1982). Encoding of target-range information and its representation in the auditory cortex of the mustached bat. *J. Neurosci.* **47:** 225–255.

O'Neill, W. E., Holt, J. R. and Gordon, M. (1992). Responses of neurons in the intermediate and ventral nuclei of the lateral lemniscus of the mustached bat to sinusoidal and pseudorandom amplitude modulations. *Abstr. Assoc. Res. Otolaryngol.* Vol. 140, p. 418.

Park, T. J. and Pollak, G. D. (1993). GABA shapes a topographic organization of response latency in the mustached bat's inferior colliculus. *J. Neurosci.* **13:** 5172–5187.

Saitoh, I. and Suga, N. (1992). Effects of inhibitory amino acid antagonists on delay tuning of FM–FM neurons of the mustached bat. *Abstr. Assoc. Res. Otolaryngol.* Vol. 141, p. 421.

Suga, N. (1964a). Recovery cycles and responses to frequency modulated tone pulses in auditory neurons of echolocating bats. *J. Physiol.* **175:** 50–80.

Suga, N. (1964b). Single unit activity in cochlear nucleus and inferior colliculus of echolocating bats. *J. Physiol.* **172:** 449–474.

Suga, N. (1965a). Responses of cortical auditory neurons to frequency modulated sounds in echolocating bats. *Nature* **206:** 890–891.

Suga, N. (1965b). Analysis of frequency modulated sounds by neurons of echolocating bats. *J. Physiol.* **179:** 26–53.

Suga, N. (1965c). Functional properties of auditory neurons in the cortex of echolocating bats. *J. Physiol.* **181:** 671–700.

Suga, N. (1968). Analysis of frequency-modulated and complex sounds by single auditory neurons of bats. *J. Physiol.* **198:** 51–80.

Suga, N. (1969). Classification of inferior collicular neurons of bats in terms of responses to pure tones, FM sounds and noise bursts. *J. Physiol.* **200:** 555–574.

Suga, N. (1990). Cortical computational maps for auditory imaging. *Neural Networks* **3:** 3–21.

Suga, N. and Horikawa, J. (1986). Multiple time axes for representation of echo delays in the auditory cortex of the mustached bat. *J. Neurophysiol.* **55:** 776–805.

Suga, N. and O'Neill, W. E. (1979). Neural axis representing target range in the auditory cortex of the mustached bat. *Science* **206:** 351–353.

Suga, N. and Schlegel, P. (1973). Coding and processing in the auditory systems of FM-Signal-producing bats. *J. Acoust. Soc. Am.* **54:** 174–190.

Suga, N., O'Neill, W. E. and Manabe, T. (1978). Cortical neurons sensitive to combininations of information-bearing elements of biosonar signals in the mustache bat. *Science* **200**: 778–781.

Suga, N., O'Neill, W. E., Kujirai, K. and Manabe, T. (1983). Specificity of combination-sensitive neurons for processing of complex biosonar signals in the auditory cortex of the mustached bat. *J. Neurophysiol.* **49**: 1573–1626.

Suga, N., Olsen, J. F. and Butman, J. A. (1990). Specialized subsystems for processing biologically important complex sounds: cross-correlation analysis for ranging in bat's brain. *Cold Spring Harbor Symp. Quant. Biol.* **55**: 585–597.

Sullivan, W. E., III (1982a). Neural representation of target distance in auditory cortex of the echolocating bat Myotis Lucifigus. *J. Neurophysiol.* **48**: 1011–1032.

Sullivan, W. E., III (1982b). Possible neural mechanisms of target distance coding in auditory system of the echolocating bat *Myotis lucifigus*. *J. Neurophysiol.* **48**: 1033–1047.

Sun, X., Jen, P. H. S., Sun, D. and Zhang, S. (1989). Corticofugal influences on the responses of bat inferior collicular neurons to sound stimulation. *Brain Res.* **495**: 1–8.

Teng, H. and Suga, N. (1994). Cortico-thalamic control: Effect of electrical stimulation of the cortical FM–FM area on responses of thalamic FM–FM neurons in the mustached bat. *Abstr. Assoc. Res. Otolaryngol.* p. 86.

Winer, J. A., Wenstrup, J. J. and Larue, D. T. (1992). Patterns of GABAergic immunoreactivity define subdivisions of the mustached bat's medial geniculate body. *J. Comp. Neurol.* **319**: 172–190.

Wenstrup, J. J. and Grose, C. D. (1993). Inputs to comination-sensitive neurons in the medial geniculate body of the mustached bat. *Neurosci. Abstr.* Vol. 19, p. 1426.

Wong, D. (1991). Cellular organization of the cat's auditory cortex. In *Neurobiology of Hearing: The Central Auditory System* (R. A. Altschuler *et al.*, eds), pp. 367–387. Raven Press, New York.

Wong, D., Maekawa, M. and Tanaka, H. (1992). The effect of pulse repetition rate on the delay sensitivity of neurons in the auditory cortex of the FM bat *Myotis lucifugus. J. Comp. Physiol.* **A170**: 393–402.

Yan, J. and Suga, N. (1995). Differences in response properties between collicular and thalamic delay-tuned neurons of the mustached bat. *J. Neurophysiol.* Submitted.

The Olivocochlear Feedback Gain Control Subsystem: Ascending Input from the Small Cell Cap of the Cochlear Nucleus?

D. O. KIM, K. PARHAM, H. ZHAO AND S. GHOSHAL

Division of Otolaryngology, Surgical Research Center, Center for Neurological Sciences, University of Connecticut Health Center, Farmington, CT 06030-1110, USA

Introduction

Cochlear mechanics was postulated to be active by Gold in 1948. In more recent years, there have been considerable controversies about cochlear mechanics being linear (e.g. Békésy, 1960; Evans and Wilson, 1975) vs. non-linear (Rhode, 1971), passive vs. active (e.g. Kemp, 1978; Zwicker, 1979; Kim *et al.*, 1980), and exhibiting broad tuning vs. sharp tuning (e.g. Khanna and Leonard, 1982; Sellick *et al.*, 1982; Robles *et al.*, 1985). A current view accepted by many investigators is that the outer hair cells (OHCs) of the mammalian cochlea underlie cochlear biomechanical amplification (e.g. Mountain, 1980; Siegel and Kim, 1982; Davis, 1983; Neely and Kim, 1983), rendering the cochlear mechanics active, nonlinear and sharply tuned. The motility of the OHCs (Brownell *et al.*, 1985; Zenner *et al.*, 1985) is postulated to underlie the mechanically amplifying function of the OHCs. The existence of descending projections of olivocochlear (OC) neurons located in the brain stem to the cochlea (Rasmussen, 1960), more specifically the medial and lateral OC (MOC and LOC) neurons to the OHCs and type I afferent fibers beneath the inner hair cells (IHCs), respectively (e.g. Warr and Guinan, 1979; Warr, 1992), is generally accepted.

It was hypothesized that the OHC-MOC and IHC-LOC neurons (together with their associated intervening neurons) form two distinct

31

parallel subsystems, each of which comprises a brain-stem reflex circuit subserving a separate feedback control subsystem (Kim, 1984). It was further hypothesized that the function of the OHC subsystem is *not* to transmit the usual auditory information to the brain but to provide gain control for the active non-linear cochlear biomechanics, and that the function of type II cochlear afferent neurons is to transmit information about the operating point associated with the motor function of the OHCs. The ascending portions of such reflex circuits, regarding the regions and cell types of the cochlear nucleus (CN) projecting to the OC neurons, are presently much more poorly understood than the descending portions. The former information is only beginning to be elucidated (e.g. Thompson and Thompson, 1991).

One candidate for a source of the ascending input to the MOC neurons is the small cell cap (SCC) of the CN. The SCC, described by Osen (1969), is also called the juxtagranular area (Liberman, 1993). The rationale for hypothesizing this is as follows. The MOC neurons are postulated to provide an automatic gain control on the cochlear biomechanics through a negative feedback on the OHCs' mechanical amplifying action (e.g. Kim, 1984) as shown schematically in Fig. 1. In this model, the OHCs' action provides a high gain at low stimulus intensity and, as the intensity is increased, the negative feedback from the MOC neurons decreases the gain of the cochlear biomechanical amplifier such that the cochlear output amplitude is kept in a relatively narrow range. In order for the MOC neurons to achieve such a feedback-control role, they need accurate information about stimulus intensity. Therefore, a requirement for the ascending input from the CN to the MOC neurons is an ability to convey accurate stimulus intensity information as depicted in Fig. 1.

Most of the CN neurons are believed to receive inputs from the auditory nerve (AN) fibers without MOC collateral inputs. A certain small population of CN neurons, i.e. cells closely associated with the granule cell layer (GCL) (Osen *et al.*, 1984), such as multipolar cells beneath the GCL extending their dendrites into the GCL and possibly granule cells (GCs) (Brown *et al.*, 1988b; Benson and Brown, 1990; Brown, 1993); are believed to receive input from MOC collaterals. Liberman (1991, 1993) observed that the SCC receives input preferentially from low spontaneous rate (SR) AN fibers. Since the SCC is in general subjacent and closely associated with the GCL, we postulate that the SCC neurons receive both MOC collaterals and low-SR AN inputs. Figure 2 depicts these pathways schematically. One consequence of the automatic gain control of the cochlear biomechanics is that the cochlear output is not an accurate indicator of the absolute stimulus intensity. The range of the cochlear output amplitude is

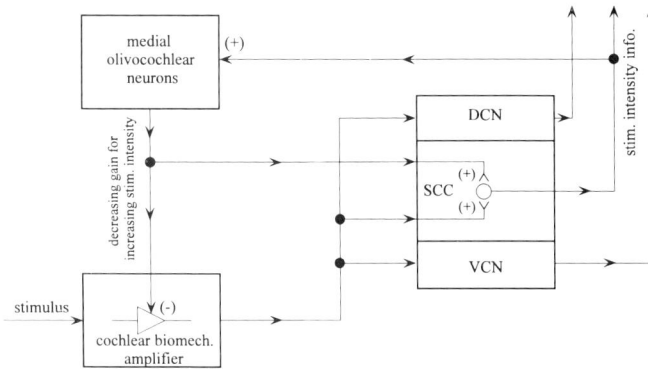

Fig. 1. A block diagram of the MOC feedback system and automatic gain control of the cochlear biomechanics.

reduced to a value much smaller than that of the stimulus by the gain-control action. The MOC collateral inputs to the VCN are suggested to be excitatory based on the synaptic morphology (Benson and Brown, 1990). Physiologically, the effects of acetylcholine (ACh) on VCN neurons, when detectable, are mainly excitatory (Caspary *et al.*, 1983). Reported observations of the effects of electrically stimulating the OC neurons on CN neurons are variable (e.g. Comis and Whitfield, 1968; Starr and Wernick, 1968). Since the collateral fibers of OC neurons entering the VCN are positive for acetylcholinesterase (Osen and Roth, 1969), and since MOC neurons are believed to use ACh as a neurotransmitter (e.g. Altschuler and Fex, 1986), Caspary *et al.*'s (1983) ACh observation also supports the view that the MOC collateral inputs to the VCN are mainly excitatory. From a theoretical view point, we believe that *excitatory* inputs from MOC collateral to the VCN (as depicted in Figs 1 and 2) are *necessary*. We expect that certain CN neurons, such as SCC neurons, that receive MOC collateral and AN inputs are better able to convey stimulus intensity information than the bulk of the CN neurons that do not receive MOC collateral input. Benson and Brown (1990) made a similar suggestion.

About 95% of the AN fibers, called type I, are believed to originate from the IHCs and 5%, called type II, from the OHCs (Spoendlin, 1972). The type I fibers are subdivided into two or three spontaneous rate (SR) groups (e.g. Liberman, 1978; Kim and Molnar, 1979). If they are grouped into two with a border at 20 spikes/sec, which corresponds to the middle of a broad trough in the bimodal distribution of SR, the

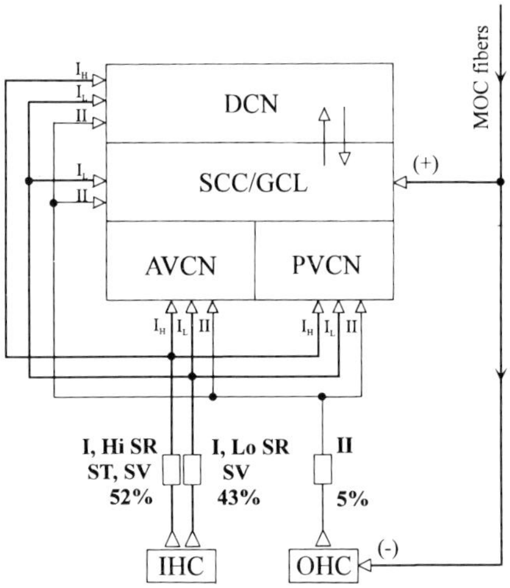

Fig. 2. A schematic wiring diagram of the two types of cochlear hair cells, three types of ganglion cells, subdivisions of the CN and MOC fibers. 'SV' and 'ST', which stand for scala vestibuli and tympani, respectively, are intended to denote that low-SR ganglion cells tend to be on the SV side and high-SR cells on both ST and SV sides.

proportions of low- and high-SR groups among the total AN fibers are about 43 and 52%, respectively. Figure 2 depicts these AN fiber groups schematically.

The expectation that the SCC neurons receive preferentially low-SR AN fiber input would also favor the predicted role of SCC neurons in conveying stimulus intensity information over a wider dynamic range, because the low-SR AN fibers exhibit a wider dynamic range than high-SR fibers (Sachs and Abbas, 1974). The GCL and the adjoining area (which we associate with the SCC) are targets of the type II AN fibers originating from the OHCs (Brown *et al.*, 1988a; Berglund and Brown, 1994) as depicted in Fig. 2. The type II AN input to the SCC further implicates the SCC in an OHC-MOC reflex circuit.

The projections from the principal or large-cell *core* areas of the CN have been extensively studied. The SCC *shell* area surrounding the VCN core, by contrast, has not received as much attention in the literature so far. We believe that the SCC is emerging as an important region of the CN and needs to be further investigated. The goal of this paper is to examine theoretical implications of the SCC neurons that are postulated to take part in the OHC-MOC reflex circuit and to investigate the projections from and to the SCC using neuroanatomical tract-tracing methods.

Methods for Neuroanatomical Study

Focal injections of tracer were made using a Picospritzer (several pulses of 30–100 PSI, 10–100 msec) and a micropipette (tip, about 20 mm) positioned in a superficial area (depth, about 200 μm) in the left cat anteroventral cochlear nucleus (AVCN) including the GCL and SCC. The tracer was biotinylated dextran (BD), 10–20%, alone or in a mixture with tritiated [3H]leucine (50 μCi/μl) and/or dextran tetramethyl rhodamine (10%). The volume of the tracer injected was about 0.1 μl. Prior to placing an injection pipette, compound neural responses were recorded with another pipette filled with a common electrolyte (e.g. 1 M NaCl) to guide the placement of the injection pipette. The characteristic frequency (CF) of the injection site was determined by recording compound neural responses through the injection pippette.

The tracer was allowed to be transported over a survival period of 5–10 days. Whereas [³H]leucine is known to be anterogradely transported (Cowan *et al.*, 1972), we found that BD was transported both antero- and retrogradely. About 24 hr prior to perfusing the cat, horseradish peroxidase (HRP) was injected either into incisions at the floor of the fourth ventricle or into the cochlea through the round window for the purpose of labeling OC neurons. The cats were perfused transcardially with a fixative (4% paraformaldehyde, 0.2% gluteraldehyde). Brain stem tissue was frozen-sectioned at 50–75 μm. The sections were initially processed for HRP using diaminobenzidine (DAB) (Mesulam, 1982) without metal intensification, yielding reddish-brown reaction product. The same sections were subsequently processed for BD using an avidin–biotin complex (ABC) procedure and DAB with metal intensification (Veenman *et al.*, 1992), yielding black-gray reaction product. [³H]Leucine labels were examined using autoradiography with an exposure time of 4–9 weeks.

Results

In results from several cats with focal injections of BD restricted in the superficial AVCN, we have found BD-labeled fibers and swellings in a number of auditory regions of the brain stem (Kim *et al.*, 1994a). In cases where BD was injected in a mixture with [³H]leucine, some of the fibers in the brain stem auditory pathways were double labeled with BD and [³H]leucine. Since [³H]leucine is known to be a specific anterograde tracer, the double label is regarded as an anterograde transport from the injection site. These results suggest that the SCC projects to the following regions: DCN, ipsilaterally; the periolivary areas including

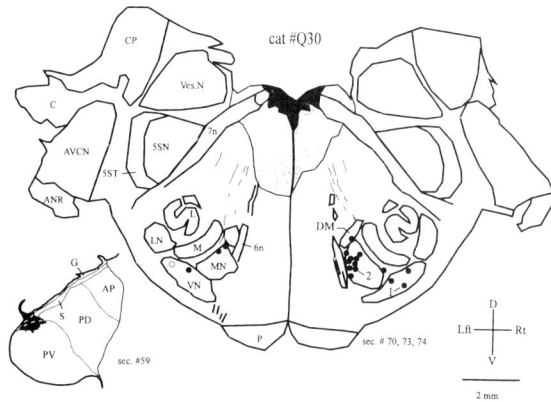

Fig. 3. Transverse sections of the brain stem of cat (#Q30). BD injection was in superficial AVCN as shown by a black area in the lower left inset. HRP was injected into incisions made at the midline and slightly off the midline in an area bordering the floor of the fourth ventricle; the black and gray areas represent the HRP injection site and an adjoining fringe. The small closed circles represent MOC neurons. One small open circle in the left VN is a BD-filled cell. These cells were from three sections (#70, 73 and 74), 50 μm each. Two MOC cells marked '1' and '2' are illustrated at a higher magnification in Fig. 4. 5ST, spinal trigeminal tract; 5SN, spinal trigeminal nucleus; 6n, 6th (abducens) nerve root; 7n, 7th (facial) nerve root; ANR, auditory nerve root; C, cerebellum; CP, cerebellar peduncle; D, dorsal; DM, dorsomedial periolivary nucleus; G, granule cell layer; L, LSO; Lft, left; LN, LNTB; M, MSO; MN, MNTB. Rt, right; S, small cell cap; V, ventral; Ves. N, vestibular nuclei; VN, VNTB.

ventral and medial nuclei of trapezoid body (VNTB and MNTB) which are known to contain MOC neurons, bilaterally; ventral and dorsal nuclei of lateral lemniscus (VNLL and DNLL), mainly contralaterally; and inferior colliculus (IC), mainly contralaterally. BD label was found in fibers and swellings as well as in cell bodies in an overlapping area within the DCN of these cats. Since the BD label in the cell bodies indicates a retrograde transport, the results suggest that the SCC and DCN are reciprocally connected (depicted by the double arrows in Fig. 2).

Fig. 4 (opposite). High-magnification descriptions of two MOC cells (Marked '1' and '2' in Fig. 3). The color plates are photographs obtained with a ×63 oil objective lens and the line drawings are camera lucida drawings obtained with a ×100 oil objective lens. In the color plates, the arrowheads point to BD-labeled swellings in close apposition with the HRP-labeled MOC cell soma (a) or dendrite (d); the arrows point to faint BD-labeled fibers connecting the swellings. The BD and HRP labels are distinguished by the color and darkness. In (d), the HRP-filled thick fiber running across the MOC cell 2 is interpreted to be an axon of another MOC cell.

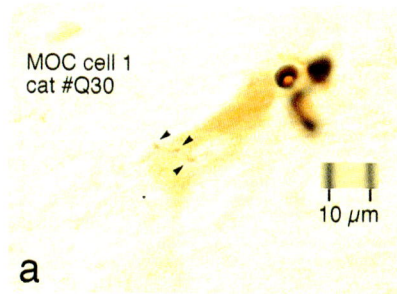

MOC cell 1
cat #Q30

10 μm

a

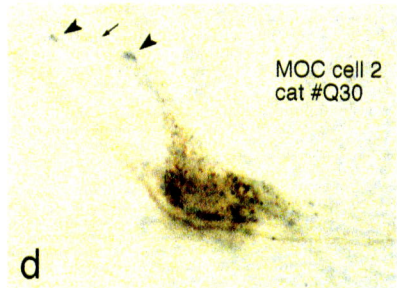

MOC cell 2
cat #Q30

d

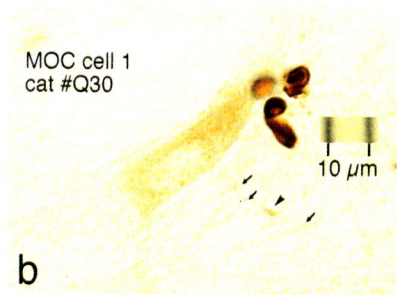

MOC cell 1
cat #Q30

10 μm

b

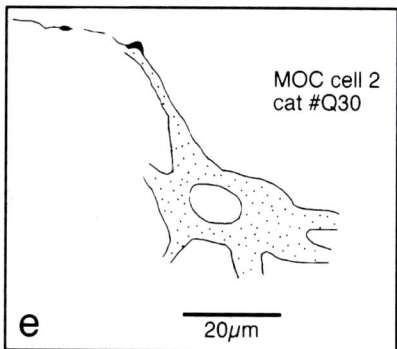

MOC cell 2
cat #Q30

20μm

e

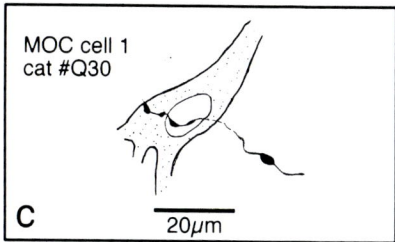

MOC cell 1
cat #Q30

20μm

c

Figures 3 and 4 describe the results from one cat, #Q30. The inset in the lower left part of Fig. 3 describes the injection site in a transverse section of the left AVCN where BD was injected. The injection site was relatively restricted in a superficial area of the AVCN that included GCL, SCC and a slight encroachment in the extreme dorsolateral edge of the posteroventral (PV) part of the AVCN. The terminology of the CN subdivisions is from Brawer *et al.* (1974). The main part of Fig. 3 represents a transverse section of the same cat's brain stem describing the area of HRP injection bordering the floor of the fourth ventricle and labeled MOC neurons (depicted by small solid circles). The identity of the MOC neurons is based on the cells being labeled with HRP and being located in an area ventromedial to the medial superior olive (MSO), marked by 'M'. Occasionally, BD-labeled cells were found in the periolivary region; one such example is shown by a small open circle in the left VNTB (marked 'VN') in Fig. 3. It is possible that some of these BD-labeled cells in the periolivary regions are MOC neurons that were labeled through their collateral endings at the injection site in the SCC/GCL area. Thin line segments in Fig. 3 between the HRP injection site and MOC neurons represent HRP-filled axons, which are interpreted to be axons of MOC neurons.

We have observed that HRP-filled (reddish-brown) MOC cells and BD-filled (black-grey) axons/swellings come in close apposition. Examples of such double-labeled cases, two MOC cells marked '1' and '2' in Fig. 3(b), are shown in Fig. 4. Figures 4(a) and (b) are color photomicrographs showing MOC cell 1. Three BD-labeled swellings (arrowheads) are in close apposition with the MOC cell soma (4a) and, at a different focal plane (4b), another BD-labeled swelling (arrow head) is present nearby. Figure 4(c) is a camera lucida drawing which represents superimposed images of the cell and the BD-labeled swellings and a connecting fiber (arrows in Fig. 4b) viewed at various focal planes. Figures 4(d) and (e) are descriptions for MOC cell 2 where the arrowheads point to the two BD-labeled swellings, one of which is in close apposition with a dendrite of the MOC cell; the arrow points to a fiber connecting the swellings. These findings support the hypothesis that the SCC projects to the MOC neurons. Caution should be taken, however, because the source of some of the BD-labeled swellings in cat #Q30 may be neurons located in the PV subdivision, because the injection site encroached into PV slightly.

In cats where the injection site of BD was well restricted in the superficial GCL and SCC areas, most of the BD-labeled fibers were quite thin, ranging from 0.3 to 1.5 µm in diameter. If the SCC neurons convey stimulus intensity information as postulated in the Introduction, the thin diameters of the fibers imply that the SCC neurons convey the intensity information slowly with a long time constant, perhaps tens (or hundreds) of milliseconds. We believe that the long time constant is desirable in the gain-control operation of the OHC-MOC feedback subsystem, because a relatively slow adjustment of the cochlear amplifier gain over a number of cycles of a signal is more desirable than an adjustment made cycle by cycle. MOC efferent units were observed to exhibit a broad range of latencies, about 5–40 msec (Liberman, 1988). The shorter latencies of the MOC fibers might imply that the MOC neurons receive some fast ascending inputs which might arise from non-SCC CN neurons.

In cats injected with BD in the superficial AVCN, such as that represented in Figs 3 and 4, we found BD-labeled fibers in the vestibular nerve and its root (VN and VNR) ventral to the anterior part of the AVCN. It is possible that some of these fibers (with diameters of about 2.5 μm) are MOC fibers that picked up the BD label via their collateral endings at the injection site. This possibility is consistent with the MOC fibers' known diameters (Arnesen and Osen, 1984) and trajectories that include the VNR *en route* to the cochlea (Liberman and Brown, 1986). We also found BD label in granule cells (GCs) and small stellate cells (SSCs) embedded in the middle of the VNR, indicating that these cells project to the SCC/GCL area. In addition, there were numerous thin BD-labeled fibers in the VNR. These fibers and cells appeared continuous with a *shell* of SCC/GC area in the AVCN containing BD-labeled thin fibers, GCs and SSCs. This finding raises a question of whether (or how) these cells of the VNR may mediate an integration of vestibular and auditory information.

We have examined the cochleas of some of the cats injected with BD in the superficial AVCN and found that BD label was present in central and peripheral axons and somata of cochlear ganglion cells and a few fibers in the intraganglionic spiral bundle. The latter are interpreted to be MOC fibers based on the previous finding of MOC fibers in that location (Rassmussen, 1953; Liberman and Brown, 1986). BD-labeled ganglion cells were considered mostly type I cells judging by the shape, size and the ratio of the diameters of the peripheral and central processes near the soma (Kiang *et al.*, 1982). BD-labeled ganglion cells did include type II cells, some of which were recognized by the pseudomonopolar shape (Kiang *et al.*, 1982). The presence of BD label in type II ganglion cells and putative MOC fibers observed in our study supports the notion that the SCC/GCL receives inputs from the two sources as depicted in Fig. 2. Since the SCC receives preferential input from low-SR AN fibers, and since low-SR cochlear ganglion cells are preferentially on the scala vestibuli side (Kawase and Liberman, 1992), it is predicted that the BD label should be preferentially on the SV side.

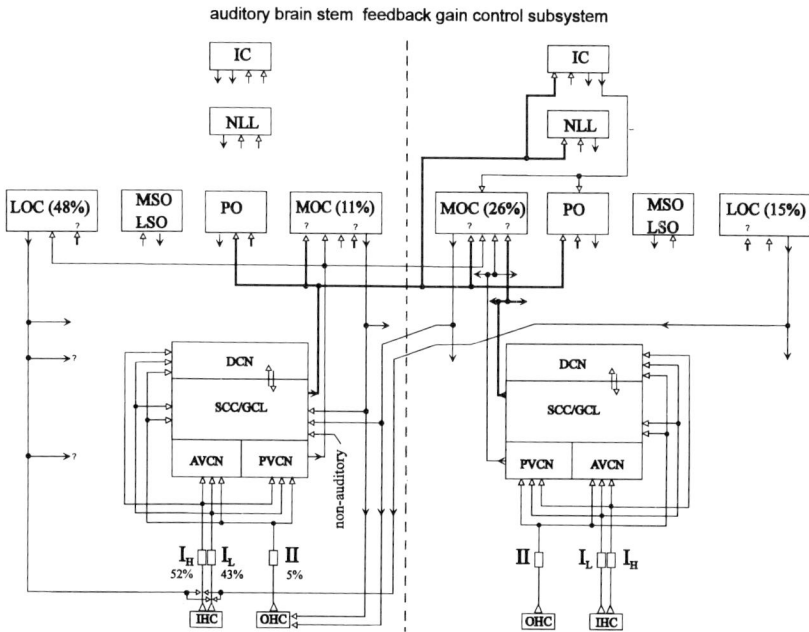

Fig. 5. A block diagram of the feedback gain control subsystem of the auditory brain stem between the cochlea and the midbrain.

Preliminary results of our study on this subject were supported by Kim *et al.* (1994b).

We have begun single-unit physiological recordings in the SCC region of the cat. One finding is that the units recorded in a superficial region (depth 0.4 mm) of the AVCN tended to exhibit low SR, typically less than 5 spikes/sec. Physiological response properties of these units, e.g. poststimulus histogram type, excitatory–inhibitory response area type, discharge rate vs. stimulus level behavior, are currently under investigation.

Discussion

We are interested in gaining an overall understanding of the relationships among many parts of the feedback gain control subsystem of the auditory brain stem. Figure 5 is a schematic wiring diagram depicting the auditory system between the cochlea and the midbrain representing the present hypotheses. In this diagram, we have chosen to omit the connections of the large-cell core parts of the auditory pathways in order to describe details of the parts involving the SCC and other parts postulated to take part in the feedback gain control subsystem. Figure 5 retains a hypothesis proposed by Kim (1984) that the OHCs, type II AN fibers, a part of type I AN fibers and a special subset of CN neurons provide stimulus intensity information to the MOC neurons, forming a biomechanical gain control feedback subsystem. Figure 5 is an updated version which specifies that the SCC/GCL and low-SR AN fibers (marked 'I_L' in Fig. 5) are postulated to be the corresponding parts of the OHC-MOC subsystem. The projections from the SCC, highlighted with bold lines, are based on our observations in cats as described above. The percentages of the four groups of OC neurons are based on Warren and Liberman (1989). Some elements of the hypotheses depicted in Fig. 5 were suggested by other investigators. For example, Warr (1992, Fig. 7.4) showed a schematic diagram describing a hypothesis that CN multipolar cells beneath the GCL, which extend dendrites into the GCL and receive inputs from type II AN fibers, MOC collaterals and low-SR type I AN fibers, may project to the MOC and LOC neurons. The present hypothesis (Figs 2 and 5) is similar to that of Warr.

We consider that the following are possible mechanisms whereby MOC neurons can modulate the cochlear amplifier. Discharges of MOC neurons would release ACh as neurotransmitter at the MOC-OHC synapse (e.g. Altschuler and Fex, 1986). Under such a condition, the OHC is generally expected to be hyperpolarized inferring from *in vitro* observations (Housley and Ashmore, 1991). From the polarity of the OHC motility, which is well agreed by many studies reported, a hyperpolarization is expected to be accompanied by an extension of the OHC length or no detectable change if the membrane potential is in a extension-saturation region. Actual *in vitro* observations of ACh effects on the OHC length are variable: a contraction (Brownell *et al.*, 1985); a contraction followed by a return to, or an extension beyond the original length (Slepecky *et al.*, 1988; Plinkert *et al.*, 1990). The reason for these variable effects is not clear now. Perhaps, an initial Ca^{2+} influx through an ACh-activated cation channel followed by a K^+ efflux through a Ca^{2+}-activated K^+ channel (Fuchs and Murrow, 1992; Doi and Ohmori, 1993) underlie the variable and sometimes biphasic length change of the OHC. The polarities and time courses of OHC length change by ACh should be further clarified in future studies. Since the MOC signal to the OHC is not expected to change rapidly, e.g. over one period of the characteristic frequency (CF), but slowly, e.g. over many periods of CF, the change in OHC potential/length arising from MOC discharges is expected to be determined by a slow motile mechanism of the OHC.

The OHC 'motility slope gain' (in nm/mV) changes as a function of the operating point of the OHC potential/length since the length vs. voltage function is non-linear and sigmoidal (Evans *et al.*, 1991; Santos-Sacchi, 1991) such that the motility gain is reduced when the OHC resting potential is hyperpolarized. In this regard, an increase in the discharges of MOC neurons is expected to decrease the OHC motility gain. The resting values of the membrane potential, cell length and turgor (or osmotic) pressure of the OHC may be interdependent. The OHC turgor pressure influences the OHC motility gain (Santos-Sacchi, 1991). The MOC neurons may control the OHC turgor pressure (e.g. Patuzzi, 1995) besides the membrane potential and cell length, and through these variables control the gains of the OHC motility and the cochlear amplifier. The observations that electrical stimulation of OC fibers affected distortion-product otoacoustic emissions (Mountain, 1980; Siegel and Kim, 1982) and basilar-membrane motion (Dolan and Nuttal, 1992) provide a general support for the present notion.

Regarding Fig. 5, the projections from the PVCN to the MOC (bilateral) and LOC (ipsilateral) are based on the finding of Thompson and Thompson (1991) in the guinea-pig. The cell type and the specific location within the PVCN that give rise to the projections to the OC neurons in the Thompson and Thompson study are not known. The concept of ascending projection from the SCC to the IC is based on our present anterograde observations as well as the retrograde observations of Adams (1979) and Oliver (1987) following HRP injection into the IC. The projections from the SCC to the IC and NLL, depicted in Fig. 5, are thought to convey stimulus intensity information to these upper brain-stem nuclei which is ultimately expected to be conveyed to the cerebral cortex. Compared with the MOC subsystem, possible functional roles played by the LOC subsystem and the ascending inputs to LOC neurons are much less understood.

A descending projection from the IC to the MOC neurons was suggested by a physiological study of Rajan (1990), where stimulation of the IC was observed to reduce a temporary threshold shift of the cochlea exposed to a loud sound. The projection from the IC to the MOC neurons was anatomically confirmed by Thompson and Thompson (1993) and Vetter *et al.* (1993). The IC-to-MOC projection is expected to allow the upper brain-stem and more rostral areas of the brain to modify the cochlear amplifier via the MOC neurons. The observations that the acoustic startle reflex amplitude was increased (Davis *et al.*, 1986) and the N_1 compound action potential of the AN was increased (Meloni and Davis, 1993) when the animal was injected with a dopamine agonist are interpreted to be a result of increased cochlear response arising from inhibitions of MOC neurons by dopamine-sensitive neurons in higher brain centers. Thus, these observations are possible examples where higher centers of the central nervous system modify the gain of the cochlear biomechanical amplifier via the MOC feedback pathway.

Besides the projections to MOC neurons, another way that the central nervous system can influence the auditory pathways is through projections to the CN. Non-auditory structures are known to project to the CN. For example, somatosensory dorsal column nuclei and spinal trigeminal nuclei project to the CN (Itoh *et al.*, 1987; Weinberg and Rustioni, 1987; Young *et al.*, 1993; Wright and Ryugo, 1994). The GCL and adjacent areas of the VCN are a target of the somatosensory input. As depicted in Fig. 5, such non-auditory inputs may allow the central nervous system to modify auditory signal processing, including a change in the cochlear amplifier gain via the SCC/GCL-MOC pathway. Some non-auditory inputs may be initially onto the GCs, which may subsequently influence the SCC neurons.

The OC feedback gain control of the cochlear amplifier postulated here has significant implications on the binaural processing that is believed to take place in the superior olivary complex (SOC). For example, as discernible in Figs 1 and 5, the intensity information carried in the ascending pathways of the large-cell CN core subdivisions (i.e. DCN, AVCN and PVCN) is affected by the amount of feedback adjustment on the cochlear amplifier gain. Thus, the inputs to the lateral and medial superior olive (LSO and MSO) from each ear would not be totally determined by the stimulus to one ear alone, but would be substantially modified by the stimulus applied to the opposite ear through binaural actions of MOC neurons (Liberman, 1988), which in general would be changing dynamically. Thus, the binaural processing system should be viewed to include not only the LSO/MSO but also the MOC–cochlear amplifier–CN.

Another implication of the feedback gain control on the binaural localization processing is that, if this gain adjustment were made unilaterally in each ear independent of the opposite ear, the interaural intensity information supplied to the LSO would be disrupted, leading to an error in localizing a sound source. Therefore, the cochlear amplifier gain control *must* be coordinated in the two ears. A possible design that satisfies such a requirement is one where the same amount of gain change is applied to the two ears by ensuring that the total firing rates of MOC neurons projecting to the two ears are the same at any given time. Perhaps such a requirement for equalizing the cochlear amplifier gains in the two ears is the functional reason why most MOC neurons exhibit binaural facilitation and/or excitation (Liberman, 1988), implying that they receive binaural inputs (Thompson and Thompson, 1991), and why some MOC neurons project to the two cochleas (e.g. Thompson and Thompson, 1986).

Acknowledgements

This study was supported in part by the National Institute on Deafness and Other Communicative Disorders, National Institutes of Health of the US. We thank Drs D. K. Morest, D. L. Oliver and E. M. Ostapoff for advising us on the neuroanatomical methods for the brain stem, Dr M. C. Liberman and Ms L. Dodds for advising us on the methods of processing cochlear histology and Messrs L. LaBrie and J. Paparello for making comments on the manuscript.

References

Adams, J. C. (1979). Ascending projections to the inferior colliculus. *J. Comp. Neurol.* **183:** 519–538.

Altschuler, R. A. and Fex, J. (1986). Efferent neurotransmitters. In *Neurobiology of Hearing: The Cochlea* (R. A. Altschuler, R. P. Bobbin and D. W. Hoffman eds), pp. 383–396. Raven Press, New York.

Arnesen, A. R. and Osen, K. K. (1984). Fibre population of the vestibulocochlear anastomosis in the cat. *Acta Otolaryngol.* **98:** 255–269.

Békésy, G. von (1960). *Experiments in Hearing.* McGraw-Hill, New York.

Benson, T. E. and Brown, M. C. (1990). Synapses formed by olivocochlear axon branches in the mouse cochlear nucleus. *J. Comp. Neurol.* **295:** 52–70.

Berglund, A. M. and Brown, M. C. (1994). Central trajectories of type II spiral ganglion cells from various cochlear regions in mice. *Hear. Res.* **75:** 121–130.

Brawer, J. R., Morest, D. K. and Kane, E. C. (1974). The neuronal architecture of the cochlear nucleus of the cat. *J. Comp. Neurol.* **155:** 251–300.

Brown, M. C. (1993). Fiber pathways and branching patterns of biocytin-labeled olivocochlear neurons in the mouse brainstem. *J. Comp. Neurol.* **337:** 600–613.

Brown, M. C., Berglund, A. M., Kiang, N. Y. S. and Ryugo, D. K. (1988a).

Central trajectories of type II spiral ganglion neurons. *J. Comp. Neurol.* **278**: 581–590.

Brown, M. C., Liberman, M. C., T. E. Benson and Ryugo, D. K. (1988b). Brainstem branches from olivocochlear axons in cats and rodents. *J. Comp. Neurol.* **278**: 591–603.

Brownell, W. E., Bader, C. R., Bertrand, D. and de Ribaupierre, Y. (1985). Evoked mechanical responses of isolated cochlear outer hair cells. *Science.* **227**: 194–196.

Caspary, D. M., Havey, D. C., and Faingold, C. L. (1983). Effects of acetylcholine on cochlear nucleus neurons. *Exp. Neurol.* **82**: 491–498.

Comis, S. D. and Whitfield, I. C. (1968). Influence of centrifugal pathways on unit activity in the cochlear nucleus. *J. Neurophysiol.* **31**: 62–68.

Cowan, W. M., Gottlieb, D. I., Hendrickson, A., Price, J. L. and Woolsey, T. A. (1972). The autoradiographic demonstration of axonal connections in the central nervous system. *Brain Res.* **37**: 21–51.

Davis, H. (1983). An active process in cochlear mechanics. *Hear. Res.* **9**: 79–90.

Davis, M., Commissaris, R. L., Cassella, J. V., Yang, S., Dember, L. and Harty, T. P. (1986). Differential effects of dopamine agonists on acoustically and electrically elicited startle responses: Comparison to effects of strychnine. *Brain Res.* **371**: 58–69.

Doi, T. and Ohmori, H. (1993). Acetylcholine increases intracellular Ca^{2+} concentration and hyperpolarizes the guinea-pig outer hair cell. *Hearing Res.* **67**: 179–188.

Dolan, D. F. and Nuttal, A. L. (1992). Laser Doppler vibrometer measurements of guinea pig middle ear and basilar membrane motion. *Assoc. Res. Otolar. Meeting Abstr.* Vol. 15, p. 98.

Evans, B. N., Hallworth, R. and Dallos, P. (1991). Outer hair cell electromotility: the sensitivity and vulnerability of the DC component. *Hearing Res.* **52**: 288–304.

Evans, E. F. and Wilson, J. P. (1975). Cochlear tuning properties: concurrent basilar membrane and single nerve fiber measurements. *Science.* **190**: 1218–1221.

Fuchs, P. A. and Murrow, B. W. (1992). Cholinergic inhibition of short (outer) hair cells of the chick's cochlea. *J. Neurosci.* **12**: 800–809.

Gold, T. (1948). Hearing II. The physical basis of the action of the cochlea. *Proc. R. Soc. Lond. B* **135:** 492–498.

Housley, G. D. and Ashmore, J. F. (1991). Direct measurement of the action of acetylcholine on isolated outer hair cells of the guinea pig cochlea. *Proc. R. Soc. Lond. B* **244:** 161–167.

Itoh, K., Kamiya, H., Mitani, A., Yasui, Y., Takada, M. and Noburo, M. (1987). Direct projections from the dorsal column nuclei and the spinal trigeminal nuclei to the cochlear nuclei in the cat. *Brain Res.* **400:** 145–150.

Kawase, T. and Liberman, M. C. (1992). Spatial organization of the auditory nerve according to spontaneous discharge rate. *J. Comp. Neurol.* **319:** 312–318.

Kemp, D. T. (1978). Stimulated acoustic emissions from within the human auditory system. *J. Acoust. Soc. Am.* **64:** 1386–1391.

Khanna, S. M. and Leonard, D. G. B. (1982). Basilar membrane tuning in the cat cochlea. *Science* **215:** 305–306.

Kiang, N. Y. S., Rho, J. M., Northrop, C. C., Liberman, M. C. and Ryugo, D. K. (1982). Hair-cell innervation by spiral ganglion cells in adult cats. *Science.* **217:** 175–177.

Kim, D. O. (1984). Functional roles of the inner- and outer-hair-cell subsystems in the cochlea and brainstem. In *Hearing Science: Recent Advances* (C. I. Berlin, ed.), pp. 241–262. College Hill Press, San Diego, CA.

Kim, D. O. and Molnar, C. E. (1979). A population study of cochlear nerve fibers: comparison of spatial distributions of average-rate and phase-locking measures of responses to single tones. *J. Neurophys.* **42:** 16–30.

Kim, D. O., Neely, S. T., Molnar, C. E. and Matthews, J. W. (1980). An active cochlear model with negative damping in the partition: Comparison with Rhode's ante- and post-mortem observations. In *Psychophysical, physiological and behavioural studies in hearing* (G. van den Brink, and F. A. Bilsen, eds). Delft University Press, Noordwijkerhout, The Netherlands.

Kim, D. O., Parham, K., Sirianni, J. G. and Chang, S. O. (1991). Spatial response profiles of posteroventral cochlear nucleus neurons and auditory-nerve fibers in unanesthetized decerebrate cats: response to pure tones. *J. Acoust. Soc. Am.* **89:** 2804–2817.

Kim, D. O., Ghoshal, S., Zhao, H., Parham, K., Morest, D. K. and Oliver, D. L. (1994a). Anterograde/retrograde labeling of connections from/to the small cell cap region of the cat cochlear nucleus. *Assoc. Res. Otolaryngol. Meeting Abstr.* Vol. 17, p. 27.

Kim, D. O., Parham, K., Zhao, H. and Ghoshal, S. (1994b). Retrograde labeling of cochlear ganglion cells with injection of biotinylated dextran (BD) into granule-cell (GC) and small-cell-cap (SCC) regions of the cat anteroventral cochlear nucleus (AVCN). *Soc. Neurosci. Abstr.* Vol. 20.

Liberman, M. C. (1978). Auditory-nerve response from cats raised in a low-noise chamber. *J. Acoust. Soc. Am.* **63:** 442–455.

Liberman, M. C. (1988). Response properties of cochlear efferent neurons: monaural vs. binaural stimulation and the effects of noise. *J. Neurophysiol.* **60:** 1779–1798.

Liberman, M. C. (1991). Central projections of auditory-nerve fibers of differing spontaneous rate. I. Anteroventral cochlear nucleus. *J. Comp. Neurol.* **313:** 240–258.

Liberman, M. C. (1993). Central projections of auditory nerve fibers of differing spontaneous rate, II: Posteroventral and dorsal cochlear nuclei. *J. Comp. Neurol.* **327:** 17–36.

Liberman, M. C. and Brown, M. C. (1986). Physiology and anatomy of single olivocochlear neurons in the cat. *Hearing Res.* **24:** 17–36.

Meloni, E. G. and Davis, M. (1993). Enhancement of the compound action potential recorded from the cochlear nucleus of the rat following administration of D-amphetamine and apomorphine. *Soc. Neurosci. Abstr.* Vol. 19, p. 822.

Mesulam, M. (1982). *Tracing Neural Connections with Horseradish Peroxidase.* Wiley, New York.

Mountain, D. C. (1980). Changes in endolymphatic potential and crossed olivocochlear bundle stimulation alter cochlear mechanics. *Science* **210:** 71–72.

Neely, S. T. and Kim, D. O. (1983). An active cochlear model showing sharp tuning and high sensitivity. *Hearing Res.* **9:** 123–130.

Oliver, D. L. (1987). Projections to the inferior colliculus from the anteroventral cochlear nucleus in the cat: possible substrates for binaural interaction. *J. Comp. Neurol.* **264:** 24–46.

Osen, K. K. (1969). Cytoarchitecture of the cochlear nuclei in the cat. *J. Comp. Neurol.* **136:** 453–483.

Osen, K. K. and Roth, K. (1969). Histochemical localization of cholinesterases in the cochlear nuclei of the cat, with notes on the origin of acetylcholinesterase-positive afferents and superior olive. *Brain Res.* **16:** 165–185.

Osen, K. K., Mugnaini, E., Dahl, A. L. and Christiansen, A. H. (1984). Histochemical localization of acetylcholinesterase in the cochlear and superior olivary nuclei. A reappraisal with emphasis on the cochlear granule cell system. *Arch. Ital. Biol.* **122:** 169–212.

Patuzzi, R. (1995). Monitoring cochlear homeostasis with automatic analysis of the low-frequency cochlear microphonic. In *Active Hearing* (Å. Flock, D. Ottoson and M. Ulfendahl, eds), pp. 141–153 (this volume). Elsevier, Oxford.

Plinkert, P. K., Gitter, A. H., Zimmerman, U., Kirchner, T., Tzqartos, S. and Zenner, H. P. (1990). Visualization and functional testing of acetylcholine receptor-like molecules in cochlear outer hair cells. *Hearing Res.* **44:** 25–34.

Rajan, R. (1990). Electrical stimulation of the inferior colliculus at low rates protects the cochlea from auditory desensitization. *Brain Res.* **506:** 192–204.

Rasmussen, G. L. (1953). Further observations of the efferent cochlear bundle. *J. Comp. Neurol.* **99:** 61–74.

Rasmussen, G. L. (1960). Efferent fibers of the cochlear nerve and cochlear nucleus. In *Neural Mechanisms of the Auditory and Vestibular Systems* (G. L. Rasmussen and W. F. Windle, eds), pp. 105–115. Charles C. Thomas, Springfield, IL.

Rhode, W. S. (1971). Observations of the vibration of the basilar membrane in squirrel monkeys using the Mössbauer technique. *J. Acoust. Soc. Am.* **49:** 1218–1231.

Robles, L., Ruggero, M. A. and Rich, N. C. (1985). Mössbauer measurements of the mechanical response to single-tone and two-tone stimuli at the base of the chinchilla cochlea. In *Peripheral Auditory Mechanisms, Proceedings* (J. B. Allen, J. L. Hall, A. Hubbard, S. T. Neely and A. Tubis, eds), Vol. 64, pp. 121–128. Springer-Verlag, Berlin.

Sachs, M. B. and Abbas, P. J. (1974). Rate versus level functions for auditory-nerve fibers in cats: tone-burst stimuli. *J. Acoust. Soc. Am.* **56:** 1835–1847.

Santos-Sacchi, J. (1991). Reversible inhibition of voltage-dependent outer hair cell motility and capacitance. *J. Neurosci.* **11:** 3096–3110.

Sellick, P. M., Patuzzi, R. and Johnstone, B. M. (1982). Measurement of basilar membrane motion in the guinea pig using the Mössbauer technique. *J. Acoust. Soc. Am.* **72:** 131–141.

Siegel, J. H. and Kim, D. O. (1982). Efferent neural control of cochlear mechanics? Olivocochlear bundle stimulation affects cochlear biomechanical nonlinearity. *Hearing Res.* **6:** 171–182.

Slepecky, N., Ulfendahl, M. and Flock, A. (1988). Shortening and elongation of isolated outer hair cells in response to applications of potassium gluconate, acetylcholine and cationized ferritin. *Hearing Res.* **34:** 119–126.

Spoendlin, H. (1972). Innervation densities of the cochlea. *Acta Otolaryngol.* **73:** 235–248.

Starr, A. and Wernick, J. S. (1968). Olivocochlear bundle stimulation: effects on spontaneous and tone-evoked activities of single units in cat cochlear nucleus. *J. Neurophysiol.* **31:** 549–564.

Thompson, A. M. and Thompson, G. C. (1991). Posteroventral cochlear nucleus projections to olivocochlear neurons. *J. Comp. Neurol.* **303:** 267–285.

Thompson, A. M. and Thompson, G. C. (1993). Relationship of descending inferior colliculus projections to olivocochlear neurons. *J. Comp. Neurol.* **335:** 402–412.

Thompson, G. C. and Thompson, A. M. (1986). Olivocochlear neurons in the squirrel monkey brainstem. *J. Comp. Neurol.* **254:** 246–258.

Veenman, C. L., Reiner, A. and Honig, M. G. (1992). Biotinylated dextran amine as an anterograde tracer for single- and double-labeling studies. *J. Neurosci. Methods* **41:** 239–254.

Vetter, D. E., Saldaña, E. and Mugnaini, E. (1993). Input from the inferior colliculus to medial olivocochlear neurons in the rat: a double label study with PHA-L and cholera toxin. *Hearing Res.* **70:** 173–186.

Warr, W. B. (1992). Organization of olivocochlear efferent systems in mammals. In *The Mammalian Auditory Pathway: Neuroanatomy* (D. B. Webster, A. N. Popper and R. R. Fay, eds), pp. 410–448. Springer-Verlag, New York.

Warr, W. B. and Guinan, J. J. (1979). Efferent innervation of the organ of corti: two separate systems. *Brain Res.* **173:** 152–155.

Warren, E. H., III and Liberman, M. C. (1989). Effects of contralateral sound on auditory-nerve responses. I. Contributions of cochlear efferents. *Hearing Res.* **37:** 89–104.

Weinberg, R. J. and Rustioni, A. (1987). A cuneocochlear pathway in the rat. *Neuroscience* **20:** 209–219.

Wright, D. D. and Ryugo, D. K. (1994). Projections from dorsal column nuclei neurons form mossy fiber-like terminals in the dorsal cochlear nucleus of rat. *Assoc. Res. Otolar. Meeting Abstr.* Vol. 17, p. 28.

Young, E. D., Nelken, I. and Conley, R. A. (1993). Cochlear nuclear responses evoked from dorsal column (somatosensory) nuclei. *Assoc. Res. Otolar. Meeting Abstr.* Vol. 16: p. 124.

Zenner, H. P., Zimmermann, U. and Schmitt, U. (1985). Reversible contraction of isolated mammalian cochlear hair cells. *Hearing Res.* **18:** 127–133.

Zwicker E. (1979). A model describing nonlinearities in hearing by active processes with saturation at 40 dB. *Biol. Cyber.* **35:** 243–250.

Afferent Synaptic Transmission from Hair Cells

HARUNORI OHMORI

Department of Physiology, Faculty of Medicine, Kyoto University, Kyoto, 606-01 Japan

Hair cells are sensory transducer cells in the acoustic and vestibular organs. Their transducing activities have been studied extensively in several species (reviewed by Flock, 1971; Crawford and Fettiplace, 1980; Hudspeth, 1983; Russell and Sellick, 1983; Ohmori, 1985). It has been demonstrated that transducer channels have defined unitary conductance and broad cation selectivity with fast gating kinetics (Corey and Hudspeth, 1983; Ohmori, 1985; Crawford *et al.*, 1991). Depolarization of the hair cell membrane activates calcium channels and increases the intracellular calcium concentration (Ohmori, 1984, 1988; Roberts *et al.*, 1990), and should release neurotransmitter at the synapse between hair cell and afferent nerve terminal. We have developed three types of preparation to investigate the nature of afferent synapse transmission. The first method involved the cut-out preparation of ampulla of the semi-circular canal organ of a chick (Yamashita and Ohmori, 1990). Both bouton type and calyx type afferent nerve fibers were identified by morphology and the nature of synaptic transmission was studied. The second preparation was the primary culture of the cochlear ganglion neuron of a chick (Yamaguchi and Ohmori, 1990). The excitability of the cochlear ganglion neuron was investigated and the glutamate sensitivity was identified. The last preparation used a pairing of hair cells and cultured cerebellar granule cells; the cerebellar granule cell culture was used as a detector for neurotransmitter released from hair cells by membrane depolarization (Kataoka and Ohmori, 1994).

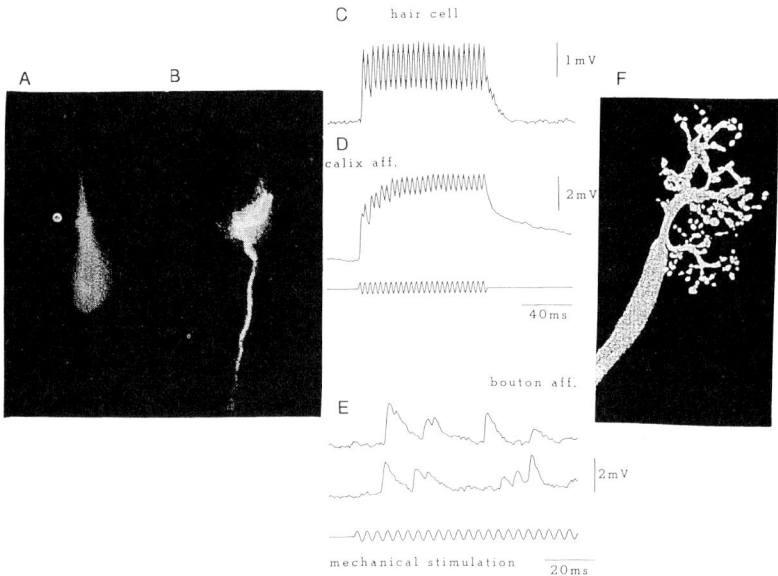

Fig. 1. Lucifer Yellow fluorescence images of a hair cell, and two types of afferent nerve terminals and electrical responses recorded in the isolated ampulla preparation. Lucifer Yellow was injected into hair cells and nerve terminals through the recording microelectrode. In (D) and (E) the trace at the bottom indicates the timing and waveform of mechanical stimulation applied to the bathing medium. (Modified from Yamashita and Ohmori, 1990.)

Synapse Potentials in Bouton-Type and Calyx-Type Afferent Nerves in the Ampulla of Semicircular Canal Organ

Electrical responses were recorded from hair cells and nerve terminals in the isolated preparation of ampulla of the semicircular canal organ of a chick (Fig. 1). Both bouton-type afferent terminals and calyx-type afferent terminals were identified from fluorescence images by penetrating the cell with a Lucifer Yellow CH filled microelectrode (Yamashita and Ohmori, 1990). The recorded nerve terminal was identified as the afferent terminal when action potentials were generated in response to the mechanical stimulation applied diffusely in the bathig medium. Oscillatory mechanical stimuli of about 250 Hz were applied for 100–150 msec.

When recorded from the bouton-type nerve terminals (Fig. 1E,F), mechanical stimulation produced fast-rise/slow-decay synaptic potentials, similar to the appearance of synaptic potentials observed in the other chemical synapses. The amplitude and frequency of synaptic potentials were decreased during mechanical stimulation. This is reflected in the rapid adaptation of spike generation during mechanical stimulation in the bouton-type afferent nerve fibers (Yamashita and Ohmori, 1990).

The calyx-type nerve terminal showed slowly rising post-synaptic potentials (Fig. 1B,D). An individual cycle of mechanical stimulation was followed by the synaptic potential and the rough appearance was similar to the transduction potential in the hair cell. However, the post-synaptic potential in the calyx-type afferent increased gradually during stimulation and showed a prolonged depolarization (for about 50 msec), even after the suspension of mechanical stimulation. On the contrary, the hair cell transduction potential was returned to the baseline level within 20 msec after the suspension of mechanical stimulation (Fig. 1A,C). Moreover, the amplitude of the post-synaptic potentials was increased by membrane hyperpolarization of calyx afferents, and there was no dye coupling between hair cells and calyx afferents. Therefore, the calyx-type synapse most probably functions via chemical transmission.

If the neurotransmitter is released and taken up by the same mechanism in these two types of synapses, the difference in the post-synaptic potential might indicate the difference in the lifetime of the released neurotransmitter between these two types of synapses. The lifetime of the neurotransmitter could be prolonged in the calyx-type synapse probably by preventing diffusion and uptake of the neurotransmitter by encapsulating hair cells.

Glutamate-Induced Current in the Primary Cultured Cochlear Ganglion Neuron

A cochlear ganglion neuron was cultured from a chick embryo (Yamaguchi and Ohmori, 1990). The electrical activities of the cultured neuron were studied by the whole-cell patch recording technique. The cochlear ganglion neuron exhibited a sodium channel, two types of calcium channel (the low threshold inactivating type and the high threshold non-inactivating type), and both an inward rectifying potassium channel and an outward rectifying potassium channel. These channels should contribute to the generation of action potentials. Although we have recorded only from the cell body of cochlear ganglion neuron, and it is not certain whether similar chemosensitivity existed in the afferent nerve terminal on one of its pseudo-bipolar processes, the most important finding in this cultured neuron should be the observation of glutamate sensitivity (Fig. 2). When glutamate (30

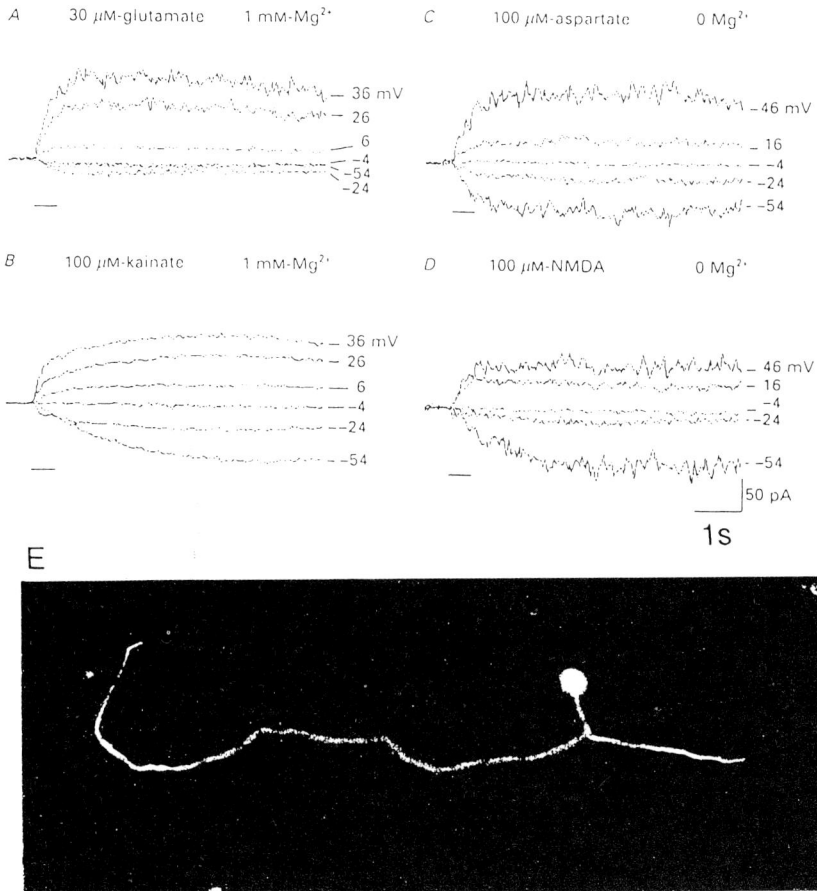

Fig. 2. Sensitivity to glutamate and related amino acids in cultured cochlear ganglion neuron. A known concentration of agonist was puff-applied to the cell. (E) shows the immuno-staining to neurofilament. (Modified from Yamaguchi and Ohmori, 1990.)

μM) was puff-applied to the cell body, ionic currents were induced. The amplitude of the outward current was increased with positive shift of the membrane potential, while the amplitude was decreased at membrane potentials more negative than –25 mV in the normal Mg^{2+}-containing medium. The overall current–voltage relationship was J-shaped. This suppression of the inward current at relatively large negative membrane potentials was due to the block of the glutamate receptor channel by extracellular Mg^{2+} ions (Fig. 2A). In the Mg^{2+}-free extracellular medium, the glutamate (or aspartate)-induced current was linear vs. membrane potentials (Fig. 2C), and NMDA could induce the current (Fig. 2D). Most (84%) of the glutamate-activated current was suppressed by APV (100 μM, the antagonist of the NMDA receptor channel). Therefore, the NMDA subtype of glutamate receptor was found in the cochlear ganglion neuron. As observed in other preparations (Ascher and Nowak, 1988), this channel was permeable to Ca^{2+}, Ba^{2+} and Sr^{2+}.

Activation of Glutamate Receptor on the Granule Cell by Membrane Depolarization of Closely Placed Hair Cells

Our finding of glutamate sensitivity in the cultured cochlear ganglion neuron strongly suggests that glutamate is the most likely candidate as the neurotransmitter released from hair cells. We have therefore performed an experiment to investigate directly the release of glutamate from hair cells. As has already been established in the retina (Copenhagen and Jahr, 1989), we have utilized a glutamate-sensitive neuron as a detector and have formed a bioassay system for the glutamate that is presumably released from cochlear hair cells.

As a detector neuron we have adopted a rat primary cultured cerebellar granule cell. We could identify the huge Purkinje cell and the small granular cell in the cerebellar culture. Glutamate receptors are expressed in both Purkinje cells and granule cells; however, the non-NMDA-type glutamate receptor is exclusively expressed in the mature Purkinje cell, and the NMDA-type glutamate receptor in the granule cell. The sensitivity to glutamate is quite different between these two types of glutamate receptors, and the sensitivity is about an order of concentration higher in the NMDA-type glutamate receptor than in the non-NMDA-type glutamate receptor (Mayer and Westbrook, 1987).

A single hair cell was transported and was placed close to the single granule cell. The distance between these two cells must be about 1 μm and sometimes could be much larger as we could occasionally identify a clear space between two cells. Two closely apposed cells were whole-cell voltage clamped, and the granule cell was generally held at +55 mV. This potential was selected in order to eliminate the Mg^{2+} block of the NMDA receptor channel. Hair cells were held at −65 mV and were step depolarized for 500–1000 msec. Depolarization for less than 20 msec could not induce any current responses in the granule cell.

The whole-cell patch recording is notorious in washing the soluble component out of the cell, and seems to induce run-down of calcium channels even in hair cells (Ohmori, 1984). Since we were trying to induce transmitter release from hair cells by membrane depolarization, namely by the influx of calcium ions through the voltage-gated calcium channels, we have adopted the Nystatin perforated-patch recording technique of Horn and Marty (1988).

The hair cell depolarization induced a large outward current in the granule cell held at +55 mV. The size of outward current was dependent on the level of depolarization of hair cells, and peaked at about +10 mV. When the holding potential of the granule cell was changed from +55 mV to +25, −25 and −55 mV, the overall voltage dependence looked J-shaped in the Mg^{2+} (1 mM) containing external medium; the current was suppressed at −55 mV.

We used a 10 mM Ca^{2+} and 1 mM Mg^{2+} containing external medium, and a CsCl, EGTA (0.1 mM) based internal medium. The presence of magnesium ions in the external medium should make the current–voltage relationship of the NMDA receptor channel J-shaped. However, the NMDA receptor channel is not the only chemically activated ionic channel in the cerebellar granule cell; there are at least $GABA_A$ type receptor channels. The activation of $GABA_A$ receptor should produce an outward current at depolarized potentials and an inward current at hyperpolarized potentials in the granule cell under our recording conditions. We have therefore discriminated these two types of currents pharmacologically by applying both APV (200 μM) and bicuculine (200 μM, the antagonist of $GABA_A$-type receptor). Only APV could suppress the hair-cell-induced current responses in the granule cell (Fig. 3).

Glutamate is the Most Likely Candidate as the Neurotransmitter Released from Hair Cells

Pharmacological experiments utilizing the NMDA receptor antagonist APV suggest that the granule cell response induced by the hair cell depolarization is mediated by NMDA receptor. This is confirmed by the absence of responses in the Purkinje cell which has the non-NMDA-subtype of glutamate receptor and is less sensitive to glutamate. These observations support the idea that glutamate or related excitatory amino acids are the most likely candidates for the neurotransmitter released from hair cells. However, we have not yet excluded the possibility that neurotransmitters other than glutamate are released from hair cells, since we have not yet demonstrated a direct release of glutamate from hair cells (in photoreceptor cell – see Ayoub *et al.*, 1989). Since our preparation is not a pure hair cell/ granule cell pair, and there seem to be some synaptic inputs to the detector granule cell, it

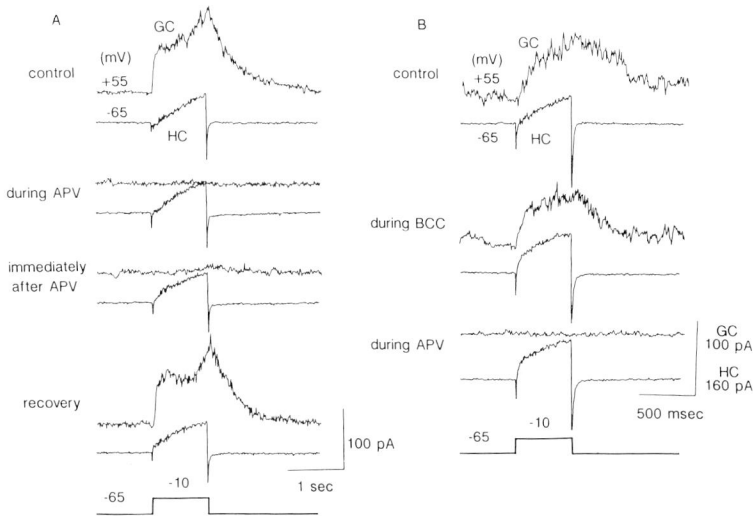

Fig. 3. Hair cell depolarizations induced currents in closely placed cerebellar granule cell. A pair of traces are shown for each experiment. The top traces in the pair were recorded from a granule cell (labeled GC) and the bottom traces were from a hair cell (labeled HC). The trace at the bottom in each panel indicates the pulse protocol applied to the hair cell. (A) shows the block of the granule cell response by APV (200 µM), and (B) shows the insensitivity of the granule cell response to bicucullin (200 mM). (Kataoka and Ohmori, 1994).

can be presumed that fine processes of the third neuron might release glutamate to the detector granule cell after receiving a non-glutamate neurotransmitter from a depolarized hair cell. If this could occur, the released substance from hair cell must activate the presynaptic terminals making contacts on the detector granule cell because the conduction of action potentials was blocked by TTX (1 µM) in this experiment. If the granule cell response was induced through such a synapse, the response would show a rapid rise and a slower decay, as is the case of ordinary synaptic transmission. However, this was not observed in the granule cell response recorded here. The granule cell generated outward currents in a graded appearance. Another possibility would be an extracellular accumulation of K^+ about the granule cell due to hair cell depolarization. We have utilized CsCl-based intracellular medium in the patch pippette, but the hair cell interior did not seem to be dialyzed completely with this medium, as large outward currents were observed by hair cell depolarization. Since the granule cell response was blocked by APV, local accumulation of K^+ would have again depolarized the third neuron. However, this could not be the case either, since the granule cell response was bell-shaped when the level of hair cell depolarization was changed and amplitudes of granule cell responses were plotted against the hair cell membrane potential. If K^+ ions were carried by outward currents and accumulated around the cell, such accumulation would become larger with the positive shift of hair cell membrane potential, and this would have produced a progressive increase or at least a saturation of the granule cell response, but not the observed decrease of it at large positive membrane potentials.

Estimation of the Glutamate Concentration at the Granule Cell

The size of outward current recorded in the granule cell at +55 mV was 2–100 pA (27 ± 30 pA, mean ± SD; n = 20 cells) in response to the hair cell depolarization to –10 mV. Glutamate (1 μM) was applied to the same granule cell at the end of each experiment, and generated outward currents of 30–350 pA (160 ± 88 pA, n = 13) at +55 mV. The ratio of granule cell response induced by hair cell depolarization to that induced by 1 μM glutamate was 0.04–0.8 (0.2 ± 0.2, n = 13). This might indicate that glutamate released from the hair cell would be raised to about 0.2–0.8 μM (0.5 ± 0.5 μM, n = 13 cells; calculated from the dose–response relationship of glutamate-activated current in granule cells) around the granule cell, if glutamate is released from the hair cell (Kataoka and Ohmori, 1994).

References

Ascher, P. and Nowak, L. (1988). The role of divalent cations in *N*-methyl-D-aspartate responses of mouse central neurones in culture. *J. Physiol.* **399**: 247–266.

Ayoub, G. S., Korenbrot, J. I. and Copenhagen, D. R. (1989). Release of endogenous glutamate from isolated cone photoreceptors of the lizard. *Neurosci. Res., Suppl.* **10**: S47–S55.

Copenhagen, D. R. and Jahr, C. E. (1989). Release of endogenous excitatory amino acids from turtle photoreceptors. *Nature* **341**: 536–539.

Corey, D. P. and Hudspeth, A. J. (1983). Kinetics of the receptor current in bullfrog saccular hair cells. *J. Neurosci.* **3**: 962–976.

Crawford, A. C. and Fettiplace, R. (1980). The frequency selectivity of auditory nerve fibers and hair cells in the cochlear of the turtle. *J. Physiol.* **306**: 79–125.

Crawford, A. C., Evans, M. G. and Fettiplace, R. (1991). The actions of calcium on the mechano-electrical transducer current of turtle hair cells. *J. Physiol.* **434:** 369–398.

Flock, A. (1971). Sensory transduction in hair cells. In *Handbook of Sensory Physiology* (W. R. Lowenstein ed.), Vol. 1, pp. 396–441. Springer, Berlin.

Horn, R. and Marty, A. (1988). Muscarinic activation of ionic currents measured by a new whole cell recording method. *J. Gen. Physiol.* **92:** 145–159.

Hudspeth, A. J. (1983). Mechanoelectrical transduction by hair cells in the acousticolateralis sensory system. *Ann. Rev. Neurosci.* **6:** 187–215.

Kataoka, Y. and Ohmori, H. (1994). Activation of glutamate receptors in response to membrane depolarization of hair cells isolated from chick cochlear. *J. Physiol.* **477:** 403–414.

Mayer, M. L. and Westbrook, G. L. (1987). The physiology of excitatory amino acids in the vertebrate central nervous system. *Prog. Neurobiol.* **28:** 197–276.

Ohmori, H. (1984). Studies of ionic currents in the isolated vestibular hair cell of the chick. *J. Physiol.* **350:** 561–581.

Ohmori, H. (1985). Mechano-electrical transduction currents in isolated vestibular hair cell of the chick. *J. Physiol.* **359:** 189–217.

Roberts, W. M., Jacobs, R. A. and Hudspeth, A. J. (1990). Colocalization of ion channels involved in frequency selectivity and synaptic transmission at presynaptic active zones of hair cells. *J. Neurosci.* **10:** 3664–3684.

Russell, I. J. and Sellick, P. M. (1983). Low-frequency characteristics of intracellularly recorded receptor potentials in guinea-pig cochlear hair cells. *J. Physiol.* **338:** 179–206.

Yamaguchi, K. and Ohmori, H. (1990). Voltage-gated and chemically gated ionic channels in the cultured cochlear ganglion neurone of the chick. *J. Physiol.* **420:** 185–206.

Yamashita, M. and Ohmori, H. (1990). Synaptic responses to mechanical stimulation in calyceal and bouton type vestibular afferents studied in an isolated preparation of semicircular canal ampullae of chicken. *Exp. Brain Res.* **80:** 475–488.

Synaptic Transmission by Auditory Hair Cells

FERNÁN JARAMILLO

Department of Physiology, Emory University School of Medicine, Atlanta, GA 30322, USA

Introduction

The exquisite ability of the hair cell to transduce hair bundle displacements into electrical signals is rivaled by its capacity to faithfully relay these electrical signals to the brain. This demanding task is entrusted to the synapses that the hair cell forms with the terminals of auditory afferent fibers. Whereas our understanding of mechano-electrical transduction has grown tremendously during the past decade, a comparable increase in our knowledge of the hair cell synapse has been lacking. Here current understanding of the hair cell's synaptic function is briefly summarized. Although we are mainly interested in synaptic function in the auditory system, the progress that has been made in our understanding of vestibular hair cell synapses is not ignored here. This discussion, however, does not include the specialized calyceal synapses that type I hair cells form with afferent fibers in the vestibular system.

Synapses are highly specialized to relay specific types of information. At the neuromuscular junction a premium is placed on reliability. Thus, an individual nerve terminal can contain from a few to a large array of release sites. Although a particular site might fail to release a quantum of neurotransmitter in response to an invading action potential, the collective response is one that ensures the release of a large number of quanta. Therefore, under normal circumstances, the neuromuscular synapse is fail-safe. In contrast, synapses in the central nervous system appear to be quite different. Individual synapses can comprise an individual release site that often fails to release transmitter in response to an arriving action potential (Stevens, 1993).

Hair cell synapses, like their counterparts in the visual system's photoreceptors, operate according to a different set of rules. These sensory receptors do not, for the most part, exhibit regenerative electrical responses. Instead, they respond to graded receptor potentials, allowing them to encode continuously varying stimuli. As in the case of photoreceptors, there is little doubt that the hair cell synapse is glutamatergic (Klinke, 1981; Annoni *et al.*, 1984; Sewell and Mroz, 1987, 1990; Ehrenberger and Felix, 1991; Starr and Sewell, 1991; Zucca *et al.*, 1991). A mounting body of evidence indicates that both common classes of glutamate receptors, the NMDA type and the ampa/kainate types, are present at the hair cell synapse. However, the precise role that each conductance plays remains unclear.

Until recently it has not been possible to ascertain the voltage-dependent release of glutamate from hair cells. However, in this volume Ohmori describes the likely release of glutamate from vestibular hair cells. In these experiments Ohmori uses glutamate-sensitive cultured neurons as the detector for transmitter release. This approach, which has also been successfully employed to assay the release of transmitter by advancing growth cones as well as by retinal photoreceptors, could significantly improve our understanding of the hair cell's synaptic function.

Hair Cell Synaptic Function

Information about the function of hair cell synapses is extremely sparse. The early work of Furukawa and his colleagues demonstrated the release of quanta of neurotransmitter by hair cells in the goldfish sacculus (Furukawa *et al.*, 1978). These synapses are characterized by a marked adaptive rundown in the number of quanta released in response to successive cycles of stimulation. For reasons that are discussed below, it is possible that a similar rundown in the auditory synapses of higher vertebrates might turn out to be undesirable. Unfortunately, our ignorance about synaptic transmission in these species forces us to infer the behavior of auditory hair cell synapses from the cadence of firing of auditory fibers, which appears to be random: when an auditory nerve fiber fires in response to a particular cycle of stimulation, the probability that it will respond to any successive cycle (provided it has not fired in response to any intermediate cycle) remains constant. This constancy holds regardless of how many cycles are ineffective, suggesting a process that has no 'memory' (Hind *et al.*, 1966; Crawford and Fettiplace, 1980; also see

Gummer, 1991). It is possible that memoryless transmission is simply a consequence of the low probability of transmitter release at individual release sites. If this probability is very low, there might be no trace of it by the time the next cycle of stimulation arrives. However, this appears not to be the case, because the memoryless nature of firing is well established at frequencies at which the probability of a fiber responding to a specific cycle of stimulation is well above 0.5 (Hind *et al.*, 1966).

The ability of auditory fibers to preserve timing information is characterized by phase locking, i.e. by the tendency of auditory fibers to fire at a preferred time (phase) of the stimulus cycle. This ability extends to several kilohertz in the squirrel monkey (Hind *et al.*, 1966) and the chick (Warchol and Dallos, 1990), to frequencies near 10 kHz in the barn owl (Sullivan and Konishi, 1984), and to frequencies as high as 18 kHz in the cat (Teich *et al.*, 1993). At these frequencies most cycles of stimulation are ineffective, due to the fibers' refractoriness. However, it seems inescapable that every cycle of stimulation involves a rise and a decline in the probability of firing. In general, and even at these high frequencies, firing in response to a cycle of stimulation (or the lack thereof), appears not to have an effect on future activity.

The history-independent transmission of information bears significantly on the likely specialization of auditory hair cell synapses. A universal feature of non-sensory synapses is the use-dependent modification of their efficacy. At the neuromuscular synapse this tendency is manifested in paired-pulse facilitation, augmentation and post-tetanic potentiation, as well as in synaptic fatigue. In the central nervous system the processes responsible for long-term potentiation and depression provide the means for permanently maintained changes in synaptic efficacy (reviewed in Stevens, 1993). There is not, to my knowledge, any synapse capable of responding in a way that is not affected by its previous activity. It is quite feasible that auditory hair cell synapses have developed specific specializations that enable the history-independent transmission of sensory information. At present we have no information regarding the molecular steps that mediate exocytosis in hair cells. Similarly, we have very little information about the kinetics of neurotransmitter release and binding to receptors, the kinetics of neurotrasmitter clearance, and the details of action potential generation in auditory fiber boutons. Careful study of these questions is needed if we are to understand the basis of the hair cell synapse's unparalleled ability to relay timing information.

Synaptic Organization

Morphological specializations

The synapses of hair cells, like those of photoreceptors and retinal bipolar cells, are characterized by the presence of osmiophilic bodies which vary in size and form, depending on the species and hair cell location (Hama and Saito, 1977; Bagger-Sjoback and Gulley, 1979; Swetlitschkin and Vollrath, 1988). These bodies, which range from flattened ribbons or bars to near spheroids, are surrounded by numerous clear neurotransmitter vesicles, which appear to be linked to the dense bodies by fine filaments. The plasma membranes of both hair cell and nerve terminal appear thickened in transmission electron micrographs, and although the biochemical composition of these densities has not been characterized in hair cells, it is possible that they have elements in common with some of the components found in the densities of central synapses (Walsh and Kuruc, 1992).

Ideas concerning the function of the dense bodies are largely speculative, and are limited by a near complete ignorance about their biochemical composition. Proposed functional roles include the idea of the dense body acting as a transporter in charge of delivering transmitter vesicles to the release site, as well as a transmitter loading system, put in place to provide for the rapid loading of transmitter vesicles (Osborne, 1977). An antibody that recognizes synaptic ribbons in the retina (Balkema, 1991) might provide the means for some progress in our knowledge of the composition of dense bodies. Recently Balkema has reported evidence indicating that an antibody that recognizes photoreceptor-synapse ribbons, cross-reacts with the tricarboxylic acid (TCA) cycle enzyme aconitase (1994 Soc. Neurosci. Abstr., personal communication). One possibility is that the dense bodies contain aconitase as well as other TCA cycle enzymes in order to ensure an elevated concentration of the intermediaries required for the synthesis of amino acids such as glutamate. It will be interesting to determine whether this or similar antibodies cross-react with hair cell dense bodies.

Ionic conductances

The organization of presynaptic ion channels is best understood in the amphibian hair cell, where ~40 Ca^{2+} channels are clustered with ~90 Ca^{2+}-activated K^+ channels, allowing for the interactions that mediate electrical resonance (Roberts *et al.*, 1988). This work has been recently confirmed as well as furthered by the description of the means to locate these release sites in live, dissociated hair cells (Issa and Hudspeth, 1994). This is a significant technical development that should facilitate the study of synaptic transmission in dissociated cells.

The activation of these voltage-dependent Ca^{2+} channels appears to be extraordinarily rapid in comparison to those of other cells (Hudspeth and Lewis, 1988; Fuchs and Evans, 1990b). The activation in the hair cells of the chick cochlea, perhaps the fastest measured so far, has a time constant for activation of 350 µsec at room temperature, corresponding to a bandwidth of ~450 Hz. This cutoff frequency is likely to be significantly higher at physiological temperatures. If we use a conservative Q10 of 2 for the kinetics of gating, then at 40 deg C the cutoff frequency would be ~2 kHz, in line with the degree of phase locking observed in the chick (Warchol and Dallos, 1990).

The pharmacology of the neurotransmitter receptors is glutamatergic, although the small size of afferent terminals has precluded a thorough pharmacological characterization. It is quite possible that these receptors are endowed with unusually fast kinetics, as has been demonstrated for those mediating transmission in the nucleus *magnocellularis* of the chick (Raman and Trussel, 1992)

Ca²⁺ buffering

The short-term, use-dependent modification of synaptic efficacy has often been explained in terms of residual presynaptic Ca^{2+} following transmitter release (Katz and Miledi, 1968; Delaney *et al.*, 1989; Yamada and Zucker, 1992). The hair cell's ability to rapidly buffer Ca^{2+} following transmitter release is likely to be a requisite for the normal transfer of timing information, particularly at high frequencies. There are numerous intracellular components, both mobile and immobile, that might account for the hair cell's ability to buffer Ca^{2+}. A mobile buffer, with a efficacy comparable to that of 1 mM of the Ca^{2+} chelator BAPTA, has been described (Roberts, 1993), although its molecular identity has not been established.

Exocytosis and membrane recycling

Until recently exocytosis has not been directly monitored in hair cells. However, a report at the 1994 meeting of the US Society for Neuroscience by T. Parsons and W. M. Roberts (Parsons, Lenzi, Almers and

Roberts, in preparation) provides evidence for an increase in membrane area, measured as an increase in the cell's capacitance, following long-lasting depolarizations. This result is interpreted as being due to the exocytosis of a large number of synaptic vesicles. Interestingly, this membrane can be reinternalized with a remarkably fast time constant of 12 sec, provided that the cytoplasmic contents of the cell are not dialyzed during recording, suggesting that perhaps a mobile factor is required for endocytosis.

The visualization of individual release sites at dissociated hair cells (Issa and Hudspeth, 1994) opens the possibility of exocytosis measurements using the cell-attached patch configuration. Although these studies pose formidable signal-to-noise difficulties, the potential reward renders them worthy of undertaking.

Summary and Future Directions

Auditory hair cells are adapted to cope with the demands imposed by the need to transfer high-frequency information. Unlike many other synapses, they are capable of continuously responding in a history-independent fashion. The main challenge for future research is to gain a better understanding of the biochemical machinery that enables the hair cell to operate in this fashion. A better characterization of the molecular composition of dense bodies and of the presynaptic densities associated with the plasma membrane is needed.

We will also have to develop better assays for transmitter release by hair cells. This question is difficult to approach in intact systems because a precise stimulus cannot be delivered to the hair cell's mechanosensory organelle, the hair bundle. The reason for this difficulty lies in the complicated, and poorly understood relationship between acoustic stimulation and hair bundle displacement. Numerous factors contribute to this complicated relationship. Most prominent among these are the mechanical properties of the cochlea and its active frequency tuning mechanism. Furthermore, because of the inaccessibility of postsynaptic terminals, transmitter release must be inferred from the pattern of auditory nerve fiber firing. In order to avoid these complications it will be important to study the release of neurotransmitter by isolated hair cells directly.

Acknowledgements

Supported by the Emory University Research Council and by an Alfred. P. Sloan Fellowship.

References

Annoni, J. M. , Cochran, S. L. and Precht, W. (1984). Pharmacology of the vestibular hair cell-afferent fiber synapse in the frog. *J. Neurosci.* **4**: 2106–2116.

Bagger-Sjoback, D. and Gulley, R. L. (1979). Synaptic structures in the type II hair cell in the vestibular system of the guinea pig. *Acta Otolaryngol.* **88**: 401–411.

Balkema, G. W. (1991). A synaptic antigen (B16) is localized in retinal synaptic ribbons. *J. Comp. Neurol.* **312**: 573–583.

Crawford, A. C. and Fettiplace, R. (1980). The frequency response of auditory nerve fibres and hair cells in the cochlea of the turtle. *J. Physiol.* **306**: 79–125.

Delaney, K. R., Zucker, R. S. and Tank, D. W. (1989). Calcium in motor nerve terminals associated with posttetanic potentiation. *J. Neurosci.* **10**: 3558–3367.

Ehrenberger, K. and Felix, D. (1991). Glutamate receptors in afferent cochlear neurotransmission in guinea pigs. *Hearing Res.* **52**: 73–80.

Evans, E. F. (1975). Cochlear nerve and cochlear nucleus. In *Handbook of Sensory Physiology* (W. Keidel and W. D. Neff, eds), Vol. V/2, pp. 1–108. Springer-Verlag, Berlin.

Fuchs, P. A. and Evans, M. G. (1990). Calcium currents in hair cells isolated from the cochlea of the chick. *J. Physiol.* **429**: 553–568.

Furukawa, T. and Matsuura, S. (1978). Adaptive rundown of excitatory post-synaptic potentials at synapses between hair cells and eight nerve fibers in the goldfish. *J. Physiol.* **276**: 193–209.

Furukawa, T., Hayashida, Y. and Matsuura, S. (1978). Quantal analysis of the size of excitatory post-synaptic potentials at synapses between hair cells and afferent nerve fibres in goldfish. *J. Physiol.* **276**: 211–226.

Gummer, A. W. (1991). Post-synaptic inhibition can explain the concentration of short inter-spike-intervals in avian auditory nerve fibers. *Hearing Res.* **55**: 231–243.

Hama, K. and Saito, K. (1977). Fine structure of the afferent synapse of the hair cells in the saccular macula of the goldfish, with special reference to the anastomosing tubes. *J. Neurocytol.* **6**: 361–373.

Hind, J. E., Anderson, D. J., Brugge, J. F. and Rose, J. E. (1966). Phase-locked response to low frequency tones in single auditory nerve fibers of the squirrel monkey. *J. Neurophysiol.* **30**: 769–793.

Issa, N. and Hudspeth, A. J. (1994). Clustering of Ca^{2+} channels and Ca^{2+}-activated K+ channels at fluorescently labeled presynaptic active zones of hair cells. *Proc. Natl. Acad. Sci. USA* (in press).

Katz, B. and Miledi, R. (1968). The role of calcium in neuromuscular facilitation. *J. Physiol.* **195**: 81–492.

Klinke, R. (1981). Neurotransmitter in the cochlea and the cochlear nucleus. *Acta Otoryn.* **91**: 541–554.

Konishi, M. (1991). The neural algorithm for sound localization in the owl. *The Harvey Lectures*, Series 86, pp. 47–64.

Osborne, M. P. (1977). Role of vesicles with some observations on vertebrate sensory cells. In *Synapsis* (G. Cotrell and P. Usherwood, eds), pp. 40–63. Academic Press, New York.

Parsons, T. D., Lenzi, D., Alwers, W. and Roberts, W. M. (1994). Calcium-triggered exocytosis and endocytosis in an isolated presynaptic cell: capacitance measurements in saccular hair cells. *Neuron* **13**: 875–883.

Raman, I. M. and Trussel, L. O. (1992). The kinetics of the response to glutamate and kainate in neurons of the avian cochlear nucleus. *Neuron* **9**: 173–186.

Roberts, W. M. (1993). Spatial calcium buffering in hair cells. *Nature* 363, 74–76.

Roberts, W. M., Jacobs, R. A. and Hudspeth A. J. (1990). Colocalization of ion channels involved in frequency selectivity and synaptic transmission at presynaptic active zones of hair cells. *J. Neurosci.* **10**: 3664–3684.

Sewell, W. F. and Mroz, E. A. (1987). Neuroactive substances in inner ear extracts. *J. Neurosci.* **7**: 2465–2475.

Sewell, W. F. and Mroz, E. A. (1990). Purification of a low-molecular-weight excitatory substance from the inner ears of goldfish. *Hearing Res.* **50**: 127–137.

Starr, P. A. and Sewell, W. F. (1991). Neurotransmitter release from hair cells and its blockade by glutamate-receptor antagonists. *Hearing Res.* **52:** 23–42.

Stevens, C. F. (1993). Quantal release of neurotransmitter and long-term 9. *Cell* **72**/*Neuron* **10:**(Suppl), 55–63.

Sullivan, W. E. and Konishi M. (1984). Segregation of stimulus phase and intensity coding in the cochlear nucleus of the barn owl. *J. Neurosci.* **4:** 1787–1799.

Swetlitschkin, R. and Vollrath, L. (1988). Synaptic bodies in the different rows of outer hair cells in the guinea pig cochlea. *Ann. Otol. Rhinol. Laryngol.* **97:** 308–312.

Takahashi, T., Moiseff, A. and Konishi, M. (1984). Time and intensity cues are processed independently in the auditory system of the owl. *J. Neurosci.* **4:** 1781–1786.

Teich M. C., Khanna, S. M. and Guiney, P. C. (1993). Spectral characteristics and synchrony in primary auditory nerve fibers in response to pure acoustic stimuli. *J. Stat. Physics* **70:** 257–279.

Walsh, M. J. and Kuruc, N. (1992). The postsynaptic density: constituent and associated proteins characterized by electrophoresis, immunoblotting, and peptide sequencing. *J. Neurochem.* **59:** 667–78.

Warchol, M. E. and Dallos, P. (1990). Neural coding in the chick cochlear nucleus. *J. Comp. Physiol.* **166:** 721–734.

Yamada, W. M. and Zucker, R. S. (1992). Time course of transmitter release calculated from simulations of a calcium diffusion model. *Biophys. J.* **61:** 671–682.

Zucca, G., Botta, L., Milesi, V., Dagani., F. and Valli, P. (1992). Evidence for L-glutamate release in frog vestibular organs. *Hearing Res.* **63:** 52–56.

Molecular Mechanisms Involving Neurotransmitters after Excitotoxicity at the Inner Hair Cell–Auditory Nerve Synapse

MICHEL EYBALIN, SAAID SAFIEDDINE, CHRISTINE GERVAIS D'ALDIN, RÉMY PUJOL AND JEAN-LUC PUEL

INSERM U. 254, Neurobiologie de l'Audition – Plasticité Synaptique, CHU Hôpital St Charles, 34295 Montpellier Cedex 5, France

An excitatory amino acid, probably glutamate, is likely used as a fast transmitter at the synapse between the cochlear inner hair cells (IHCs) and the radial dendrites of the type I primary auditory neurons (see for a review, Eybalin, 1993). The effects of this transmitter are mediated primarily by α–amino-3-hydroxy-5-methyl-4-isoxazole propionic acid (AMPA)/kainate receptors but also by N-methyl-D-aspartate (NMDA) receptors (see Eybalin, 1993). Metabotropic receptors (mGluR1) linked to the inositol phosphate (IP) turnover have also been detected in the cochlea (Eybalin *et al.*, 1993a) but their function remains unknown.

In addition to its excitatory effect, it has been shown in the central nervous system that an excess of extracellular glutamate, resulting from an excess release or an insufficient reuptake, leads to prolonged depolarization resulting in an osmotic imbalance and a massive swelling. Cell death can occur later, possibly due to a Ca^{2+} homeostasis defect (for review see Choi and Rothman, 1990). This excitotoxicity can be mimicked by glutamate agonists. In the cochlea, we have previously shown that non-NMDA glutamate agonists induced an immediate swelling of radial dendrites which culminates in a membrane rupture and a complete loss of the intradendritic content (Puel *et al.*, 1991, 1994a; Pujol *et al.*, 1985). This swelling was dose dependent (Puel *et al.*, 1994a) and underlaid a dose-dependent loss of auditory nerve response (Puel *et al.*, 1991, 1994a).

It has been hypothesized (Eybalin, 1993; Pujol, 1991) that the acute radial dendrite swelling observed in the organ of Corti after noise trauma or ischemia (Beagley, 1965; Spoendlin, 1971, 1976; Robertson, 1983; Billett *et al.*, 1989; Pujol, 1991) occurred through the same mechanism as the swelling induced by glutamate agonists. Therefore, they should be considered as excitotoxic events, at least concerning the damages on primary auditory neurons and their peripheral processes. Indeed, when a prior intracochlear administration of glutamate antagonists was performed before the onset of ischemia (Pujol *et al.*, 1992, 1993; Puel *et al.*, 1994a) or noise trauma (Puel *et al.*, 1994b), no dendritic swelling occurred. As in the electrophysiological experiments in the normal cochlea, a predominant implication of the AMPA/kainate receptors in the excitotoxic phenomenon was recognized since non-NMDA antagonists blocked most of the dendritic swelling, while NMDA antagonists showed weak efficacy. However, the combination of NMDA and non-NMDA antagonists was necessary to block the swelling of a minor population of dendrites.

Our current investigations concern the events which occur after the immediate dendritic swelling induced by AMPA excitotoxicity or noise trauma (6 kHz, 130 dB SPL, 30 min). These events include neuronal death (Spoendlin, 1976, 1984; Juiz *et al.*, 1989) and also a regrowth of new dendritic processes by surviving neurons up to the IHCs. This regrowth (Eybalin, 1993) may explain the apparently normal innervation of the IHC basal pole 10 days after the kainate administration in the experiments by Juiz *et al.* (1989). Accordingly, in the case of the AMPA excitotoxicity, we observed a dendrite regeneration whose time-course was relatively fast (Fig. 1). Within 1 day, the newly formed dendrites had reached the IHC base and already formed some rare synapses, still immature in their morphology (Pujol *et al.*, 1994). This regrowth process was concomitant to a partial recovery of ABRs (about 30% 1 day after the AMPA administration; Puel *et al.*, 1993). At 5 days post exposure, the ABRs returned to normal values while typical synaptic differentiations were observed at contacts between terminals of the radial dendrites and IHCs. In addition, the regenerated dendrites were also normally contacted by efferent endings. Similar observations were noted when using noise trauma instead of AMPA as an excitotoxic event (Puel *et al.*, 1994b). No difference was observed concerning the dendrite regeneration and neosynaptogenesis process at the IHC level between the two excitotoxic models. However, the ABR-thresholds did not completely recover after the noise trauma (about 40 dB recovery over an immediate 60 dB loss). The remaining threshold loss could be accounted for by permanent hair cell damage.

Among the variety of molecular events underlying the recovery of

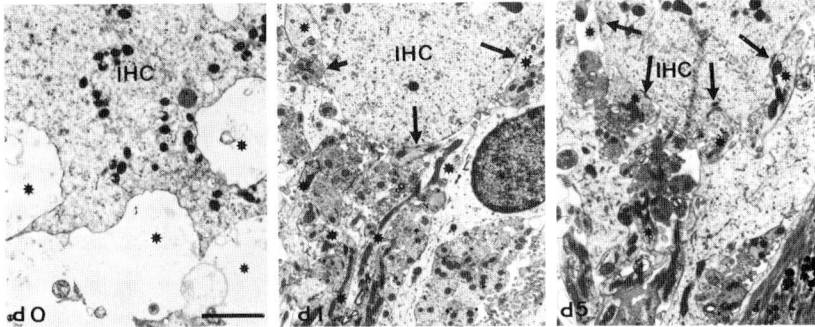

Fig. 1. The synaptic basal pole of an inner hair cell (IHC) immediately (d0); 24 hr (d1) and 5 days (d5) after an intrascalar perfusion of 200 μM AMPA. The acute effect of AMPA (d0) is a dramatic swelling of the radial dendrites with membrane disruption (asterisks). Twenty-four hours after the AMPA perfusion (d1), the dendritic swelling is no longer seen at the IHC basal pole which has recovered a normal shape. New auditory dendrites (arrows) and efferent terminal (short arrow) contact the IHC. Five days after AMPA (d5), the IHC basal pole is apparently normal, with synaptic contacts with regenerated radial dendrites (arrows). asterisks: afferent dendrites. The scale bar (2.5 mm) applies for all the figures.

the auditory nerve responses and the IHC neosynaptogenesis after an excitotoxicity, one should consider the expression of glutamate receptors in primary auditory neurons since some of them have been involved in trophic events in nervous tissue *in vivo* and *in vitro* (e.g. Cline, 1991; Burgoyne *et al.*, 1993). Clues about the implications of these receptors in the neosynaptogenesis process at the IHC level may then be obtained. Similarly, one should also consider possible trophic functions of lateral efferent neurotransmitters and their receptors in this plastic process (see Zilles, 1992). The present chapter summarizes the data we have acquired using *in situ* hybridization concerning these various molecules, and their putative role in the afferent and efferent neosynaptogenesis of IHCs. Biotin- or digoxigenin-labeled oligoprobes were used as previously described (Safieddine and Eybalin, 1992). The intensity of labeling was quantitated using an IM 640 SAMBA image analysis system (Alcatel-TITN Answare, Grenoble, France).

Transient Increase of Glutamate Receptor mRNAs in Primary Auditory Neurons

Previous *in situ* hybridization studies have shown that type I primary auditory neurons contain mRNAs encoding for the GluR2 and GluR3

subunits of AMPA/kainate receptors (Ryan *et al.*, 1991; Safieddine and
Eybalin, 1992), the NMDAR1 subunit of NMDA receptors (Safieddine
and Eybalin, 1992; Kuriyama *et al.*, 1993) and the mGluR1 subtype of
metabotropic receptors (Eybalin *et al.*, 1993a). Consistent results were
retrieved in control cochleas, i.e. cochleas contralateral to AMPA-treated
cochleas, perfused with artificial perilymph only, or from non-
traumatized guinea-pigs.

Whatever the excitotoxic event was, no variation in the expression of
the GluR2 (Fig. 3) and the GluR3 mRNAs was seen in the primary
auditory neurons at any stage studied (1–5 days). On the contrary, the
expression of NMDAR1 (Fig. 3) and mGluR1 (Fig. 2) mRNAs was
increased 24 hr after excitotoxicity when compared to that of the control
cochleas. A return to a normal level after 5 days was then observed
(Eybalin *et al.*, 1993b; Safieddine *et al.*, 1993).

mRNA synthesis of glutamate receptors after excitotoxicity most
likely underlies a replacement of lost receptors on postsynaptic radial
dendrites. However, the transient increased expression of NMDAR1

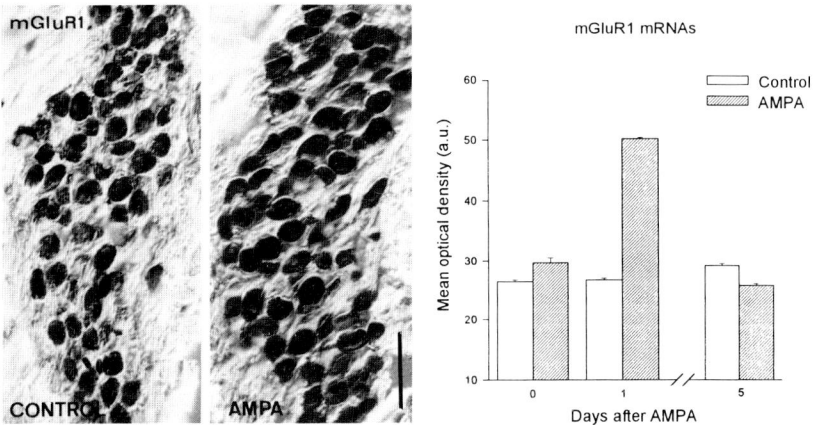

Fig. 2. mGluR1 mRNAs in primary auditory neurons in the second turn of guinea-pig
cochleas detected using a digoxigenin-labeled antisense oligoprobe (scale bar: 50 μm).
Primary auditory neurons from a control cochlea (contralateral unperfused cochlea) and 24
hr after an intrascalar 200 μM AMPA perfusion (d1). Note the higher intensity of the
labeling of primary auditory neurons one day after the AMPA perfusion. The right side of
the figure shows a quantitative analysis as a function of time of the mGluR1 mRNAs
hybridization signal in primary auditory neurons from AMPA-treated cochleas (AMPA)
and control cochleas. The analysis was performed on two non-serial sections per cochlea (*n*
= 3 per stage). The mGluR1 mRNA level was increased in the AMPA-treated cochleas by
±60% after 24 hr. After 5 days, mRNA levels equivalent to those before the AMPA perfusion
were observed. The data are expressed as means ± SD. a.u., arbitrary unit.

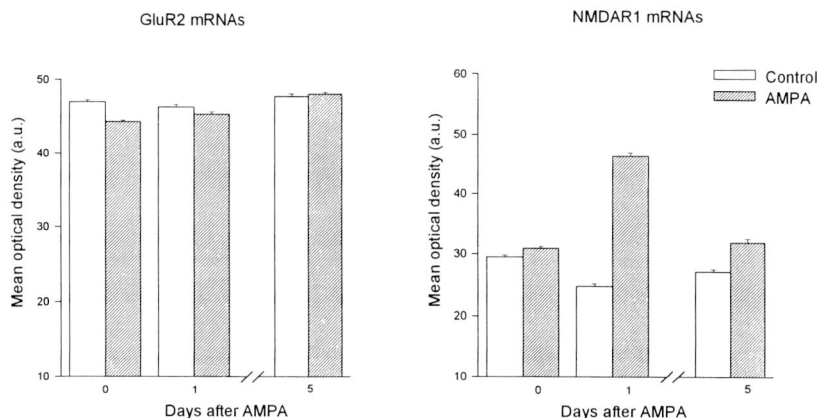

Fig. 3. Quantitative analysis as a function of time of the GluR2 and NMDAR1 mRNAs hybridization signal in spiral ganglion neurons from AMPA-treated cochleas (AMPA) and contralateral unperfused cochleas (control). The conditions of the analysis were the same as in the case of the mGluR1 mRNAs analysis. The GluR2 mRNA level was not affected by the AMPA treatment. In contrast, the NMDAR1 mRNA level was increased in the AMPA-treated cochleas by æ60% after 24 hr, reaching a mean optical density equivalent to that of the GluR2 mRNAs. After 5 days, mRNA levels equivalent to those before the AMPA perfusion were observed. The data are expressed as means ± SD. a.u., arbitrary unit.

and mGluR1 mRNAs suggest other functional implications. During the period of ABR recovery, NMDA receptors, IP-linked metabotropic receptors and AMPA/kainate receptors may well be involved in plastic events such as the guidance of newly formed dendrites and/or the stabilization of the IHC synapses. Accordingly, NMDA receptors have been implicated in trophic actions involving axon and dendrite guidance (see Burgoyne *et al.*, 1993; Komuro and Rakic, 1993; Rossi and Slater, 1993), during periods of synaptogenesis (see McDonald and Johnston, 1990; Rossi and Slater, 1993), experience-dependent synaptic plasticity (Cline, 1991; Kalb *et al.*, 1992) and after excitotoxicity (Hori *et al.*, 1991; Carmignoto and Vicini, 1992). Similarly, metabotropic glutamate receptors are involved in postsynaptic protein synthesis in synaptoneurosomes (Weiler and Greenough, 1993) and mediate the increased glutamate-triggered IP synthesis in various brain regions after ischemia (Chen *et al.*, 1988; Seren *et al.*, 1989), or after kainate-induced epilepsy during the sprouting of mossy fibers (Mayat *et al.*, 1994). Finally, AMPA/kainate receptors promote neurite guidance and synaptogenesis (Peichl and Bolz, 1984; Mattson *et al.*, 1988). Thus the transient increase in the expression of NMDAR1 and mGluR1 mRNAs, as well as the high level of expression of GluR2 and GluR3 mRNAs,

observed in primary auditory neurons may reflect an active role of
NMDA receptors and IP-coupled metabotropic receptors, and AMPA/
kainate receptors in the regeneration and neo-synaptogenesis in the
cochlea.

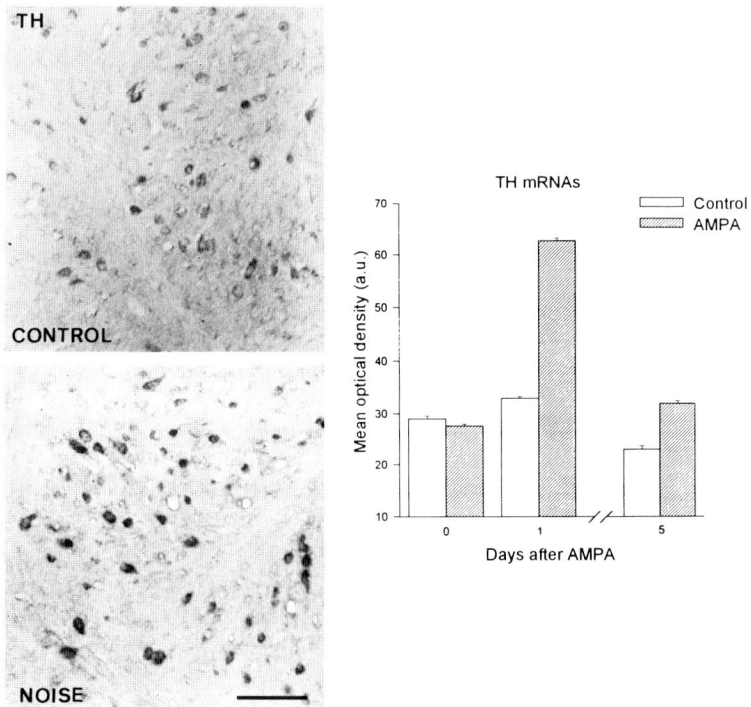

Fig. 4. Tyrosine hydroxylase (TH) mRNAs in lateral superior olives (LSO) from an
unstimulated guinea-pig (control) and from a guinea-pig exposed 24 hr previously to a
noise trauma (130 dB SPL, 6 kHz, 1 hr). Biotin-labeled TH antisense oligoprobe. Note the
increased intensity of the TH mRNA labeling in the olive from the traumatized guinea-pig.
Scale bar: 100 μm. (C). On the left, a quantitative analysis as a function of time of the TH
mRNAs hybridization signal in LSO neurons homolateral (AMPA) and contralateral
(control) to AMPA-injected cochleas. The analysis was performed using the same
conditions as in the case of glutamate receptor mRNAs. Twenty-four hours after the AMPA
perfusion, the TH mRNA level was increased by ±100% in ipsilateral LSO neurons. After 5
days, a return to a mean optical density equivalent to the levels before the AMPA perfusion
was observed. The data are expressed as means ±SD. a.u., arbitrary unit.

Transient Increase in mRNAs Encoding Synthetic Enzymes and Receptors for Lateral Efferent Neurotransmitters

According to our electron microscopic investigations (Pujol *et al.*, 1994), efferent varicosities showed very early signs of activation (increased vesicular content). In addition, there were signs that the mechanism of dendrite regeneration and neo-synaptogenesis after excitotoxicity mimicked the normal synaptogenesis of the IHCs. One day after exitotoxicity, efferent synapses were found at the IHC basal pole besides the newly established afferent synapses, similar to the innervation pattern seen in the first postnatal days in rat (Lenoir *et al.*, 1980) or mouse (Shnerson *et al.*, 1982). Thus, we investigated the involvement of the neurotransmitters/modulators of the lateral efferent system during the dendrite regrowth and neosynaptogenesis.

The mRNA levels of tyrosine hydroxylase (TH; Fig. 4), choline acetyltransferase (ChAT), glutamate decarboxylase (GAD), prepro-enkephalin (PPE) and calcitonin gene-related peptide (CGRP) were evaluated in lateral efferent neurons up to 5 days after either treatment with 200 µM AMPA (Safieddine *et al.*, 1993) or noise trauma. Although the CGRP mRNA levels did not change, for all the other molecules, expression was increased 24 hr after the excitotoxic event and returned to control level (control animal and contralateral lateral superior olive in the case of AMPA excitotoxicity) by 5 days. The time-course of the variations in mRNA levels was the same for both types of excitotoxicities and was similar to that observed for glutamatergic receptors mRNAs in primary auditory neurons.

Recent *in situ* hybridization data have shown that primary auditory neurons contain mRNAs coding for the D1 and D2 subtypes of dopaminergic receptors and for m3 muscarinic receptors (Safieddine and Eybalin, 1994). Therefore, we also evaluated the variations in the expression of these mRNAs after an acoustic trauma. While the expression of the D1 mRNAs in primary auditory neurons did not change, the expression of mRNAs encoding for D2 receptors (Fig. 5) and m3 receptors was increased 24 hr after acoustic trauma.

An upregulation of mRNAs encoding for neurotransmitter-related molecules of lateral efferent neurons is not surprising due to the need for a renewal of neurotransmitter stores. Indeed, GABA, met-enkephalin and dopamine are released in the cochlea under noise stimulation (Drescher *et al.*, 1983; Eybalin *et al.*, 1987; Gil-Loyzaga *et al.*, 1993). They could limit the extent of the dendritic swelling as suggested by the protective effects of D2 dopamine agonists on radial dendrites during ischemia (Pujol *et al.*, 1993) and noise trauma (Puel *et al.*, 1994b). Similarly, the loss of the postsynaptic receptors on radial dendrites

Active hearing

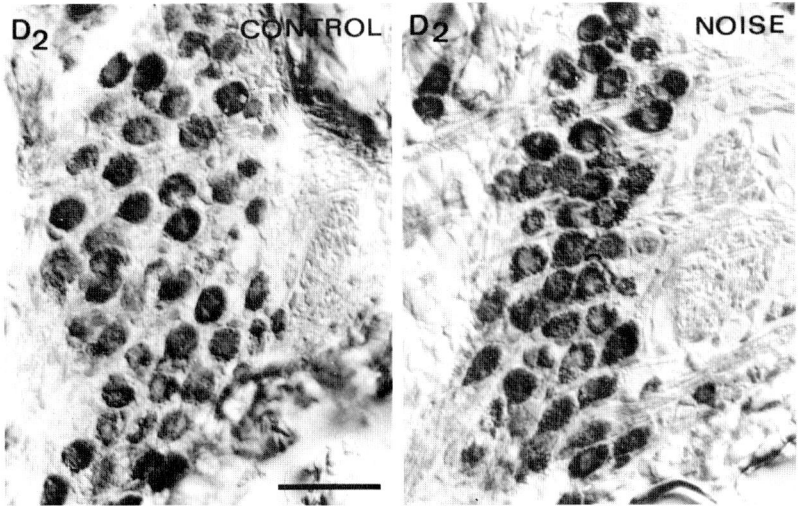

Fig. 5. D2 mRNAs in primary auditory neurons from the basal turn of an unstimulated guinea-pig (control) and from a guinea-pig exposed 24 hr previously to a noise trauma (130 dB SPL, 6 kHz, 1 hr). Digoxigenin-labeled D2 antisense oligoprobe. Increased labeling intensity can be noted in primary auditory neurons from the traumatized cochlea. Scale bar: 50 μm.

could account for the increased expression of D2 and m3 receptors in primary auditory neurons. However, the fact that not all mRNA vary after excitotoxicity (e.g. CGRP, D1 receptors in lateral efferent neurons and primary auditory neurons, respectively) suggests that transient increased mRNA synthesis may be related to trophic actions of some lateral efferent neurotransmitters. For example, the GABA trophic effects on neurite growth and synaptogenesis are well known (Wolff *et al.*, 1978; Spoerri and Wolff, 1981; Dames *et al.*, 1985). Conversely, acetylcholine, dopamine and enkephalins block neuritic and dendritic growth in nervous tissues (Lankford *et al.*, 1987, 1988; Hauser *et al.*, 1988; Mattson, 1988; Lipton and Kater 1989). It would be worth investigating whether the dendritic growth in the cochlea is under the balanced influences of lateral efferent neurotransmitters and glutamate released by IHCs. Some molecules may block these effects by stopping dendrite growth before aberrant pathways or synapses occur.

Conclusions

Primary auditory neurons can regenerate their dendrites, repair their synapses with the IHCs and lateral efferent varicosities, and recover functionally after excitotoxic injuries. The mechanisms involved in this reinnervation process likely involve ionotropic and metabotropic glutamate receptors (mainly NMDA and mGluR1?) and lateral efferent neurotransmitters. In human cochlear pathophysiology, this mechanism might well account for some of the functional recovery after ischemia-related sudden deafness or acoustic trauma. Considering the relative rapidity of the IHC neosynaptogenesis process (5 days), a first rapid phase of the functional recovery after sudden deafness or acoustic trauma may depend on the establishment of new synapses (within the 5 days), whereas a second slower phase (longer than 5 days) may depend on another mechanism, perhaps hair cell repair.

A last point should be raised concerning possible neuronal death. Ten days after a kainate injury, 34% of the primary auditory neurons had degenerated (Juiz *et al.*, 1989). Although we do not know if neuronal loss occurs in our excitotoxic models, it is conceivable that, in the case of successive excitotoxic injuries, additive effects irreversibly damage neurons and, consequently, lower or stop beneficial effects of the reinnervation. The neural presbycusis is characterized by a significant loss of primary auditory neurons (Spoendlin and Schrott, 1988) and thus may serve as a good example of this irreversible mechanism.

Acknowledgements

The authors wish to thank J. Fessenden for editorial assistance, R. Leducq, N. Renard and R. Assié for art work, and P. Sibleyras for photographic work.

References

Beagley, H. A. (1965). Acoustic trauma in the guinea pig. II. Electron microscopy. Including the morphology of cell junctions in the organ of Corti. *Acta Otolaryngol.* **60:** 479–495.

Billett, T. E., Thorne, P. and Gavin, J. B. (1989). The nature and progression of injury in the organ of Corti during ischemia. *Hearing Res.* **41:** 189–198.

Carmignoto, G. and Vicini, S. (1992). Activity-dependent decrease in NMDA receptor responses during development of the visual cortex. *Science* **258:** 1007–1011.

Chen, C.-K., Silverstein, F., Fisher, S. K., Statman, D. and Johnston, M. V. (1988). Perinatal hypoxic–ischemic brain injury enhances quisqualic acid-stimulated phosphoinositide turnover. *J. Neurochem.* **51:** 353–359.

Choi, D. W. and Rothman, S. M. (1990). The role of glutamate neurotoxicity in hypoxic–ischemic neuronal death. *Ann. Rev. Neurosci.* **13:** 171–182.

Cline, H. T. (1991). Activity-dependent plasticity in the visual system of frog and fish. *Trends Neurosci.* **14:** 104–111.

Dames, W., Joo, F., Feher, O., Toldi, J. and Wolff, J. R. (1985). Gamma-aminobutyric acid enables synaptogenesis in the intact superior cervical ganglion of the adult rat. *Neurosci. Lett.* **54:** 159–164.

Drescher, M. J., Drescher, D. G. and Medina, J. E. (1983). Effects of sound stimulation at several levels on concentrations of primary amines, including neurotransmitter candidates in perilymph of the guinea pig inner ear. *J. Neurochem.* **41:** 309–320.

Eybalin, M. (1993). Neurotransmitters and neuromodulators of the mammalian cochlea. *Physiol. Rev.* **73:** 309–373.

Eybalin, M., Rebillard, G., Jarry, T. and Cupo, A. (1987). Effect of noise level on the met-enkephalin content of guinea pig cochlea. *Brain Res.* **418:** 189–192.

Eybalin, M., Safieddine, S., Gervais d'Aldin, C., Puel, J.-L., Bartolami, S. and Pujol, R., (1993a). Cochlear neurotransmitters: neurochemical anatomy and physiological correlates. XIVth meeting of the International Society of Neurochemistry, Montpellier. *J. Neurochem.* **61**(Suppl.): S233.

Eybalin, M., Safieddine, S., Puel, J.-L., Gervais d'Aldin, C. and Pujol, R. (1993b). Acute and long term effects of AMPA on cochlear auditory neurons. Satellite meeting of the XIVth meeting of the International Society of Neurochemistry on Excitatory Amino Acid Neurotransmission, Marseilles. *J. Neurochem.* **61**(Suppl.): S273.

Gil-Loyzaga, P., Fernandez-Mateos, P., Vicente-Torres, A., Remezal, M., Cousillas, H., Arce, A. and Esquifino, A. (1993). Effects of noise stimulation on cochlear dopamine metabolism. *Brain Res.* **623:** 177–180.

Hauser, K. F., McLaughlin, P. J. and Zagon, I. S. (1987). Endogenous opioids regulate dendritic growth and spine formation in developing rat brain. *Brain Res.* **416:** 157–161.

Hori, N., Doi, N., Miyahara, S., Shinoda Y and Carpenter, D. O. (1991). Appearance of NMDA receptors triggered by anoxia independent of voltage *in vivo* and *in vitro*. *Exp. Neurol.* **112:** 304–311.

Juiz, J. M., Rueda, J., Merchan, J. A. and Sala, M. L. (1989). The effects of kainic acid on the cochlear ganglion of the rat. *Hearing Res.* **40:** 65–74.

Kalb, R. G., Lidow, M. S., Halsted, M. J. and Hockfield, S. (1992). N-Methyl-D-aspartate receptors are transiently expressed in the developing spinal cord ventral horn. *Proc. Natl. Acad. Sci. USA* **89:** 8502–8506.

Komuro, H. and Rakic, P. (1993). Modulation of neuronal migration by NMDA receptors. *Science* **260:** 95–97.

Kuriyama, H., Albin, R. L. and Altschuler, R. A. (1993). Expression of NMDA-receptor messenger RNA in the rat cochlea. *Hearing Res.* **69:** 215–220.

Lankford, K. L., Demello, F. G. and Klein, W. L. (1987). A transient embryonic dopamine receptor inhibits growth cone motility and neurite outgrowth in a subset of avian retina neurons. *Neurosci. Lett.* **75:** 169–174.

Lankford, K. L., Klein, W. L. and Demello, F. G. (1988). D1-type dopamine receptors inhibit growth cone motility in cultured retina neurons: evidence that neurotransmitters act as morphogenetic growth regulators in the developing central nervous system. *Proc. Natl. Acad. Sci. USA* **85:** 2839–2843.

Lenoir, M., Shnerson, A. and Pujol, R. (1980). Cochlear receptor development in the rat with emphasis on synaptogenesis. *Anat. Embryol.* **160:** 253–262.

Lipton, S. A. and Kater, S. B. (1989). Neurotransmitter regulation of neuronal outgrowth, plasticity and survival. *Trends Neurosci.* **7:** 265–270.

MacDonald, J. W. and Johnston, M. V. (1990). Physiological and pathophysiological roles of excitatory amino acids during central nervous system development. *Brain Res. Rev.* **15:** 41–70.

Mattson, M. P. (1988). Neurotransmitters in the regulation of neuronal cytoarchitecture. *Brain Res. Rev.* **13:** 179–212.

Mattson, M. P., Lee, R. E., Adams, M. E., Guthrie, P. B. and Kater, S. B. (1988). Interactions between entorhinal axons and target hippocampal neurons: a role for glutamate in the development of hippocampal circuitry. *Neuron* **1:** 865–876.

Mayat, E., Lerner-Natoli, M., Rondouin, G., Lebrun, F., Sassetti, I. and Récasens, M. (1994). Kainate-induced status epilepticus leads to a delayed increase in various specific glutamate metabotropic receptor responses in the hippocampus. *Brain Res.* **645:** 186–200.

Peichl, L. and Bolz, J. (1984). Kainic acid induces sprouting of retinal neurons. *Science* **223:** 503–504.

Puel, J.-L., Pujol, R., Ladrech, S. and Eybalin, M. (1991). α-Amino-3-hydroxy-5-methyl-4-isoxazole propionic acid (AMPA) electrophysiological and neurotoxic effects in the guinea pig cochlea. *Neuroscience* **45:** 63–72.

Puel, J.-L., Gervais d'Aldin, C., Safieddine, S., Pujol, R. and Eybalin, M. (1993). Synaptic plasticity after excitotoxic events in the cochlea. *Inner Ear Biol. Workshop* **30:** 19.

Puel, J.-L., Pujol, R., Tribillac, F., Ladrech, S. and Eybalin, M. (1994a). Excitatory amino acid antagonists protect cochlear auditory neurons from excitotoxicity. *J. Comp. Neurol.* **341:** 241–256.

Puel, J.-L., Gervais d'Aldin, C., Safieddine, S., Eybalin, M. and Pujol, R. (1994b). Excitotoxicity and plasticity of the IHC-auditory nerve contribute to both temporary and permanent threshold shift. In *Noise-Induced Hearing Loss* (A. Axelsson, H. Borchgrevink, D. Henderson, R. P. Hamernik and R. S. Salvi, eds). Thieme, New York, (in press).

Pujol, R. (1991). Sensitive developmental period and acoustic trauma: Facts and hypotheses. In *Noise-Induced Hearing Loss* (A. L. Dancer, D. Henderson, R. J. Salvi and R. P. Hamernik, eds), pp. 196–203. Mosby Year Book, St Louis, MO.

Pujol, R., Lenoir, M., Robertson, D., Eybalin, M. and Johnstone, B. M. (1985). Kainic acid selectivity alters auditory dendrites connected with cochlear inner hair cells. *Hearing Res.* **18:** 145–151.

Pujol, R., Puel, J.-L. and Eybalin, M. (1992). Implication of non-NMDA and NMDA receptors in cochlear ischemia. *NeuroReport* **3:** 299–302.

Pujol, R., Puel, J.-L., Gervais d'Aldin, C. and Eybalin M (1993). Pathophysiology of the glutamatergic synapses in the guinea pig cochlea. *Acta Otolaryngol.* **113:** 330–334.

Pujol, R., Gervais d'Aldin, C., Leducq, R. and Puel, J.-L. (1994). Auditory-nerve regeneration and neo-synaptogenesis below inner hair cells after excitotoxic damage. *Assoc. Res. Otolaryngol. Abstr.* Vol. 17, p. 116.

Robertson, D. (1983). Functional significance of dendritic swelling after loud sounds in the guinea pig cochlea. *Hearing Res.* **9:** 263–278.

Rossi, D. J. and Slater, N. T. (1993). The developmental onset of NMDA receptor-channel activity during neuronal migration. *Neuropharmacology* **32:** 1239–1248.

Ryan, A. F., Brumm, D. and Kraft, M. (1991). Occurrence and distribution of non-NMDA glutamate receptor mRNAs in the cochlea. *NeuroReport* **2:** 643–646.

Safieddine, S. and Eybalin, M. (1992). Co-expression of NMDA and AMPA/kainate receptor mRNAs in cochlear neurones. *NeuroReport* **3:** 1145–1148.

Safieddine, S., Puel, J.-L., Pujol, R. and Eybalin, M. (1993). Recovery of ABR responses and transitory increase of transmitter-related mRNA expression after intracochlear AMPA. *Assoc. Res. Otolaryngol.* **16:** 32.

Seren, M. S., Aldinio, C., Zanoni, R., Leon, A. and Nicoletti, F. (1989). Stimulation of inositol phospholipid hydrolysis by excitatory amino acids is enhanced in brain slices from vulnerable regions after transient global ischemia. *J. Neurochem.* **53:** 1700–1705.

Shnerson, A., Devigne, C. and Pujol, R. (1982). Age-related changes in the C57BL/6J mouse cochlea. II. Ultrastructural findings. *Dev. Brain Res.* **2:** 77–88.

Spoendlin, H. (1971). Primary structural changes in the organ of Corti after acoustic overstimulation. *Acta Otolaryngol.* **71:** 166–176.

Spoendlin, H. (1976). Anatomical changes following various noise exposures. In *Effects of Noise on Hearing* (D. Henderson, R. P. Hamernik, D. S. Dosanjh and J. H. Mills, eds), pp. 69–89. Raven Press, New York.

Spoendlin, H. (1984). Factors inducing retrograde degeneration of the cochlear nerve. *Ann. Otol. Rhinol. Laryngol.* **93**(Suppl. 112): 76–81.

Spoendlin, H. and Schrott, A. (1988). The spiral ganglion and the innervation of the human organ of Corti. *Acta Otolaryngol.* **105:** 403–410.

Spoerri, P. E. and Wolff, J. R. (1981). Effect of GABA administration on murine neuroblastoma cells in culture I: Increased membrane dynamics and formation of specialized contacts. *Cell Tissue Res.* **218:** 567–579.

Weiler, I. J. and Greenough, W. T. (1993). Metrabotropic glutamate receptors trigger postsynaptic protein synthesis. *Proc. Natl. Acad. Sci. USA* **90:** 7168–7171.

Wolff, J. R., Joo, F. and Dames, W. W. (1978). Plasticity in dendrites shown by continuous GABA administration in superior cervical ganglion of adult rat. *Nature* **274:** 72–74.

Zilles, K. (1992). Neuronal plasticity as adaptative property of the central nervous system. *Ann. Anat.* **174:** 383–391.

Sensory and Supporting Cells in the Organ of Corti: Cytoskeletal Organization Related to Cellular Function

NORMA B. SLEPECKY

Department of Bioengineering and Neuroscience, Institute for
Sensory Research, Syracuse University, Syracuse, NY 13244-5290,
USA

Introduction

Many of the presentations in this symposium have discussed the physiological and morphological changes that occur in cells of the cochlea. It is also important to ask the question, what is it about the make-up of these cells that permits them to undergo such changes or provides them with the active mechanisms to produce them? The cells of the guinea-pig cochlea have been most often studied in this regard since they are large and can be obtained from an easily accessible sensory epithelium. However, if cells from other species are studied, it will be possible to compare results across species with the long-range goal of understanding structure–function relationships and the basic mechanisms used by specific cell types with regard to sensory transduction.

In mammalian cells, the three major proteins that provide the structural basis for cell shape are actin, tubulin and intermediate filament proteins. Two of these, actin and tubulin, provide the mechanisms for cell division, cell movement, changes in tension within cells and changes in cell shape. All three have subunits which can be present as monomers or assembled into filaments and tubules, but their roles have been most often studied when they are in their polymerized forms – microfilaments, microtubules and intermediate filaments. Not much is known about intermediate filaments. They are intermediate in size between microfilaments and microtubules and they are thought to be responsible for maintaining constant cell shape over long periods of time. Other than that, they are cell type specific and it is not possible to generalize further about their biochemical and functional properties

On the other hand, actin and tubulin have been extensively studied. When present as microfilaments (actin) and microtubules (tubulin), their number, length, organization and location are determined by inter-action with specific actin-binding and microtubule-associated proteins. Calcium is thought to act as a signal which modulates these events. In resting cells, free calcium ions are maintained at low levels, bound to proteins which act as calcium buffers. On stimulation, calcium is released and interacts with calcium-dependent regulatory proteins which ultimately control metabolism, cytoskeletal integrity and cell shape.

Using fluorescently labeled antibodies, it is now possible to identify specific proteins on tissue sections. The view of cells provided by this technique shows structures which can be related to functions based on the type of protein present. Moreover, interactions at a subcellular level can be suggested based on co-localization studies of the different proteins. Thus, the immunocytochemical study of cyto-skeletal proteins, actin-binding and microtubule-associated proteins, and calcium-binding and calcium-dependent regulatory proteins has allowed us to examine individual cells from different sensory epithelia and from similar regions in different species. It provides a powerful tool to view cells and to identify areas and mechanisms of functional specialization.

Actin and Actin-Binding Proteins

Immunocytochemical studies in the mammalian cochlea have shown that actin is present predominantly as filaments in at least five different regions of the sensory hair cells (Flock and Cheung, 1977; Slepecky and Chamberlain, 1982; Arima *et al.*, 1987; Thorne *et al.*, 1987; Slepecky, 1989; Weaver *et al.*, 1993). Most of these regions are well stained by fluorescently labeled antibodies applied to tissue sections (Fig. 1). In the stereocilia, actin filaments are present in bundles, some of which continue down through the rootlet into the cuticular plate. They are present in the cuticular plate as a dense meshwork and as a circum-ferential ring at the periphery where the cell comes in contact with supporting cells at the reticular lamina. Along the lateral wall, actin filaments are organized into a lattice just inside the cell membrane, and

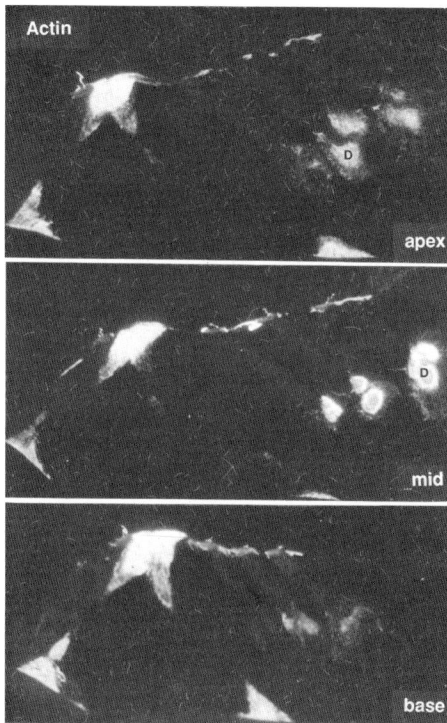

Fig. 1. In the sensory cells, actin is present in the stereocilia, the cuticular plate, the circumferential ring at the level of the cuticular plate, the cortical network and the cytoplasm. In the supporting cells, actin is present in the bundles which span the distance between the reticular lamina and the basilar membrane. The actin-containing meshwork in the Deiters cells under the OHCs are prominent in the apical and middle turns of the cochlea, but less frequent and smaller in the base.

in the cytoplasm of the hair cells there are isolated actin filaments which course throughout the cell.

Actin is also one of the major components of supporting cells (Fig. 1). Like the hair cells, there is a dense network of actin filaments at the apical portion of the pillar and Deiters cells, and similar actin-containing dense networks are found at the base of these cells where they rest on the basilar membrane (Flock *et al.*, 1982; Slepecky and Chamberlain, 1985; Arima *et al.*, 1986). Actin-containing structures have also been identified in the mid-region of the Deiters cells, just below the cup on which the outer hair cells (OHCs) rest. These structures are present in guinea pig (Holtham and Slepecky, 1994) as well as gerbil supporting cells (Fig. 1), where the structures display longitudinal gradients along the length of the sensory epithelium and are larger in the apical and middle turns of the cochlea than they are in the base. In the parts of the cell which span the distance between the reticular lamina and the basilar membrane, organized bundles of filaments and tubules are found and appear to be anchored in the actin-containing structures at both ends of the cell.

In each of these regions of the sensory and supporting cells, there are different types of actin-binding proteins present. This results in localized differences in the rate of filament assembly or disassembly and the amount and orientation of filament crosslinking—ultimately affecting the mechanical properties of the filaments, networks and bundles. In the stereocilia, fimbrin cross-links filaments into bundles (Flock *et al.*, 1982; Slepecky and Chamberlain, 1985). Tropomyosin is found in the rootlets (Slepecky and Chamberlain, 1985), where it is thought to wrap around single actin filaments. Both of these proteins may contribute to the stiffness of the stereocilia.

In the cuticular plate, actin filaments are cross-linked by α-actinin, spectrin and fimbrin (Drenckhahn *et al.*, 1982; Flock *et al.*, 1982; Slepecky and Chamberlain, 1985; Ylikoski *et al.*, 1990; Slepecky and Ulfendahl, 1992). Profilin, an actin-binding protein which regulates filament formation shows weak staining (Slepecky and Ulfendahl, 1992). Myosin present in the cuticular plate is thought to interact with actin filaments to provide tension in the cuticular plate (Drenckhahn *et al.*, 1982; Steyger *et al.*, 1989; Slepecky and Ulfendahl, 1992). Almost all of the actin-binding proteins found in or near the cuticular plate have properties which could modulate the stiffness or position of the stereocilia—ultimately affecting transduction.

The actin filaments in the lateral wall forming the cortical lattice are cross-linked by spectrin (Holley and Ashmore, 1990; Ylikoski *et al.*, 1990; Slepecky and Ulfendahl, 1992; Nishida *et al.*, 1993) and the linkage (like that in the cytoskeleton of red blood cells) is thought either to be involved in the fast OHC motility or to be flexible enough to follow the cell shape changes caused by both fast and slow types of OHC motility. The presence of myosin in this region (Drenckhahn *et al.*, 1982) also supports the idea that some type of active movement occurs here.

Unlike the sensory cells, supporting cells appear to be specialized for maintaining stable cell shape. This would provide the mechanical support required to hold the sensory cells up off the basilar membrane and to transmit movement of the basilar membrane to the stereocilia. Thus, supporting cells contain a different subset of actin-binding proteins (Slepecky and Chamberlain, 1987), and contain well-characterized intermediate filaments (Raphael *et al.*, 1987; Schrott *et al.*, 1988; Schulte and Spicer, 1989; Anniko and Arnold, 1990; Bauwens *et al.*, 1991;

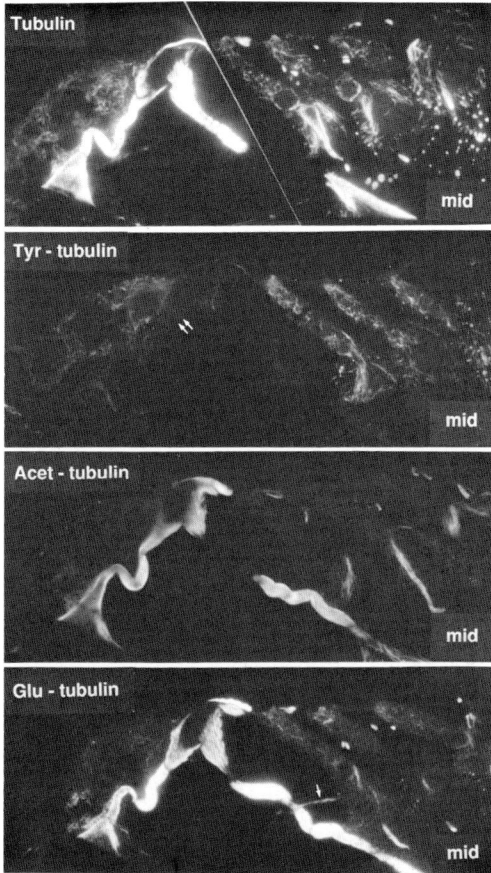

Fig. 2. Antibodies which bind to all forms of tubulin label networks of microtubules in the sensory cells and bundles of microtubules in the supporting cells. Antibodies which are specific for the newly synthesized form of tubulin containing a tyrosine residue (Tyr-tubulin) label only fine networks of microtubules predominantly in the sensory cells. Antibodies which are specific for tubulin which has been present in microtubules long enough to undergo post-translational modification such as the addition of acetyl (Acet-tubulin) or glutamyl (Glu-tubulin) residues show that these microtubules are localized almost entirely in the supporting cells.

Kuijpers *et al.*, 1991; Shi *et al.*, 1993). The apical and basal poles of the pillar and Deiters cells contain the actin cross-linking protein α-actinin and the intermediate filament protein cytokeratin. In the filament and tubule bundles spanning the length of these cells, tropomyosin co-localizes with the actin filaments. The mid-regions of the Deiters cells contain the intermediate filament protein vimentin.

Tubulin and Microtubule-Associated Proteins

Tubulin is present as microtubules, in networks throughout the cell bodies of both the inner hair cells and outer hair cells (Steyger *et al.*, 1989; Furness *et al.*, 1990; Slepecky and Ulfendahl, 1992). In both inner hair cells (IHCs) and OHCs (Fig. 2), microtubules are found in arrays below the cuticular plate and running parallel to the long axis of the cells. Recent staining with antibodies to different tubulin isoforms indicates that the microtubules in the sensory hair cells in guinea-pig (Slepecky and Ulfendahl, 1992) and gerbil (Fig. 2) are composed of the tyrosinated form of tubulin. This is the form in which newly syn-thesized tubulin first appears and its presence predominantly in the sensory cells suggests that in these cells the microtubules are dynamic structures, undergoing rapid cycles of polymerization and depolymeri-zation. This suggestion is further supported by the presence in OHCs of MAP-2, a protein thought to reversibly cross-link microfilaments with microtubules (Slepecky and Ulfendahl, 1992). These findings provide biochemical evidence for the possibility of remodeling the OHC cytoskeleton, correlated with or even causing the slow, chemically induced changes in OHC shape.

Microtubules in the supporting cells are present in conspicuous bundles which run from the apical surface at the reticular lamina to the basal surface along the basilar membrane (Angelborg and Engstrom, 1972). These microtubules differ structurally from the microtubules in the sensory hair cells in that they contain 15 rather than 13 protofibrils (Saito and Hama, 1982; Kikuchi *et al.*, 1991). Recent immuno-cyto-chemical studies indicate that the tubulin present in them has been post-translationally modified (Fig. 2) by the addition of glutamyl and/or acetyl subunits, suggesting that the microtubules are longer-lived and more stable, further supporting a role for the pillar and Deiters cells in providing mechanical support. This suggestion is reinforced by the presence in the supporting cells of intermediate filaments composed of cytokeratin and vimentin.

Calcium-Binding Proteins and Calcium-Dependent Regulatory Proteins

Calcium is thought to play a major role in cells, modulating both cell structure (through the cytoskeleton) and metabolism (through enzyme regulation). For calcium ions to be effective as a signaling mechanism, their concentration in resting cells must be maintained at low levels so that focal increases can trigger specific events. Calcium-binding

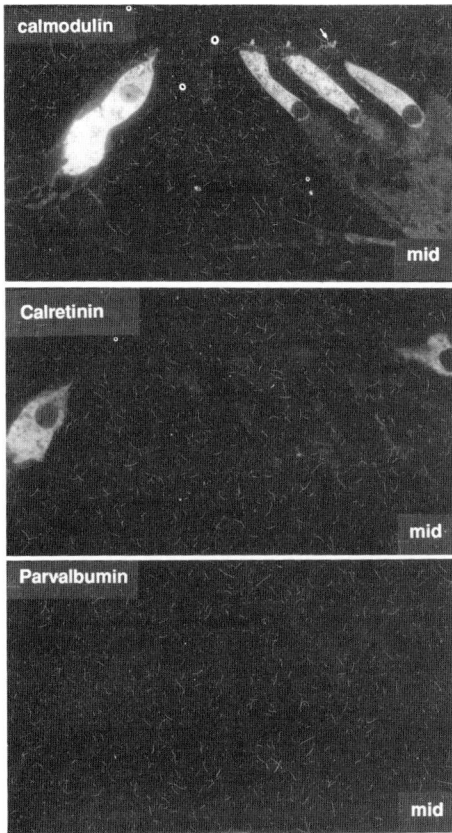

Fig. 3. The calcium-binding proteins calmodulin, calretinin and parvalbumin are present in different cell types within the organ of Corti. Calmodulin is present in IHCs and OHCs; calretinin is present in IHCs and in some of the Deiters cells (not in the Hensen cells); parvalbumin is found in IHCs only (however, staining in decalcified tissue sections is much weaker than staining in surface preparations). Staining for parvalbumin is more intense in the IHCs at the apex of the cochlea.

proteins have been suggested to play two functionally different roles in cells. Some are primarily calcium buffers (such as calbindin, parvalbumin and S-100) which are responsible for maintaining low levels of intracellular free calcium. Some (calsequestrin) bind calcium ions and release them at a critical time in response to appropriate intracellular signals. Others are calcium sensors and triggers (such as calmodulin) which bind free calcium, and in this form activate proteins and enzymes.

Calcium-binding proteins and calcium-dependent regulatory

proteins have been well studied in the adult mammalian cochlea. It is clear from these studies that most of the calcium-binding proteins studied are localized in the sensory cells. Of the proteins studied, calmodulin appears to be the most abundant. In both guinea-pig (Slepecky and Ulfendahl, 1993) and gerbil (Fig. 3), it is present in the in the stereocilia, cuticular plates and cell bodies of the IHCs and OHCs. Based on immunocytochemical labeling of tissue sections, there appears to be similar amounts along the length of the sensory epithelium.

Calsequestrin, on the other hand, appears to be OHC specific (Slepecky and Ulfendahl, 1993). This localization is of functional importance since it has been suggested that OHCs must have some specialized organelle for the release of free calcium ions which trigger length changes in OHCs. Based on its location in muscle (in sarcoplasmic reticulum), it could be expected that calsequestrin would be found in the subsurface cisterns along the lateral surface of the OHCs. However, the distribution of calsequestrin and the other calcium-binding proteins, as well as the distribution of precipitatable calcium, appears diffuse throughout the cytoplasm (Slepecky and Ulfendahl, 1993). This does not preclude membrane storage organelles for calcium ions used in intracellular signaling since the site for calcium storage and release may be more similar to the small membrane bound vesicles (calciosomes) containing a calsequestrin-like substance found in non-muscle cells and neurons.

Parvalbumin and calretinin are found in sensory hair cells in the gerbil and guinea-pig. Parvalbumin appears to be present only in IHCs in guinea-pig (Eybalin and Ripoll, 1990) and gerbil cochlea (Fig. 3). Its staining is weaker in tissue sections shown here than it is in surface preparations, and in both guinea-pig and gerbil there is more staining in the apical turns of the cochlea than there is at the base. In both species the OHCs and the supporting cells appear unstained with antibodies to parvalbumin. Calretinin is present in the IHCs and Deiters cells of the guinea-pig cochlea (Dechesne *et al.*, 1991), but in our semi-thin sections it is clearly not present in the Hensen cells of the guinea-pig cochlea (Slepecky, unpublished results) as has been previously reported. In the rat cochlea it is found in the IHCs (Dechesne *et al.*, 1991), while in the gerbil it is found in the IHCs and the third row Deiters cells (Fig. 3). In no cases have we found longitudinal gradients of this calcium-binding protein.

Calbindin is one of the most studied calcium-binding proteins in the cochlea. In guinea-pig (Slepecky and Ulfendahl, 1993), rat (Legrand *et al.*, 1988; Dechesne *et al.*, 1991), mouse (Dechesne and Thomasset, 1988) and gerbil (Fig. 4) it has been localized in the cell body and cuticular plate of the IHCs and OHCs. While parvalbumin and calretinin are only

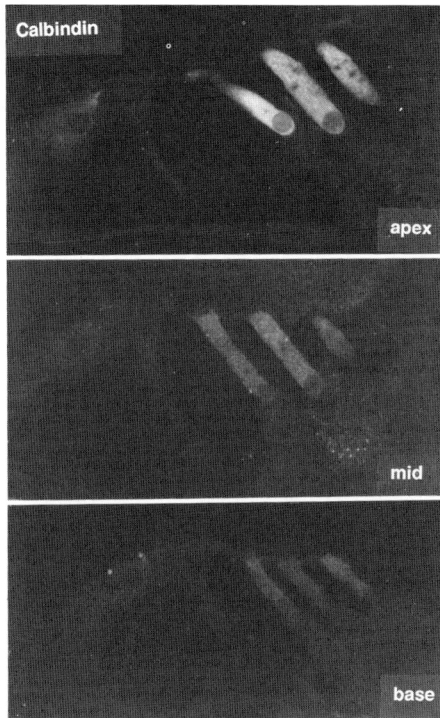

Fig. 4. Antibodies to calbindin stain IHCs and OHCs in the apex of the cochlea, but staining decreases in the middle turns and is weak or absent in the base.

in the IHCs, calbindin is conspicuously present in both the IHCs and the OHCs, but not present in supporting cells. Like parvalbumin, it displays an obvious longitudinal gradient along the length of the sensory epithelium, both in the guinea-pig (Slepecky, unpublished results) and the gerbil cochlea (Fig. 4). In the apex, it is found to stain the sensory cells intensely. IHCs hair cells in the apex of both the gerbil and guinea-pig cochlea stain brightly. IHCs in the middle and basal turns of the gerbil cochlea show decreased staining with the most basal cells showing no staining at all. IHCs in the middle and basal turns of the guinea-pig cochlea still show bright staining, with the most basal cells only showing decreased staining. OHCs in the apical turn of the gerbil cochlea show intense staining, while OHCs in the middle and basal turns show scattered loss of staining, with eventual loss of staining in the most basal cells. OHCs in the apical turn of the guinea-pig cochlea show intense staining. In the middle turns, the first row of OHCs show decreased staining, while the second and third row OHCs remain

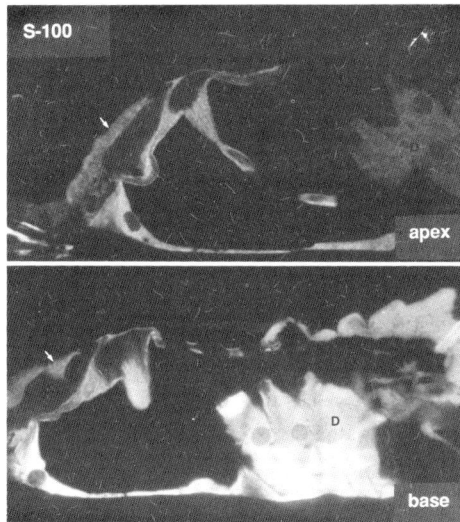

Fig. 5. S-100 appears to be a supporting cell specific protein. It is present in the IHC supporting cells (the inner border and inner phalangeal cells), the pillar cells, the Deiters cells and the Hensen cells.

brightly stained. By the basal turn, all OHCs display no staining at all. These immunocytochemical results showing longitudinal gradients for staining of calbindin have been confirmed on immunoblots where apical and basal turns have been analyzed separately.

S-100 is a calcium-binding protein which is supporting cell specific in the guinea-pig (Slepecky, unpublished results) and gerbil (Fig. 5) cochleas. The finding that in the mouse and hamster cochlea, this protein is present also in the sensory hair cells (Foster *et al.*, 1994) may represent species differences or the use of very high titer antibodies on relatively thick cryosections. Our finding that S-100 is present only in supporting cells (inner border, inner phalangeal, pillar, Deiters and Hensen cells) in two different species in 1 μm thick sections is consistent with its localization in nervous tissue where it is found in glial and Schwann cells.

Summary and Conclusions

Analysis of the type and location of various proteins within the sensory and supporting cells of the gerbil organ of Corti shows striking similarities with their location in cells of the guinea-pig cochlea. A

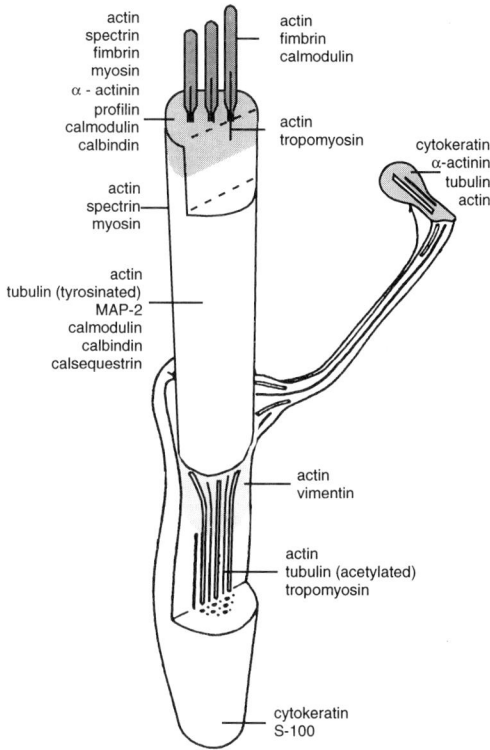

Fig. 6. A summary diagram that shows the distribution of some of the cytoskeletal and calcium-binding proteins that have been detected in the sensory and supporting cells of the cochlea to date.

summary diagram is shown in Fig. 6. In the sensory cells, actin is localized in the stereocilia, cuticular plate, circumferential ring, lateral wall and cytoplasm. The presence of various actin-binding proteins also suggests a similar organization of the actin, in cochlea across species and at various locations along the length of the sensory epithelium in the cochlea from one animal species. Actin is also present in the supporting cells, in organized bundles and in meshworks. The actin-containing specializations present in Deiters cells, just beneath the OHCs, are more pronounced in the apex of the cochlea than in the base. These structures appear similar in composition to the infracuticular network of filaments that is present in the longest hair cells at the apex of the guinea-pig cochlea, and both may be involved in providing mechanical support.

Analysis of tubulin in the sensory and supporting cells in the cochleas from different species shows that the distribution of microtubules in the individual cell types is similar along the length of the sensory epithelium. Study of post-translational modifications suggests that the microtubules in the sensory cells are dynamic, undergoing cycles of polymerization and depolymerization. This is supported by the presence in OHCs of MAP-2 (a microtubule-associated protein which reversibly cross-links actin microfilaments and microtubules) and calsequestrin (which is often found associated with membrane vesicles which sequester and release calcium upon cell stimulation).

Tubulin in the microtubule–microfilament bundles in the supporting cells has been modified by the addition of glutamyl and acetyl residues, suggesting that these microtubules are more long-lived and therefore more stable. Studies also show that the actin microfilaments in these bundles co-localize with tropomyosin and that intermediate filament proteins distribute throughout the cell. Both findings further suggest that the supporting cells are rigid and provide mechanical support for the organ of Corti, allowing movement of the basilar membrane to be transmitted to the apical surface of the sensory hair cells.

The presence of hair cell specific (parvalbumin, calmodulin and calbindin) and supporting cell specific (S-100) calcium-binding proteins suggest that these cell types have different roles within the organ of Corti and that they may be differentially regulated. Longitudinal gradients for the staining of parvalbumin (IHCs) and calbindin (IHCs and OHCs) suggest functional differences between hair cells at the apex and base of the cochlea. The finding that S-100, a glial cell specific calcium-binding protein, is found only in supporting cells suggests a metabolic support function for the non-sensory cells as well as a role in providing structural support.

References

Angelborg, C. and Engstrom, H. (1972). Supporting elements in the organ of Corti. Fibrillar structures in the supporting cells of the organ of Corti of mammals. *Acta Otolaryngol. Suppl.* **301:** 49–60.

Anniko, M. and Arnold, W. (1990). Analytical electron microscopy and monoclonal antibody techniques applied to the human inner ear. *Adv. Otorhinolaryngol.* **45:** 1–45.

Arima, T., Uemura, T. and Yamamoto, T. (1986). Cytoskeletal organization in the supporting cell of the guinea pig organ of Corti. *Hearing Res.* **24:** 169–175.

Arima, T., Uemura, T. and Yamamoto, T. (1987). Three-dimensional visualizations of the inner ear hair cell of the guinea pig. A rapid-freeze, deep-etch study of filamentous and membranous organelles. *Hearing Res.* **25:** 61–68.

Bauwens, L. J. M., Veldman, J. E., Ramaekers, F. C. S., Bouman, H. and Huizing, E. H. (1991). Expression of intermediate filament proteins in the adult human cochlea. *Ann. Otol. Rhinol. Laryngol.* **100:** 211–218.

Dechesne, C. J. and Thomasset, M. (1988). Calbindin (CaBP 28 kDa) appearance and distribution during development of the mouse inner ear. *Dev. Brain Res.* **40:** 233–242.

Dechesne, C. J., Winsky, L., Kim, H. N., Goping, G., Vu, T. D., Wenthold, R. J. and Jacobowitz, D. M. (1991). Identification and ultrastructural localization of a calretinin-like calcium-binding protein (protein 10) in the guinea pig and rat inner ear. *Brain Res.* **560:** 139–148.

Drenckhahn, D., Kellner, J., Mannherz, H. G., Groschel-Stewart, U., Kendrick-Jones, J. and Scholey, J. (1982). Absence of myosin-like immunoreactivity in stereocilia of cochlear hair cells. *Nature* **300:** 531–532.

Eybalin, M. and Ripoll, C. (1990). Immunolocalisation de la parvalbumine dans deux types de cellules glutamatergiques de la cochlée du cobaye; les cellules ciliées internes et les neurones du ganglion spiral. *C. R. Acad. Sci. Paris* **310:** 639–644.

Flock, Å. and Cheung, H. C. (1977). Actin filaments in sensory hairs of inner ear receptor cells. *J. Cell Biol.* **75:** 339–343.

Flock, Å., Bretscher, A. and Weber, K. (1982). Immunohistochemical localization of several cytoskeletal proteins in inner ear sensory and supporting cells. *Hearing Res.* **6:** 75–89.

Foster, J. D., Drescher, M. J., Hatfield, J. S. and Drescher, D. G. (1994). Immunohistochemical localization of S-100 protein in auditory and vestibular end organs of the mouse and hamster. *Hearing Res.* (in press).

Furness, D. N., Hackney, C. M. and Steyger, P. S. (1990). Organization of microtubules in cochlear hair cells. *J. Electr. Microscop. Tech.* **15:** 261–279.

Holley, M. C. and Ashmore, J. F. (1990). Spectrin, actin and the structure of the cortical lattice in mammalian cochlear outer hair cells. *J. Cell Sci.* **96:** 283–291.

Holtham, K. A. and Slepecky, N. B. (1995). A simplified method for obtaining 0.5 µm sections of small tissue specimens for use in immunocytochemistry. *J. Histochem. Cytochem.* (in press)

Kikuchi, T., Takasaka, T., Tonosaki, A., Katori, Y. and Shinkawa, H. (1991).

Microtubules of guinea pig cochlear epithelial cells. *Acta Otolaryngol.* **111:** 286–290.

Kuijpers, W., Tonnaer, E. L. G. M., Peters, T. A. and Ramaekers, F. C. S. (1991). Expression of intermediate filament proteins in the mature inner ear of the rat and guinea pig. *Hearing Res.* **52:** 133–146.

Legrand, C., Brehier, A., Clavel, M. C., Thomasset, M. and Rabie, A. (1988). Cholecalcin (28 kDa CaBP) in the rat cochlea. *Dev. Brain Res.* **38:** 121–129.

Nishida, Y., Fujimoto, T., Takagi, A., Honjo, I. and Ogawa, K. (1993). Fodrin is a constituent of the cortical lattice in outer hair cells of the guinea pig cochlea: immunocytochemical evidence. *Hearing Res.* **65:** 274–280.

Raphael, Y., Marshak, G. H., Barash, A. and Geiger, B. (1987). Modulation of intermediate-filament expression in developing cochlear epithelium. *Differentiation* **35:** 151–162.

Saito, K. and Hama, K. (1982). Structural diversity of microtubules in the the supporting cells of the sensory epithelium of guinea pig organ of Corti. *J. Electr. Microscop.* **311:** 278–281.

Schrott, A., Egg, G. and Spoendlin, H. (1988). Intermediate filaments in the cochleas of normal and mutant (w/wv, sl/sld) mice. *Arch. Otorhinolaryngol.* **245:** 250–254.

Shi, S. R., Tandon, A. K., Hausmann, R. R. M., Kalra, K. L. and Taylor, C. R. (1993). Immunohistochemical study of intermediate filament proteins on routinely processed, celloidin-embedded human temporal bone sections by using a tew technique for antigen-retrieval. *Acta Otolaryngol.* **113:** 48–54.

Slepecky, N. B. (1989). Cytoplasmic actin and cochlear outer hair cell motility. *Cell Tissue Res.* **257:** 69–75.

Slepecky, N. B. and Chamberlain, S. C. (1982). Distribution and polarity of actin in the sensory hair cells of the chinchilla cochlea. *Cell Tissue Res.* **224:** 15–24.

Slepecky, N. and Chamberlain, S. C. (1985). Immunoelectron microscopic and immunofluorescent localization of cytoskeletal and muscle-like contractile proteins in inner ear sensory hair cells. *Hearing Res.* **20:** 245–260.

Slepecky, N. B. and Chamberlain, S. C. (1987). Tropomyosin co-localizes with actin microfilaments and microtubules in supporting cells of the inner ear. *Cell Tissue Res.* **248:** 63–66.

Slepecky, N. B. and Ulfendahl, M. (1992). Actin-binding and microtubule associated proteins in the organ of Corti. *Hearing Res.* **57:** 201–215.

Slepecky, N. B. and Ulfendahl, M. (1993). Evidence for calcium-binding proteins and calcium-dependent regulatory proteins in sensory cells of the organ of Corti. *Hearing Res.* **70:** 73–84.

Steyger, P. S., Furness, D. N., Hackney, C. M. and Richardson, G. P. (1989). Tubulin and microtubules in cochlear hair cells: comparative immunocytochemistry and ultrastructure. *Hearing Res.* **42:** 1–16.

Thorne, P. R., Carlisle, L., Zajic, G., Schacht, J. and Altschuler, R. A. (1987). Differences in the distribution of F-actin in outer hair cells along the organ of Corti. *Hearing Res.* **30:** 253–266.

Weaver, S. P., Hoffpauir, J. and Schweitzer, L. (1993). Actin distribution along the lateral wall of gerbil outer hair cells. *Brain Res. Bull.* **31:** 225–228.

Ylikoski, J., Pirvora, U., Narvanen, O. and Virtanen, I. (1990). Non-erythroid spectrin (fodrin) is a prominent component of the cochlear hair cells. *Hearing Res.* **43:** 199–204.

Hair Cell Ultrastructure and Mechanotransduction: Morphological Effects of Low Extracellular Calcium Levels on Stereociliary Bundles in the Turtle Cochlea

CAROLE M. HACKNEY AND DAVID N. FURNESS

Department of Communication and Neuroscience, University of Keele, Staffordshire ST5 5BG, UK

Introduction

The basilar papilla of the turtle is homologous to the mammalian cochlea, containing about 1000 hair cells situated on an elliptical basilar membrane (Miller, 1978; Sneary, 1988). Its hair cells have narrow frequency tuning curves with characteristic frequencies ranging from about 20 Hz at the apical end of the cochlea to 600 Hz at the base (Crawford and Fettiplace, 1980; Art *et al.*, 1986). This frequency analysis is accomplished chiefly by means of an electrical resonance in the hair cells (Crawford and Fettiplace, 1981) that filters the receptor potential through the combined action of a voltage-dependent Ca^{2+}-conductance and a Ca^{2+}-activated K^+ conductance (Art and Fettiplace, 1987).

It has been proposed that the mechanotransduction channels in all vertebrate hair cells are gated by fine filamentous strands which run between the tips of the shorter stereocilia and the sides of the taller stereocilia in the next row (for review, see Pickles and Corey, 1992). These tip links have also been implicated in the adaptation that is observed when stereociliary bundles are subjected to a sustained deflection. It is thought that the adaptation represents an adjustment of

tension in the tip link specifically by sliding its attachment point along
the side of the stereocilium (Howard and Hudspeth, 1988) and this idea
has been extended by the suggestion that an active motor element might
be involved (Assad and Corey, 1992; Gillespie *et al.*, 1993; see also Corey,
1995, this volume).

However, whilst there is considerable physiological evidence to
support the view that the channels are situated near the distal ends of
the stereocilia (e.g. Hudspeth, 1982; Jaramillo and Hudspeth, 1991),
recent attempts to localize them immunocytochemically suggest that
they may be at the point at which the stereocilia of adjacent rows come
into the closest contact with each other rather than being located
precisely at one or other end of the tip link (Hackney *et al.*, 1991, 1992).
A junction-like structure is also seen at this point using transmission
electron microscopy (Hackney *et al.*, 1992). These observations have led
to the suggestion that this tip junction may also be involved in the direct
mechanical gating of the transduction channels that occurs in hair cells
(Hackney *et al.*, 1992, 1993a).

Tip links are known to be present on the stereociliary bundles of
turtle hair cells (Sneary, 1988; Hackney *et al.*, 1993b). It has also been
shown that reducing the level of extracellular calcium around turtle hair
cells (to 50 μM) results in an increase in the current flowing through the
transduction channels and also reduces or abolishes adaptation; if the
calcium level is lowered sufficiently (to 1 μM or below) the transduction
current is irreversibly abolished (Crawford *et al.*, 1991). These
observations led Crawford *et al.* (1991) to suggest a dual action of
calcium ions: an external block of the transduction channels and an
intracellular effect that could be best modelled by internal Ca^{2+}
stabilizing one of the closed states of the channel. They also attributed
the irreversible loss of sensitivity to bundle displacements that they
observed following the treatment of turtle hair cells with very low Ca^{2+}
levels to the rupture of mechanical linkages to the transduction
channels. Moreover, when the extracellular calcium level is drastically
reduced using the calcium chelator BAPTA around bullfrog saccular
hair cells, electron microscopical studies suggest that the tip links are
lost (Assad *et al.*, 1991). However, no comparable morphological
information is available on the effects of lowering the extracellular Ca^{2+}
level on the tip links or the tip junction in turtles, information which
might also assist in distinguishing between these two structures as
possible locations for the transduction channels.

The aim of this preliminary study was therefore to reduce the
extracellular calcium level around the turtle basilar papilla to levels
known to produce loss of the transduction current and to investigate the

concomitant effects on the tip links and tip junctions using scanning (SEM) and transmission electron microscopy (TEM).

Materials and Methods

The basilar papillae of 12 red-eared turtles, *Trachemys* (formerly *Pseudemys*) *scripta elegans*, with carapaces 8–9 cm in length were obtained as in the physiological studies of Crawford and Fettiplace (e.g. Crawford and Fettiplace, 1985). Animals were decapitated and the cochlear duct and lagena dissected out in artificial perilymph of composition (in mM): NaCl, 130; KCl, 4; $MgCl_2$, 2.2; $CaCl_2$, 2.8; Na-HEPES, 5, pH 7.6. The cochlear duct was then opened up and exposed for 20 min to the peptidase subtilisin BPN' (Sigma, 30 µg/ml) made up in the same solution. The tectorial membrane was then peeled from the underlying epithelium and the tissue incubated for 20 min at room temperature in an artificial perilymph solution containing either 2.8 mM Ca^{2+}, or buffered with EGTA (in mM: NaCl, 125; KCl, 4; $CaCl_2$, 4.3; NaHEPES, 5; EGTA, 5, pH 7.6) to produce a solution containing 0.1 µM Ca^{2+}. The tissue was then fixed in 2.5% glutaraldehyde in 0.1 M sodium cacodylate buffer containing 2–3 mM $CaCl_2$ (pH 7.4) for 2 hr. After washing in buffer, the cochlear duct was post-fixed in 1% OsO_4.

For SEM, the papillae were impregnated with osmium using the OTOTO technique as described previously for the mammalian cochlea (Furness and Hackney, 1986). For this, the tissue was thoroughly washed in distilled water and placed in a saturated solution of thiocarbohydrazide for 20 min followed by further thorough washing and exposure to OsO_4 for 1 hr. This procedure was repeated twice more, followed by dehydration via an alcohol series and critical point drying from dry absolute ethanol using liquid carbon dioxide as the transitional fluid. The cochlea was then attached to a scanning stub using electroconductive glue (silver DAG: Agar Scientific) and viewed using an Hitachi S-4500 field emission scanning electron microscope operated at 2–5 kV.

For TEM, the cochlear duct was fixed in the same way but following the first post-fixation step with OsO_4, was dehydrated and embedded in Spurr resin. Ultrathin sections (70 nm) were stained with uranyl acetate and lead citrate and examined in a JEOL 100CX electron microscope operated at 100 kV.

Active hearing

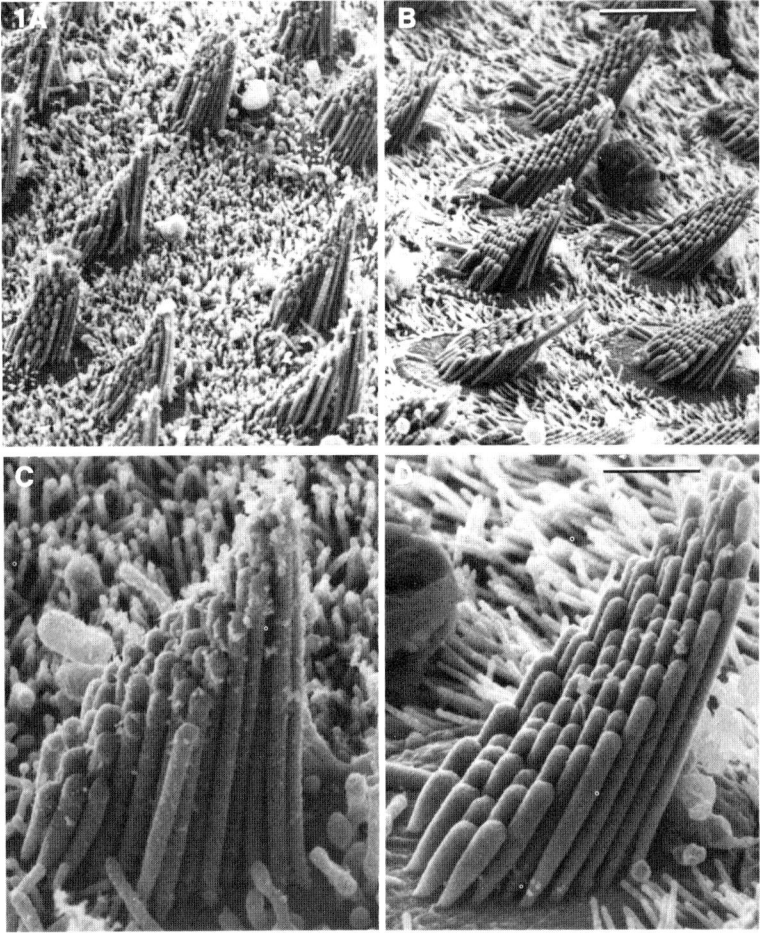

Fig. 1. (A, B) Low-power scanning electron micrographs taken at the same magnification and as far as possible from the same viewing angle showing hair bundles from the midregion of turtle papillae incubated in a solution containing 2.8 mM Ca^{2+} (A) or 0.1 μM Ca^{2+} (B). Note that the bundles corresponding to the low Ca^{2+} treatment appear to be leaning in the direction of the kinocilium more than those from the higher Ca^{2+} treatment. Scale bar = 5 μm. (C, D) Higher magnification micrographs of two hair bundles taken as above from the high Ca^{2+} treatment (C) and the low Ca treatment (D). Note that the whole bundle on the right appears to be leaning more in the positive direction (i.e. the direction in which the bundle would have to be deflected to cause depolarization of the hair cell) than the bundle on the left. Scale bar = 1 μm.

Results

At lower magnifications in the scanning electron microscope, the two most obvious differences between the papillae incubated in the 2.8 mM (high) Ca^{2+} medium compared with the 0.1 µM (low) Ca^{2+} medium are the greater amount of debris adhering to the surface of the specimen in the former and the change in the angle made by the hair bundles with the surface of the reticular lamina (Fig. 1A, B). In the low Ca^{2+} specimens, the hair bundles appear to be tilted in the positive direction (Fig. 1), i.e. the direction in which deflection produces depolarization of the hair cells under normal circumstances (e.g. Flock, 1977; Shotwell *et al.*, 1981). At higher magnifications, other changes are also apparent. Most striking is the disappearance of the majority of the tip links following the low Ca^{2+} treatment (Fig. 2), but also noticeable is the change in the tip shapes which appear more rounded and do not have the 'tenting' that is associated with the tip links. Another feature which is less obvious when the bundle is observed from above, but is more apparent in side view, is the greater spacing between the tips and sides of the adjacent stereocilia that is observed following the low Ca^{2+} treatment (Figs 2 and 3). In SEM, there is a change in the tip junction region of the stereocilia, with the membranes of adjacent stereocilia appearing to be virtually fused following the high Ca^{2+} treatment (Fig. 3A) but showing more separation in the low Ca^{2+} treatment. This impression is confirmed in TEM, which reveals both the loss of tip links and the separation of membranes (Fig. 4). This separation of membranes also reveals regularly spaced particles in the tip junction region (Fig. 4B).

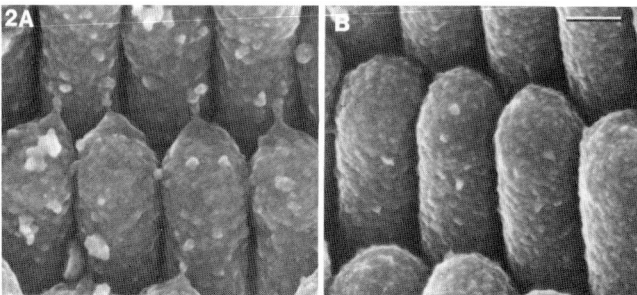

Fig. 2. (A, B) Scanning electron micrographs of the tips of stereocilia from hair bundles from the midregion of turtle papillae incubated in a solution containing 2.8 mM Ca^{2+} (A) or 0.1 µM Ca^{2+} (B). Note the presence of tip links in (A) and their absence in (B). Also notice the change in shape of the tips in (B) due to the loss of 'tenting' and the increased gaps between the rows of stereocilia. Scale bar = 200 nm.

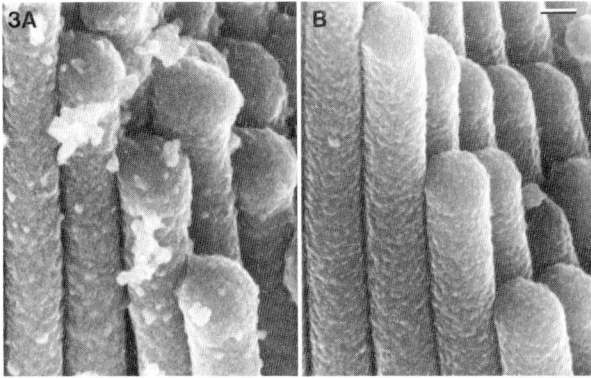

Fig. 3. (A, B) Scanning electron micrographs showing the tip junction region of stereocilia from hair bundles from the midregion of turtle papillae incubated in a solution containing 2.8 mM Ca^{2+} (A) or 0.1 μM Ca^{2+} (B). Note the virtual fusion of the tips of the shorter stereocilia to the sides of the taller stereocilia seen in (A) and the apparent gap in this location seen in (B). Scale bar = 150 nm.

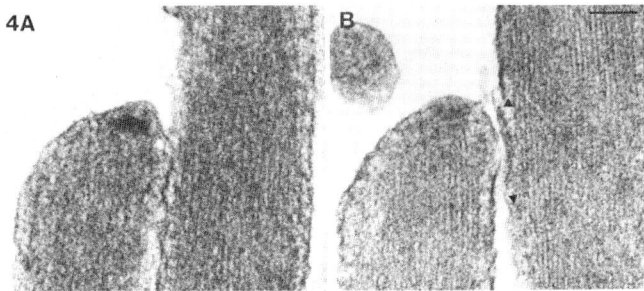

Fig. 4. (A, B) Transmission electron micrographs of the tips of stereocilia from hair bundles from the midregion of turtle papillae incubated in a solution containing 2.8 mM Ca^{2+} (A) or 0.1 μM Ca^{2+} (B). Note the virtual fusion of the apical membrane of the shorter stereocilium with the side of the taller stereocilium seen in (A) and the increased gap in this location seen in (B). A junction-like structure appears to occur where the stereocilia have separated (region delimited by arrowheads). Scale bar = 100 nm.

Discussion

The morphological changes described here for the turtle papilla are consistent with the displacement-clamp measurements of the forces made in bullfrog hair cells during the alteration of extracellular Ca^{2+} levels (Jaramillo and Hudspeth, 1993). When the extracellular Ca^{2+} level was drastically lowered in those experiments using the calcium chelator

BAPTA, the bundle ultimately moved in the positive direction, i.e. towards the kinocilium. This is observed ultrastructurally in the turtle papilla also, although the fact that the material has been subjected to fixation and drying must been borne in mind in making comparisons with the physiological experiments.

In addition, the disappearance of the tip links observed following low Ca^{2+} treatment (Assad *et al.*, 1991) is also seen in the turtle papilla. The converse of this is worth noting in that, in addition to the tip links, there appeared to be more surface material adhering to the bundles in the high Ca^{2+} treatment indicating the possible retention of glycocalyx in these specimens. In the experiments in which bullfrog saccular hair cells were subjected to low Ca^{2+} levels, Assad *et al.* (1991) reported no other systematic changes in the relationship of the stereocilia within the bundle to each other, e.g. they did not splay. In the turtle material, however, it is clear that other changes do also occur in the region of the stereociliary tips, i.e. they move apart slightly and the point of contact which we have termed here the tip junction also separates. If lowering the calcium level had removed the tip links but left the tip junctions intact, it would have provided a means of distinguishing between the two sites in terms of their possible involvement in transduction. However, as both structures appear to be affected by this treatment, this type of experiment does not preclude the possible involvement of the tip junction in mechanotransduction.

Acknowledgements

We are grateful to Professor R. Fettiplace who provided the material used in the SEM part of this study, and to Professor A. Crawford for providing the material studied using TEM. This work has been supported by the Wellcome Trust and Hearing Research Trust; the field emission electron microscope was provided by a grant to D.N.F. and C.M.H. from the Wellcome Trust.

References

Art, J. J. and Fettiplace, R. (1987). Variation of membrane properties in hair cells isolated from the turtle cochlea. *J. Physiol.* **385:** 207–242.

Art, J. J., Crawford, A. C. and Fettiplace, R. (1986). Electrical resonance and membrane currents in turtle cochlear hair cells. *Hearing Res.* **22:** 31–36.

Assad, J. A. and Corey, D. P. (1992). An active motor model for adaptation by vertebrate hair cells. *J. Neurosci.* **12:** 3291–3309.

Assad, J. A., Shepherd, G. M. G. and Corey, D. P. (1991). Tip-link integrity and mechanical transduction in vertebrate hair cells. *Neuron* **7:** 985–994.

Crawford, A. C. and Fettiplace, R. (1980). The frequency selectivity of auditory nerve fibers and hair cells in the cochlea of the turtle. *J. Physiol.* **306:** 79–125.

Crawford, A. C. and Fettiplace, R. (1981). An electrical tuning mechanism in turtle cochlear hair cells. *J. Physiol.* **312:** 377–412.

Crawford, A. C. and Fettiplace, R. (1985). The mechanical properties of ciliary bundles of turtle cochlear hair cells. *J. Physiol.* **364:** 359–379.

Crawford, A. C., Evans, M. G. and Fettiplace, R. (1991). The actions of calcium on the mechano-electrical transducer current of turtle hair cells. *J. Physiol.* **434:** 369–398.

Flock, Å. (1977). Physiological properties of sensory hairs in the ear. In *Psychophysics and Physiology of Hearing* (E. F. Evans and J. P. Wilson, eds), pp. 15–25. Academic Press, London.

Furness, D. N. and Hackney, C. M. (1986). High-resolution scanning-electron microscopy of stereocilia using the osmium-thiocarbohydrazide coating technique. *Hearing Res.* **21:** 243–249.

Gillespie, P. G., Wagner, M. C. and Hudspeth, A. J. (1993). Identification of a 120 kD hair-bundle myosin located near stereociliary tips. *Neuron* **11:** 581–594.

Hackney, C. M., Furness, D. N. and Benos, D. J. (1991). Localization of putative mechanoelectrical transducer channels in cochlear hair cells by immunoelectron microscopy. *Scanning Microsc.* **5:** 741–746.

Hackney, C. M., Furness, D. N., Benos, D. J., Woodley, J. F. and Barratt, J. (1992). Putative immunolocalization of the mechanoelectrical transduction channels in mammalian cochlear hair cells. *Proc. R. Soc. Lond. B* **248:** 215–221.

Hackney, C. M., Fettiplace, R. and Furness, D. N. (1993a). The functional morphology of stereociliary bundles on turtle cochlear hair cells. *Hearing Res.* **69:** 163–175.

Hackney, C. M., Furness, D. N. and Mahendrasingam, S. (1993b). The mechanotransduction channels in cochlear hair cells may be revealed by antibodies which recognise other amiloride-sensitive channels. In *Bio-*

physics of Hair Cell Sensory Systems (H. Duifhuis, J. W. Horst, P. van Dijk and S. M. van Netten, eds), pp. 107–115. World Scientific, Singapore.

Howard, J. and Hudspeth, A. J. (1988). Compliance of the hair bundle associated with gating of mechanoelectrical transduction channels in the bull-frog's saccular hair cell. *Neuron* **1**: 189–199.

Hudspeth, A. J. (1982). The cellular basis of hearing: the biophysics of hair cells. *Science* **230**: 745–752.

Jaramillo, F. and Hudspeth, A. J. (1991). Localization of the hair-cell's transduction channels at the hair bundle's top by iontophoretic application of a channel blocker. *Neuron* **7**: 409–420.

Jaramillo, F. and Hudspeth, A. J. (1993). Displacement-clamp measurements of the forces exerted by gating springs in the hair bundle. *Proc. Natl. Acad. Sci. USA* **90**: 1330–1334.

Miller, M. R. (1978). Scanning electron microscopic studies of the papilla basilaris of some turtles and snakes. *Am. J. Anat.* **151**: 409–435.

Pickles, J. O. and Corey, D. P. (1992). Mechanoelectrical transduction by hair cells. *Trends Neurosci.* **15**: 254–259.

Shotwell, S. L., Jacobs, R. and Hudspeth, A. J. (1981). Directional sensitivity of individual vertebrate hair cells to controlled deflection of their hair bundles. *Ann. NY Acad. Sci.* **374**: 1–10.

Sneary, M. G. (1988). Auditory receptor of the red-eared turtle: I. General ultrastructure. *J. Comp. Neurol.* **276**: 573–587.

Transducer Currents and Bundle Movements in Outer Hair Cells of Neonatal Mice

C. J. KROS, G. W. T. LENNAN AND G. P. RICHARDSON

School of Biological Sciences, The University of Sussex, Falmer, Brighton BN1 9QG, UK

Introduction

In hair cells, the transducer current forms the link between mechanical stimulation to the hair bundle and generation of the receptor potential. Transducer currents arise because of the presence of mechanosensitive ion channels in the hair bundle. Properties of the transduction mechanism can be investigated by applying either displacements (usually with a stiff glass fibre) or forces to the bundle. Displacement by a stiff fibre is suitable as a means of inducing transducer currents, but gives no information about the bundle's mechanical properties. Bundle mechanics are reported to be affected by the opening of the transducer channels (Howard and Hudspeth, 1988; Russell *et al.*, 1992), and by adaptation of the transducer currents or receptor potentials (Howard and Hudspeth, 1987; Russell *et al.*, 1989; Assad and Corey, 1992). Such mechanical changes, which may be important for the normal physiology of the cochlea (Jaramillo *et al.*, 1993), can be measured by monitoring bundle movements in response to a force stimulus (Howard and Hudspeth, 1987), or by directly measuring the force exerted by the bundle (Jaramillo and Hudspeth, 1993).

Here we report results of experiments in which a fluid jet was used to apply force steps to the hair bundles of neonatal outer hair cells (OHCs). Our aim was to investigate the nature of the bundle movements, and to correlate these bundle movements with the transducer currents.

Methods

Preparation and electrical recording

The preparation and electrical recording techniques were largely as described before (Kros *et al.*, 1992; Rüsch *et al.*, 1994). The cochlear cultures were bathed at room temperature (22–24 deg C) in a solution containing (concentrations in mM): 144 NaCl, 0.7 NaH$_2$PO$_4$, 5.8 KCl, 1.3 CaCl$_2$, 0.9 MgCl$_2$, 5.6 D-glucose, 10 HEPES–NaOH, pH 7.4. Amino acids and vitamins for Eagle's Minimal Essential Medium were added from concentrates (Gibco, UK).

Cells were voltage-clamped in the whole-cell configuration with an Axopatch 200A. Patch-pipettes drawn from soda-glass capillaries were coated with ski-wax (Astra-Gruppen, Norway) and had resistances in the bath of 2.3–2.6 MΩ. The membrane capacitance of the OHCs used in this study was 5.4 ± 0.9 pF (n = 8, mean ± SD, in the rest of this paper mean results are also presented ± SD). Residual series resistance after electronic compensation was 1–6 MΩ, the time constant of the clamp was 10–20 μsec. No corrections were made for the voltage drop across the series resistance, which was at most 2 mV. Patch-pipettes were filled with CsCl-based solutions, to minimize voltage-dependent potassium currents. Their composition was (mM): 135 CsCl, 0.1 CaCl$_2$, 5 EGTA–NaOH, 3.5 MgCl$_2$, 2.5 Na$_2$ATP, 5 HEPES–NaOH, pH 7.3. The holding potential was –84 mV or +76 mV, including a –4 mV correction for the liquid junction potential measured between intra- and extracellular solutions.

Measurement of bundle displacements

To monitor the position of the free-standing hair bundles, a 4 mW diode laser (LAS-200–670–5, LaserMax, USA), emitting plane-polarized light at 670 nm, was combined with the Nomarski optics of a Zeiss ACM top-focusing microscope (Oberkochen, Germany) to act as a localized, bright light source. The design resembles the early version of the laser differential interferometer developed by Denk and Webb (Denk *et al.*, 1989; Denk and Webb, 1990). The original Nomarski optics of the ACM (condenser NA 0.63; water-immersion objective ×40, NA 0.75) were left intact. The elliptical laser beam was circularized to form a Gaussian beam with a $1/e^2$ diameter of 0.67 mm and focused to a 9 μm spot in a pinhole. A polarization-preserving cube beamsplitter positioned between the field stop and the condenser combined the polarized

Köhler illumination (546 nm) and the laser beam, with the same vibration axis.

The first Wollaston prism and the condenser were used to split the beam and focus two diffraction-limited spots, with diameters of 2 μm, and a centre-to-centre separation of about 0.5 μm, near the top of the hair bundle. The beams were recombined by the second, movable, Wollaston prism, which was positioned so as to form an image of one wing of the bundle with a bright and a dark edge. The image of the top of the hair bundle was focused onto a pair of photodiodes (LD2–5T, Centronic, UK) through the objective and a ×25 eyepiece, at a magnification of ×700. A Leitz (Germany) binocular phototube with back-reflection was used to visualize the preparation as well as the position of the photodiodes.

A photodiode pair was used differentially, rather than a single photodiode, to reduce noise due to intensity fluctuations of the laser. The photodiodes were used in photoconductive mode, with a reverse bias of 15 V, and the amplified output signal had a bandwidth at the light intensities used for these experiments of 0–13 kHz.

To calibrate bundle displacements, the photodiode pair and the laser beams were scanned in concert over the bundle with computer-controlled piezotranslators (Physik Instrumente, Germany), and the photo-voltage was recorded as a function of displacement. The range over which the laser and the detector were scanned corresponded to bundle movements over a distance from –300 nm (inhibitory direction) to +500 nm (excitatory direction) away from the resting position. Such calibrations were performed just before and after all recordings of bundle movements, to minimize distortion in the optical recordings due to drift in the preparation. A figure of a bundle movement calibration is shown in Kros *et al.* (1993).

OHCs with the tallest bundles (3–5 μm) were selected for these experiments. Control experiments showed that the edge of the cuticular plate moved by up to 20% of the movements measured at the top of the hair bundle. This would cause a small overestimate of the bundle movement. The imperfect optical sectioning would, however, cause an underestimate of a comparable magnitude. No corrections were applied for these possible sources of error.

Mechanical stimulation

Hair bundles were mechanically stimulated by a fluid jet driven by a piezoelectric disc, as described before (Kros *et al.*, 1992; Rüsch *et al.*, 1994). Internal diameters of the tips of the jet pipettes were 5–11 μm. The pipette was placed at a distance of 10–15 µm from the hair bundle, generally on the modiolar side of the cell, so that fluid flowing out of the pipette would move the bundle in the excitatory direction. Mechanical stimuli were force steps lasting 50 msec. The driver voltage for the fluid jet was low-pass filtered at 0.5 kHz (eight-pole Bessel), to avoid excitation of the disc's first resonance, which occurred at about 4 kHz.

Figure 1(A) shows records of the movement of a glass probe which was placed 10 µm from the tip of the pipette delivering the fluid jet. Steps in driver voltage resulted with little delay (<400 µsec) in displacement steps of the probe that followed the waveform of the driver voltage. Figure 1(B) shows that the relation between driver voltage to the piezo disc and the displacement of the glass probe was linear. In particular, fluid flowing out of the pipette resulted in equally large displacements as fluid flowing into the pipette. From such experiments we concluded that the force exerted by the fluid jet was also a linear function of the driver voltage.

Data acquisition

Records of the driver voltage to the fluid jet, the output signal of the photodiodes and the transducer currents were low-pass filtered with 2.5

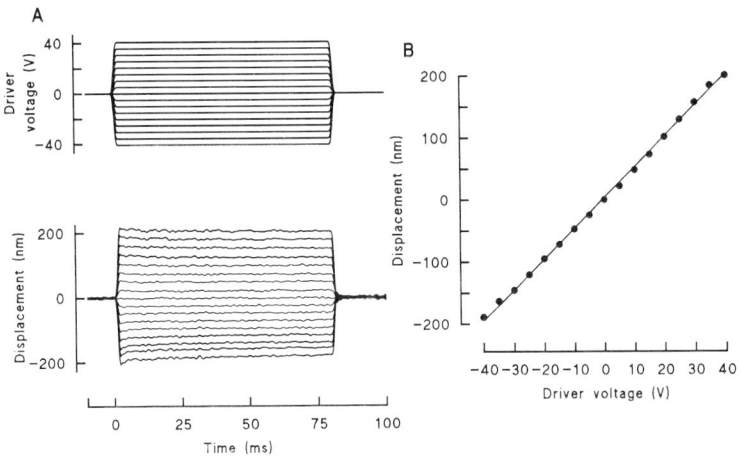

Fig. 1. Calibration of the fluid-jet stimulator. (A) Displacements of a glass probe of known stiffness (below) in response to the steps in driver voltage to the fluid-jet (above). Tip diameter of the jet-pipet 9 µm. Averages of five presentations. (B) Steady-state probe displacement as a function of driver voltage. The fitted line has a slope of 5.0 nm/V. The pressure exerted by the fluid-jet was calculated as 12 pN/µm²V. Positive displacements are away from the jet-pipette.

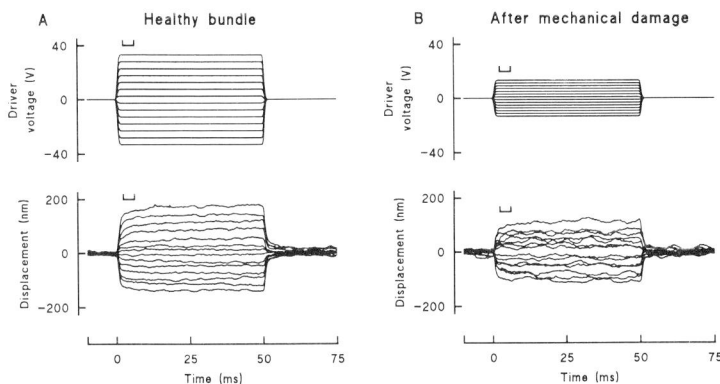

Fig. 2. Bundle movements of an OHC elicited by force steps before (A) and after (B) mechanical damage. Responses to 14 force steps with increments of 2.5 V (A) and 1.0 V (B) are shown. In this experiment, unusually, the jet-stimulator was placed on the strial side of the bundle, so fluid flowing into the pipette caused the bundle to move in the excitatory direction (plotted positive, also in the other figures). The slow component of the bundle movement contributed about 20% of the total displacement. Responses are single traces. Bundle height 4.5 μm. Markers indicate where the displacements plotted in Fig. 3 were measured.

kHz eight-pole Bessel filters and sampled at 5 kHz by a Labmaster 100 kHz DMA data acquisition board (Scientific Solutions, USA). This board was also used to generate the driver voltage to the fluid jet. The recorded 10–90% rise time of the driver-voltage steps was 880 μsec.

Results

Movements of the hair bundle

Figure 2 shows displacements of an OHC bundle in response to force steps exerted by the fluid jet. The bundle displacements had a fast component that followed the time course of the driver voltage with little delay (<400 μsec), as was observed for the movements of glass probes (Fig. 1A). In contrast to the movements of the glass probes, the bundles often displayed a further slow relaxation with time constants between 8 and 55 msec. Measurements of cuticular plate motion showed that this slow component could not be attributed to translation of the sensory epithelium. The magnitude of the slow component varied from 10 to 35% of the total displacement. It was similar in appearance to the slow relaxations of the hair bundle observed in response to stimulation with

flexible glass fibres that were originally described for frog saccular hair
cells by Howard and Hudspeth (1987), and were also seen in OHCs of
neonatal mice by Russell *et al.* (1989). These relaxations were reported to
be associated with adaptation of the transducer currents, and attributed
to a reduction in the bundle's stiffness due to slippage of the 'tip links'
down the stereocilia (Howard and Hudspeth, 1987; Assad and Corey,
1992).

In neonatal mice the slow component of hair bundle relaxation upon
stimulation with a flexible glass fibre was reported to be only present in
adapting hair cells. Non-adapting cells were approximately half as stiff
as adapting cells and did not show a slow component in their
mechanical relaxations (Russell *et al.*, 1989). Such non-adapting OHCs
were suggested to represent a pathological condition.

Since we did not encounter hair cells which spontaneously lacked a
slow component in the bundle relaxations, we attempted to simulate
this pathology by thoroughly bashing the hair bundle with a
patch-pipette. Figure 2(B) shows the responses of the bundle of Fig. 2(A)
after being damaged. Although the intensity of the 14 stimulus steps
shown was reduced by a factor of 2.5, the displacements were almost as
large as before damage was inflicted, indicating a reduction in bundle
stiffness. Nevertheless, under our experimental conditions, the slowly
relaxing component to the bundle movements remained.

The increased low-frequency noise in the bundle movements
following mechanical damage (Fig. 2B) may indicate an increase in

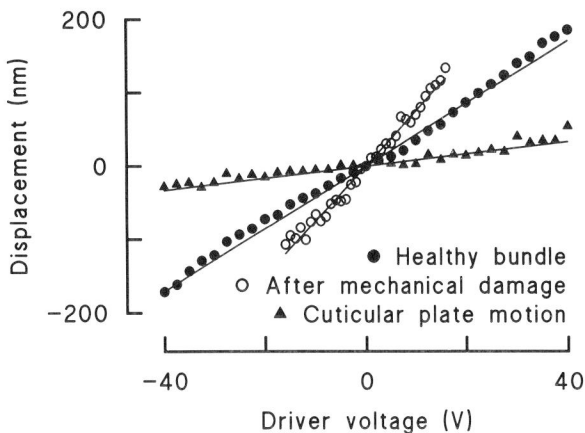

Fig. 3. Bundle displacements versus driver voltage for the OHC of Fig. 2. Slopes of the
regression lines: healthy bundle: 4.27 nm/V; after mechanical damage: 7.48 nm/V; cuticular
plate motion: 0.83 nm/V.

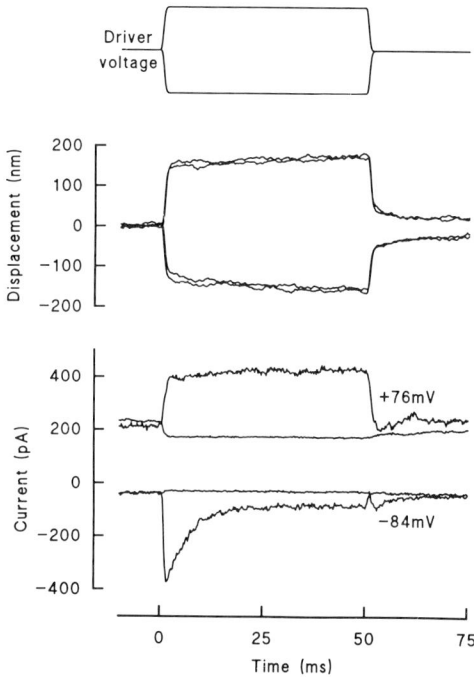

Fig. 4. OHC transducer currents and bundle movements at hyperpolarized and depolarized holding potentials. Driver-voltage steps (shown above) of 25 V caused rapid adaptation of the transducer current for excitatory bundle displacement at –84 mV (also noticeable at the offset of the inhibitory displacement step). Adaptation disappeared at +76 mV. Depolarization did not change the time course of the bundle movements, but displacements were 4% smaller than at –84 mV. Responses are single traces. Leak currents were not subtracted. Series resistance 1.6 MΩ. Bundle height 3.5 μm.

Brownian motion, commensurate with a reduced stiffness of the bundle. Figure 3 shows the bundle displacements as a function of the driver voltage (which is proportional to force on the bundle). The slopes of the regression lines indicate that the healthy bundle was 1.8 times stiffer than it was after mechanical damage. In another cell this ratio was 1.9 times. These findings make it improbable that the slow relaxations we observed in our experiments were a manifestation of the adaptation process, because the damaged bundles were extremely unlikely to transduce, let alone adapt.

A small non-linearity can be seen in the responses of the bundle before mechanical damage, which may in part be due to the gating of the transducer channels (Howard and Hudspeth, 1988; Russell *et al.*,

1992). Not enough experiments were performed to date to be confident about this though.

The slow mechanical relaxation and adaptation

To test more directly the relation between adaptation and bundle movements we measured transducer currents under voltage clamp and bundle movements simultaneously. We made use of the removal of adaptation upon depolarization of the cell (Assad *et al.*, 1989; Crawford *et al.*, 1989). Figure 4 shows the result. The cell showed strong adaptation of the transducer current at a holding potential of –84 mV, with a time constant of 5.7 msec. At a holding potential of +76 mV adaptation disappeared. In both cases the bundle movements possessed a slowly relaxing component, with time constants of about 25–40 msec, and the mechanical records were nearly overlapping. Thus, the slow mechanical relaxation had a different time constant from transducer current adaptation, and remained the same when adaptation was abolished. Similar results were observed in four other OHCs.

Transducer currents as a function of bundle displacement

Transfer functions relating transducer currents to bundle displacements of different sizes were obtained for three OHCs, at a holding potential of –84 mV. One of them is shown in Fig. 5. The shape of the transfer functions was similar to those relating transducer currents to the driver voltage to the fluid jet that we described previously (Kros *et al.*, 1992, 1993). Indeed, for the three cells used here, the transfer functions relating transducer current to displacement and driver voltage were also qualitatively similar. They were best fitted with second order Boltzmann curves. This is consistent with the transducer channel having two closed states and one open state, with displacement-dependent transitions between them:

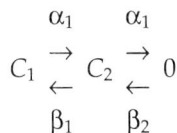

$$C_1 \underset{\beta_1}{\overset{\alpha_1}{\underset{\leftarrow}{\rightarrow}}} C_2 \underset{\beta_2}{\overset{\alpha_1}{\underset{\leftarrow}{\rightarrow}}} 0$$

The probability p_0 of the channel being open is then given by:

$$p_0 = [1 + K_2(1 + K_1)]^{-1} \qquad (1)$$

where

$$K_1 = \beta_1/\alpha_1 = \exp[a_1(x_1 - x)] \qquad (2)$$

and

$$K_2 = \beta_2/\alpha_2 = \exp[a_2(x_2 - x)] \qquad (3)$$

$K_{1,2}$ are the equilibrium constants of the two transitions, $a_{1,2}$ and $x_{1,2}$ are constants, with $x_{1,2}$ being the set points for the two displacement-dependent transitions with respect to bundle position x.

The ratio a_1/a_2 was 4.4 ± 1.5 ($n=3$ cells, as in the following estimates), indicating that the transition between the closed states was about four times more displacement sensitive than the opening transition, similar to a previous report for auditory hair cells of the turtle (Crawford *et al.*, 1991). The maximum transducer current was 594 ± 149 pA, corres-

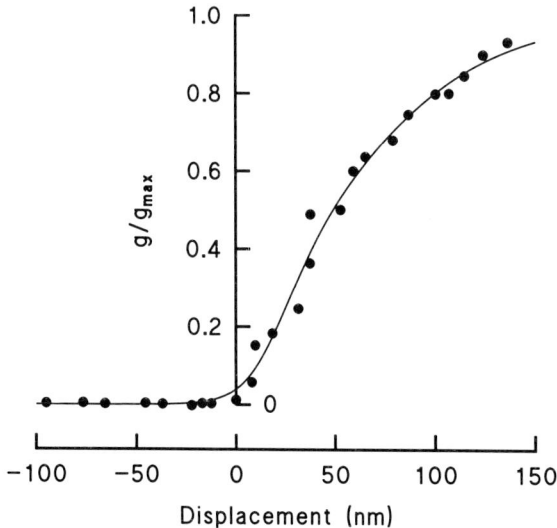

Fig. 5. Normalized OHC transducer conductance as a function of bundle displacement. This cell was stimulated with force steps. Transducer currents before the onset of adaptation and bundle displacements were measured from simultaneously recorded single traces. Maximum transducer conductance was 8.1 nS (700 pA at the holding potential of –84 mV). The curve was fitted to Eqns (1)–(3), with $a_1 = 0.082$/nm, $a_2 = 0.025$/nm, $x_1 = 24$ nm and $x_2 = 44$ nm. Series resistance 0.7 MΩ. Bundle height 4.0 μm.

ponding to a conductance of 6.8 ± 1.7 nS. At rest 5.7 ± 1.8% of the maximum available transducer conductance was activated. The sensitivity to displacement around the resting position of the bundle was 32 ± 12 pS/nm, or 2.4 ± 0.7 nS/deg of rotation. The maximum sensitivity, at a displacement in the excitatory direction of 24 ± 9 nm (0.32 ± 0.10 deg), was 75 ± 25 pS/nm, or 5.5 ± 1.3 nS/deg. Fifty per cent of the maximum available transducer current was activated at 54 ± 6 nm (0.72 ± 0.13 deg), 90% at 163 ± 28 nm (2.2 ± 0.3 deg).

Discussion

Bundle movements and adaptation

When OHC bundles of neonatal mice were stimulated with force steps from a fluid jet, the bundle movements had two kinetic components. The fast component followed the time course of the driver voltage; the slow component represented a further movement of the bundle in the direction of the applied force, with time constants ranging from 8 to 55 msec. The time course of transducer current adaptation in these cells was considerably faster (Fig. 4), with time constants of 4–7 msec (five cells). Moreover, when transducer current adaptation was abolished by depolarizing the cells, the time course of the bundle movements did not change. Mechanical damage to the bundle also did not change the time course of its displacement in response to a force step.

Therefore, the slow relaxation in the bundle movements of neonatal mouse OHCs did not represent a mechanical equivalent of adaptation as has been proposed to exist in hair bundles of the frog sacculus (Howard and Hudspeth, 1987) and indeed neonatal mouse OHCs (Russell *et al.*, 1989). A precedent for our finding was set by Crawford *et al.* (1989), who, for turtle hair cells, found no change in bundle movement upon removal of adaptation by depolarization.

If a mechanical change occurred within the bundle that was associated with adaptation, we were unable to detect it in the overall mechanics in our experiments. The physical basis of the slow relaxation that was observed in the hair bundle movements (Figs 2 and 4) is not known. It was not observed when glass fibres were used to calibrate the fluid jet (Fig. 1). It could be due to a change in the stimulus because of the micromechanics of the fluid-flow around the hair bundle, or alternatively to a slow relaxation in the hair bundle itself. At depolarized potentials, when adaptation was abolished, the transducer currents had a similar time course to the bundle movements, showing, for excitatory stimuli, a rapid increase followed by a slower further increase in current size (Fig. 4). This suggests that the observed movements were directly transferred to the gating mechanism of the transducer channels.

Sensitivity and operating range of the OHC's transduction mechanism

The relation between transducer current and bundle displacement had a similar shape to the relation between transducer current and inferred force on the bundle that we described before (Kros *et al.*, 1992). A second-order Boltzmann curve described these relations well, implying the existence of at least two closed states and one open state for the transducer channel. The transitions between the states can be described in terms of their force sensitivity (Kros *et al.*, 1992, 1993) or displacement sensitivity (Eqns 1–3).

At rest, about 6% of the total available transducer conductance was activated, towards the lower end of the range of previous estimates where the coupling of glass fibres to the bundles made estimation of the true resting conductance less certain (Ohmori, 1987; Crawford *et al.*, 1989). At the resting position, the sensitivity of the transducer was about 30 pS/nm, or 2.5 nS/deg.

The sensitivity to displacement was maximal when the bundles were moved about 25 nm away from the resting position. At 75 pS/nm (or 5.5 nS/deg), neonatal OHCs were considerably more sensitive than the most sensitive hair cells in lower vertebrates: the mean maximum sensitivity in turtle hair cells was about 7.5 pS/nm, or 0.75 nS/deg (Crawford *et al.*, 1989). Adult OHCs had a maximum sensitivity of only 1.3 pS/nm (Ashmore *et al.*, 1993).

With 50% of the total transducer conductance activated at a rotation of 0.7 deg, and 90% at 2.2 deg, the operating range of transduction was comparable to that found previously in lower-vertebrate hair cells (Crawford *et al.*, 1989). The operating range in isolated adult OHCs was much larger, with 90% of the maximum transducer current activated at a displacement of well over 500 nm (Ashmore *et al.*, 1993), as compared to 160 nm in the present study. This may point to a reduced efficiency of the coupling of bundle movement to transducer channel gating in isolated OHCs (see discussion following Ashmore *et al.*, 1993). Our measurements of sensitivity and operating range with reference to bundle displacement are in reasonable agreement with earlier predictions based on the estimated force exerted by the fluid jet on the bundles (Kros *et al.*, 1992). The measured sensitivity is about twice as large as predicted before, and the operating range is half the range predicted earlier.

Assuming a driver voltage of 150 mV in OHCs *in vivo*, and a transducer conductance which is three times larger in endolymph than in perilymph (Crawford *et al.*, 1991), our results predict a maximum OHC transducer current in the cochlea of about 3 nA, and a maximum sensitivity of forward transduction of 34 pA/nm, or 2.5 nA/deg.

Acknowledgements

This research was supported by a programme grant from the Medical Research Council to Professor I. J. Russell, C. J. K and G. P. R. Further support came from the Hearing Research Trust. C. J. K. is a Royal Society University Research Fellow. G. W. T. L. is an MRC research student.

References

Ashmore, J. F., Kolston, P. J. and Mammano, F. (1993). Dissecting components of the outer hair cell feedback loop. In *Biophysics of Hair Cell Sensory Systems* (H. Duifhuis, J. W. Horst, P. van Dijk and S. M. van Netten, eds), pp. 151–158. World Scientific, Singapore.

Assad, J. A. and Corey, D. P. (1992). An active motor model for adaptation by vertebrate hair cells. *J. Neurosci.* **12:** 3291–3309.

Assad, J. A., Hacohen, N. and Corey, D. P. (1989). Voltage dependence of adaptation and active bundle movement in bullfrog saccular hair cells. *Proc. Natl. Acad. Sci. USA* **86:** 2918–2922.

Crawford, A. C., Evans, M. G. and Fettiplace, R. (1989). Activation and adaptation of transducer currents in turtle hair cells. *J. Physiol.* **419:** 405–434.

Crawford, A. C., Evans, M. G. and Fettiplace, R. (1991). The actions of calcium on the mechano-electrical transducer current of turtle hair cells. *J. Physiol.* **434:** 369–398.

Denk, W. and Webb, W. W. (1990). Optical measurement of picometer displacements of transparent microscopic objects. *Appl. Opt.* **29:** 2382–2391.

Denk, W., Webb, W. W. and Hudspeth, A. J. (1989). Mechanical properties of sensory hair bundles are reflected in their Brownian motion measured with a laser differential interferometer. *Proc. Natl. Acad. Sci. USA* **86:** 5371–5375.

Howard, J. and Hudspeth, A. J. (1987). Mechanical relaxation of the hair bundle mediates adaptation in mechanoelectrical transduction by the bullfrog's saccular hair cell. *Proc. Natl. Acad. Sci. USA* **84:** 3064–3068.

Howard, J. and Hudspeth, A. J. (1988). Compliance of the hair bundle associated with gating of mechanoelectrical transduction channels in the bullfrog's saccular hair cell. *Neuron* **1:** 189–199.

Jaramillo, F. and Hudspeth, A. J. (1993). Displacement-clamp measurement of the forces exerted by gating springs in the hair bundle. *Proc. Natl. Acad. Sci. USA* **90:** 1330–1334.

Jaramillo, F., Markin, V. S. and Hudspeth, A. J. (1993). Auditory illusions and the single hair cell. *Nature* **364:** 527–529.

Kros, C. J., Rüsch, A., Lennan, G. W. T. and Richardson, G. P. (1993). Voltage dependence of transducer currents in outer hair cells of neonatal mice. In *Biophysics of Hair Cell Sensory Systems* (H. Duifhuis, J. W. Horst, P. van Dijk and S. M. van Netten, eds), pp. 141–150. World Scientific, Singapore.

Kros, C. J., Rüsch, A. and Richardson, G. P. (1992). Mechano-electrical transducer currents in hair cells of the cultured mouse cochlea. *Proc. R. Soc. Lond. B* **249:** 185–193.

Ohmori, H. (1987). Gating properties of the mechano-electrical transducer channel in the dissociated vestibular hair cell of the chick. *J. Physiol.* **387:** 589–609.

Rüsch, A., Kros, C. J. and Richardson, G. P. (1994). Block by amiloride and its derivatives of mechano-electrical transduction in outer hair cells of mouse cochlear cultures. *J. Physiol.* **474:** 75–86.

Russell, I. J., Kössl, M. and Richardson, G. P. (1992). Nonlinear mechanical responses of mouse cochlear hair bundles. *Proc. R. Soc. Lond. B* **250:** 217–227.

Russell, I. J., Richardson, G. P. and Kössl, M. (1989). The responses of cochlear hair cells to tonic displacements of the sensory hair bundle. *Hearing Res.* **43: 55–70.**

Cloning of Myosins from the Saccular Macula

CHARLES F. SOLC[1,2], BRUCE H. DERFLER[1,2], GEOFFREY M. DUYK[1,4] AND DAVID P. COREY[1,2,3*]

[1]Howard Hughes Medical Institute, [2]Department of Neurology, Massachusetts General Hospital, [3]Program in Neuroscience, Harvard Medical School and [4]Department of Genetics, Harvard Medical School, Massachusetts, Boston, MA 02114, USA

Introduction

Adaptation of the transduction mechanism in bullfrog saccular hair cells is associated with a rapid shift of the sensitivity curve, such that the bundle's range of maximal sensitivity moves towards the bundle's average position (Corey and Hudspeth, 1983; Eatock et al., 1987; Shepherd and Corey, 1994). A shift in the sensitivity curve, without diminution of its amplitude, is consistent with the idea that adaptation corresponds to an adjustment of the tension on transduction channels. A specific model for adaptation proposes that an adjustment in tension is brought about by a movement of the upper attachment point of each tip link along the side of that stereocilium (Howard and Hudspeth, 1987). The attachment point would slip under increased tension or would climb if tension was relaxed; a balance between slipping and climbing would normally set the resting tension at a level sufficient to open some 10–20% of transduction channels. In the model it was further speculated that the molecular motor moving the attachment point was a form of myosin climbing on the actin core of the stereocilium (Howard and Hudspeth, 1987; see Hudspeth and Gillespie, 1994, for review).

There is good evidence that adaptation does involve an adjustment of

*To whom correspondence should be addressed at: WEL414, Massachusetts General Hospital, Boston, MA 02114, USA

tension in the tip links. A relaxation of force in the hair bundle, with the same timecourse as adaptation, has been observed (Howard and Hudspeth, 1987). Manipulations that reduce calcium entry into stereocilia slow the rate of adaptation and shift the sensitivity curve in a direction corresponding to increased tension (Assad *et al.*, 1989; Crawford *et al.*, 1989; Assad and Corey, 1992). Such manipulations also cause active movement of unrestrained hair bundles (Assad and Corey, 1992). There is also some evidence that relaxation of tension in tip links may be followed by climbing of the tip-link attachment point (Shepherd *et al.*, 1991).

Evidence for the identity of the putative molecular motor is more circumstantial. Some member of the myosin family remains a favorite candidate, for a number of reasons. The actin core of the stereocilium is the likely substrate for the motor and myosins are the only molecules known to move on actin. Actin filaments in stereocilia are polarized such that myosin's power stroke would carry it towards the tip (Flock *et al.*, 1981). The climbing rate of the motor, inferred from adaptation to negative bundle deflections, is 1–2 μm/sec (Assad and Corey, 1992) and this is similar to the speed of some myosins moving on actin in *in vitro* assays. Also, glass beads coated with muscle myosin can move on the actin cores of stereocilia, at about 1–2 μm/sec (Shepherd *et al.*, 1990). The resting tension in tip links is about 14 pN, consistent with a motor consisting of a dozen or so myosin molecules (Huxley and Simmons, 1971; Assad and Corey, 1992; Hudspeth and Gillespie, 1994).

Stereocilia can be isolated for biochemical studies by several novel methods (Shepherd *et al.*, 1989; Gillespie and Hudspeth, 1991). We found that isolated stereocilia contain calmodulin and that anti-calmodulin antibodies bind at the tips of stereocilia (Shepherd *et al.*, 1989). Calmodulin may mediate the calcium dependence of adaptation, because some calmodulin inhibitors slow adaptation (Corey *et al.*, 1987) and because some myosins are regulated by calmodulin binding (e.g. Wolenski *et al.*, 1993).

More recently, Gillespie and Hudspeth (1993) have found that replacement of ATP in the cytoplasm by ADPβS, an inhibitor of some myosins, inhibits adaptation in hair cells. Biochemical studies of isolated stereocilia have revealed a protein of 120 kDa with some properties of myosins. Antibodies to a bovine adrenal myosin type I cross-react with this band and also bind to the tips of stereocilia (Gillespie *et al.*, 1993).

In seeking the molecular basis of hair cell adaptation, we have started with the assumption that the motor is indeed a myosin and have tried to identify all of the myosins expressed in the sensory epithelium of the bullfrog saccule. The myosin superfamily comprises nine myosin types with markedly differing structures, yet all have a conserved head domain. We used conserved regions within the head to generate primers for the polymerase chain reaction (PCR), in order to amplify myosin cDNAs prepared from bullfrog saccular macula. Amplified fragments were cloned, grouped by restriction digest pattern and sequenced. We found products of ten distinct myosin genes; these include five of the nine known types and a tenth, novel myosin that becomes myosin X. Antibodies raised against the cow homolog of frog myosin type Iβ label parts of the cell body and the tips of stereocilia (Gillespie *et al.*, 1993). Myosin Iβ is thus a strong candidate for the adaptation motor.

Portions of this chapter are excerpted from a research publication and reproduced with permission (Solc *et al.*, submitted).

Methods

Preparation of nucleic acid

Sacculi of adult bullfrogs (*Rana catesbeiana*) were exposed and removed whole into a dish containing an artificial perilymph (AP) solution. The otolithic membrane was removed and hair cells and supporting cells were gently scooped from the macula using a sliver of Teflon and frozen in liquid nitrogen. It is estimated that 95% of the material obtained from the macula was derived from hair cells and supporting cells. RNA was extracted using the acid guanidium method. First-strand cDNA was synthesized using SuperScript RNaseH MMLV Reverse Transcriptase (Gibco/BRL) with either random hexamers or oligo(dT) primers.

Polymerase chain reaction

Amino acid sequences corresponding to the regions between the ATP-binding site and the actin-binding domain from 17 representative myosin head sequences were aligned to determine the most commonly used amino acid at each position and the fraction of sequences that contained that amino acid. Degenerate PCR primers (five forward and four reverse) were designed based on regions of high homology. The most effective primers corresponded to the conserved LEAFGNAKT and NEKLQQ sequences, which amplified fragments ranging from 770

to 890 bp. Slightly less effective primers corresponded to the GESGA-
GKT, EQEEY and RCIKPN sequences.

Cloning

We used the T/A cloning technique, which takes advantage of the
tendency of Taq polymerase to add single 3′ 'A' overhangs to the
amplified fragments. The PCR fragments were gel-purified and cloned
into a linearized vector with single 3′ 'T' overhangs (pCR1000; Invitro-
gen).

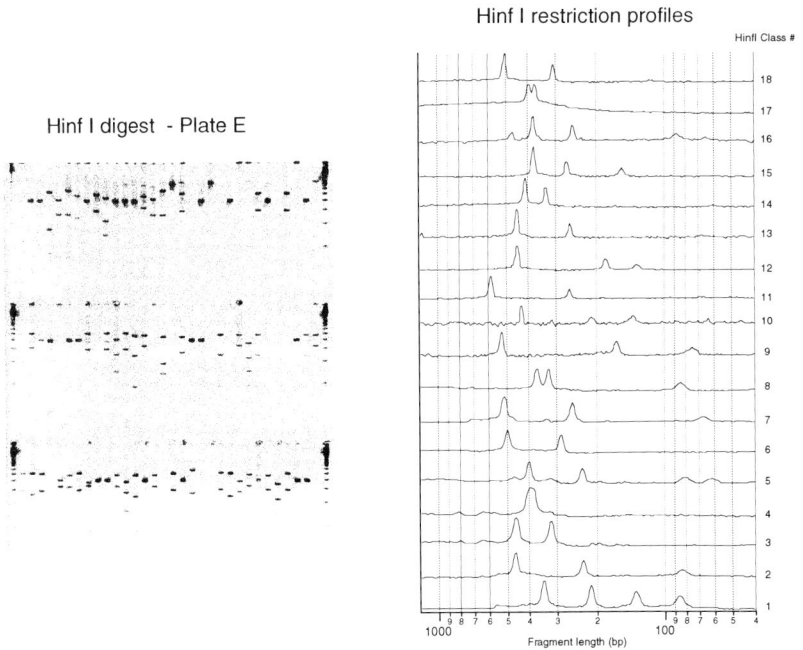

Fig. 1. *Hin*fI digests of cloned PCR products. (A) The amplified insert from each clone was
digested with the restriction enzyme *Hin*fI and the fragments from each clone were
separated on a gel. (B) Each lane was digitized with a scanner and plotted as intensity
against size (in base pairs). Twenty-two different patterns were identified among about 800
clones; shown are 18 of the 22 *Hin*fI classes.

Restriction digest fingerprint

The insert from each clone was amplified using PCR with the same primers used in the initial amplification. The PCR product was then digested with *Hin*fI by adding the enzyme directly to the PCR buffer and the fragments from each clone were separated on a gel and stained with ethidium bromide (Fig. 1a). Each lane of the gel was digitized and plotted as intensity against size (in base pairs). Novel patterns were added to a key of *Hin*fI restriction patterns (Fig. 1b).

Full-length sequence of myosin Iβ

A bullfrog kidney library was constructed in the ZAP Express vector by Stratagene. The library was screened with a probe generated by PCR from the sequence of clone 03. Positive clones were replated and the secondary plates rescreened. pBK-CMV phagemids from positives were excised and sequenced with an ABI automated sequencer. Full-length sequence was assembled in Sequencher, using at least two forward and one reverse sequences for each base.

Antibody production

Peptides corresponding to unique regions of myosin heads Iβ, VI and X were synthesized and conjugated to bovine serum albumin (HPHFVTHKLGDQKTRK, ILQNRKSPEEDEYLK and QSGCITDETIS-DEET, respectively). Three rabbits each were immunized with these antigens and boosted twice before exsanguination at 75 days (Pocono Rabbit Farm, Hazelton, Washington).

Results

Analysis of PCR products

Of all combinations of PCR primers, three produced good amplification of macular cDNA. Products had the molecular weights corresponding to approximately 770, 870 and 1350 bp. Each of these bands was subcloned into the PCR1000 cloning vector and plated. About 800 positive clones were identified. Because the head region is conserved among myosin types, we expected that each band might

Active hearing

```
            10        20        30        40        50        60        70        80        90       100
            |         |         |         |         |         |         |         |         |         |
08 PPO  LEAFGNAKT.RNDNSSPFGKIMDIEFDFKGDPLGGVISNHLLEKSPVVRQPRNERNFHIFYQMLSGASDDLLNKLKLDRDFSKYGYLSL----------
03 PPO  LEAFGNAKTLRNDNSSPFGKIMDKQFDIKGAPVGGHILNYLLEKSPVVHQNHGERNFHIFYQLLEGGEEDLLRPLGLDKNAQNYQYLIKG---------
13 PPO  -EAFGNAKTKKNDNSSPFGKFIPINFDVNGIIVGANIETYLLEKSPAIRQAKEERAFHIFYYLMSGAGEHLKNDLLLE-AYNKYRFLSN----------
27 PPO  LEAFGNAKTKKNDNSSPFGKFIPINFDVTGIVGANIETYLLEKSPAIRQAKDERTFHIFYQLLAGSGELLPSDLLLE-SVNNYRFLSN----------
23 PPO  LEAFGNAKTKKNDNSSPFGKFIPINIDKAGIIVGANIETYLLEKSPAIRQAKDERTFHIFYQILAGRGEHVKTDLLLE-GFNQYRFLSN---------
24 PPO  LEAFGNKNT.RNDNSSPFGKFIPIHFG-TGKLSSADIETYLLEKSPTFQLPNEPTYHIFYQIOSNKKPDIVEMLLITSNPYDYHYTSM----------
06 PPO  LEAFGNAKTTRNDNSSPFGKIIEIGFDKRIRIKGAHMPTILLEKSPVVFQAEEERNHIFYQLCASASLPEFNILKQG-SANSFHYTKQG---------
02 PPO  LEAFGNAKT.PNNNSSPFGKFVEIHFNEKHSVVGGFVSHVLLEKSPICVQGQDERNVHIFYPLCAGAPEEIPQKLFLN-SPDSFPYLNPGCTPYFASKET
25 PPO  LEAFGNAKT.RNNNSSPFGKFVEIHFNEKHSVVGGFVSHVLLEKSPICVQGQDERNVHIFYPLCAGAPEEIPQKLFLN-SPDSFPYLNRGCTPYFASKVT
ETB PPO LEAFGNAKT.RNNNSSPFGKFVEIHFNEKHSVVGGFVSHVLLEKSPICVQGQDERNVHIFYPLCAGAPEEIPQKLFLN-SPDSFPYLNRGCTPYFASKET
E34 PPO LEAFGNAKT.RNNNSSPFGKFVEIHFNEKHSVVGGFVSHVLLEKSPICVQGQDERNVHIFYPLCAGAPEEIPQKLFLN-SP------------------
E2E PPO LEAFGNAKTTIPINDNSSPFGKYIDIHFNKKGAIEGAKIEQILLEKSPVCROASDERNVHIFYCMLKGMPPDHKKKKLGLG-KATDFNYLCMG-------
10 PPO  LEAFGNAKT.INNNSPPFGKFIQLNICOKGHIQGGPRIVDYLLEKNRVVRQNPKERNYHIFYALLEGAGKEEREAFYLL-QTEKYHYLNQS--------

           110       120       130       140       150       160       170       180       190       200
            |         |         |         |         |         |         |         |         |         |
08 PPO  ----------------DSAKKNGIDDASNFKTVRNAMQIVGFMEHETQSVLELVATVLKLGNIEFKPESPVNGLDESKIKDKNELKEICELIGIDQS
03 PPO  ----------------OCARVSSINDKNDWKVVPRALSIINFNDDDIEELLSIVASVLHLGNVQFATDEH----GHAQVTTENQIKYLAPLLSVDST
13 PPO  ----------------GHVTIPGQODKDLFQETLEAMKIMGFPDDEQIGLLRVIAGVLQLGNIAFKKERNT---DOASMPDNTAAQKVCHLLGINVN
27 PPO  ----------------GIVPIPGQODKDNFQETMEAMHIMGFSHDEILSMLKVVSSVLQYGNIVFKKERNT---DOASMPDNTAAQKLCHLLGMNVM
23 PPO  ----------------GNIITPGQODKEIFHETMESMKIMGISHEEIMSMLRMVSAVLQFGNIVFRKERNT---DOASMPDNTAAQKLCHLLGLNVT
24 PPO  ----------------GEISVKSIDDAEELMATDEAIDILGFNOEEKMGIYKLTGAGMHYGNMKFKQKPRE---DOAESDGTEDADKVSYLMGLNSS
06 PPO  ----------------GSPVIDGVDDAKELRNTRRACALLGIGDOYOLGIFPILASILHFGNVEFKSPDADSCLISPK---HEPLIIFCDLMGVEYE
02 PPO  DKQILONRKSPEEDEYLKHGSLKDPLLDDHGDFNRMVDAMKKIGLDDTEKLDLFRVVAGVLRLGNIDFEEAGSTSGGCTLKNQSSKTLECCSKLLGLDED
25 PPO  DKQILONRKSPEEDEYPKHGSLKDPLLDD-----------TEKLDLFRVVAGVLHLGNIDFEEAGPTSGGCTLKNQSSKTLECCSKLLGLDED
ETB PPO DKQILONRKSPEEDEYLKHGSLKDPLLDDHGDFNRMVVAIKKIGLDDTEKLDLFRVVAGVLHLGNIDFEEAGSTSGGCTLKNQSSKTLECCSKLLGLDED
E34 PPO ----------------LDDHGDFNRMVVAMKKIGLDDTEKLDLFRVVAGVLHLGNIDFEEAGSTSGGCTLKNQSSKTLECCSKLLGLDED
E2E PPO ----------------KCTTCDGPDDSKEYANIRSAMKVLMFTDTENWEISPLLAAILHMGNLRYEARMYDNLDACEVVYFTSLTTAATLLEYEPOD
10 PPO  ----------------GCIITDETISDEETFQEVKTAMKVMKFTSENVREVLRLLAGILHLGNIEFIT------AGGAQICKKTALGPTAELLGLDPE

           210       220       230       240       250       260       270       280       290       300
            |         |         |         |         |         |         |         |         |         |
08 PPO  VLERAFSFRTVEA-----KOEKVSTTLNVSOAYYARDALAKNMYSRLFSWLVNRINESI-----KAQIKVPKKVMGVLDIYGFEIFEENSFEOFIINYSN
03 PPO  VLRESLIHKKIIA-----KGEELNSPLNLEOATYARDALAKAIYGRTFSWLVSKINKSLAYKGTDMHKLGSASVIGLLDIYGFEVFOHNSFEOFRINFCN
13 PPO  DFTRAILMPPIKV-----GPDYVOKAOTKEOADFAIEALAKATYERPLFRWLVHRINKAL----DKTKROGASFIGILDIAGFEIFELNSFEQLCINYTN
27 PPO  EFTRAILTPPIKV-----GPDYVOKAOTKEOADFAMEALAKATYERLFRWLVHRINKAL-----DRTKROGASFIGILDIAGFEIFELNSFEOLCINYTN
23 PPO  EFTRAILMPKIKV-----GPDYVOKAOTKEOADFAVEALRKALYERLFRSLVHRINRAL-----DRTKROGASFIGILDIAGFEIFELNSFEQLCINYTN
24 PPO  DLVKSLCNPRVKV-----GNEFVTKSONVQOVYNSIGALARSVYEKLFLWMVTRINQRL-----D-TKOSROYFIGVLDIAGFEIFDFNTFEOLCINFTN
06 PPO  EMSHWLCHRKLVT-----ATETFIKPLNRLOATNARDALSKHIYANLFNWIVCHVNKAL------LSSAKONSFIGVLDIYGFETFEINSFEOFCINYAN
02 PPO  DLRVSLTTRVMLTTAGGAKGTVIKVPLKVEOANNARDALAKAVYSRLFDHVVNRVNOCF-------PFERSSFFIGVLDIAGFEYFEHNSFEQFCINYCN
25 PPO  DLRVSLTTRVMLTTAGGAKGTVIKVPLKVEOANNARDALAKAVYSRLFDHVVNRVNOCF-------PFERSSFFFGVLDIAGFEYFEHNSFEOFCINYCN
ETB PPO DLRVSLTTRVMLTTAGGPKGTVIKVPLKY--------------SRLF-------------------EYFEHNSFEOFCINYCN
E34 PPO DLRVSLTTRVMLTTAGGAKGTVIKVPLKVEOANNARDALAKAVYSRLFDHVVNRVNOCF-------PFERSSFIGVLDIAGFEYFEHNSFEOFCINYCN
E2E PPO LMNCLTSPTIITRGETVSTPLSKEOALDVKDAFVKGIYGRLFVWIVEKINAAIYPPVSI------EPKMMPRSIGLLDIFGFENFTVNSFEOLCMNFAN
10 PPO  OLTEALTHRSMIL-----RGEEISTPLSVEOALDSPDSVAMALYSOCFAWVIKKINSRI-------KGKDDFKSIGVLDIFGFENFEVLRFEOFSINYAN

           310       320       330       340       350       360       370       380       390       400
            |         |         |         |         |         |         |         |         |         |
08 PPO  EKLOQ
03 PPO  EKLOOLFIELTLKSEQDEKESEGIAWEPVQYFNNKIICDLVEEKFKGIISILDEECLRPGEATDMTFLEKLEDTVKNHPHFVTHKLGDOKTRKVLGRDEF
13 PPO  EKLOQ
27 PPO  EKLOQ
23 PPO  EKLOQ
24 PPO  EKLOQ
06 PPO  EKLOQ
02 PPO  EKLOQ
25 PPO  EKLOQ
ETB PPO EKLOQ
E34 PPO EKLOQ
E2E PPO EKLOQ
10 PPO  EKLOQ
```

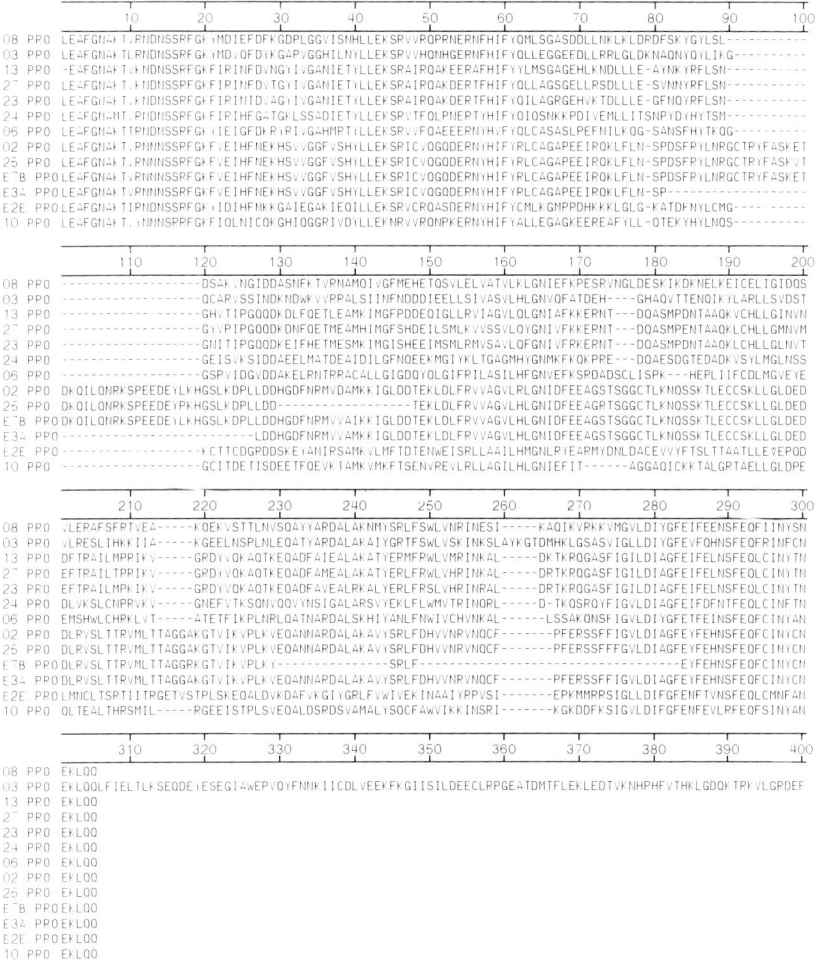

Fig. 2. Predicted amino acid sequences of cloned PCR products. Amino acid sequences were predicted from the nucleotide sequences, using the conserved LEAFGNAKT sequence to assign reading frame. Thirteen representatives out of the 21 sequenced clones are shown. Position is indicated starting at theLEAFGNAKT sequence, including insertions in myosins Iβ and VI. The last 87 amino acids of clone 03 are not shown; it ends at the actin-binding site, RCIKPN.

contain several different cDNAs. To determine the heterogeneity, we compared restriction digests of each band with digests of individual clones.

Table 1. Abundance and identification of *Hin*fI classes, isolated from 780 bp band

Clone Number	Hinf I Class	Abundance	Myosin Family	Closest Homolog	Homolog Accession
08	15	1 %	Iα	MYR1, MM Iα	B45439
03, 14	6	15 %	Iβ	MYR2,	S37146
E5B	18	1 %		MM Iβ	
13	4	38 %	IIa	NMMHC-A (nonmuscle)	A61231
I5E	19	1 %			
E9E	14	< 1 %			
27	1	5 %	IIb	MYSO_HUMAN (nonmuscle)	P35580
E2B	10	< 1 %			
17	3	9 %	IIc	MYSO_HUMAN (nonmuscle)	P35580
23	2	5 %			
18	5	1 %			
24	16	< 1 %	IId	MUSACMHCC_1 (cardiac)	M76600
04, 06	8	5 %	V	MYSD_CHICK (P190; *dilute*)	Q02440
02	11	*	VI-s1	SSUNMYOS_1	Z35331
25	7	4 %	VI-s2		
E7B	13	< 1 %	VI-s3	(pig kidney	
E3A	0	1 %	VI-s4	myosin VI)	
E2E	17	12 %	VII	pig kidney myosin VII	not in GenBank
10	12	< 1 %	X	novel	not in GenBank

About 800 clones were assigned to one or another *Hin*fI class based on digest pattern; abundance of each class is shown as a percentage of the total. Clone 02 (*) was from the 870 bp band, so is not shown as a percentage. One or two clones from each *Hin*fI class was sequenced; sequences shown in Fig. 2 are indicated in bold. Sequences were subjected to phylogeny analysis for assignment to myosin families by Dr Richard Cheney (Yale University). Myosins Iα and Iβ are assigned according to the nomenclature of Sherr *et al.* (1993). Myosin V, VI and VII families are according to Cheney *et al.* (1993). Alternative splice forms of myosin VI are denoted s1–s4. Clone 10 did not fall into any known family by phylogeny analysis. Sequences were also compared to myosin sequences in the Genbank database, and the closest homolog is noted.

To sort all 800 clones into pools representing distinct mRNAs, we did a *Hin*fI digest of each clone, separated fragments on gels and scanned the gels to generate profiles for easier comparison. Twenty-two distinct profiles were found (Fig. 1). We then sequenced one or two clones corresponding to each profile (Fig. 2). Two of the 22 *Hin*fI classes were 'junk' DNA, with the same primer on both ends, but the rest had high homology to myosins.

Assignment of myosins to families

With the nucleotide sequences we could infer amino acid sequences for each myosin head fragment. It was then possible to assign each *Hin*fI class to a myosin family (Table 1). Some were different alleles with silent mutations in *Hin*fI sites; others were closely related sequences (clones 03, 14 and E5B); still others were alternative splice forms of a single gene (clones 02, 25, E7B and E3A). Altogether, the 20 sequences represented the products of 10 myosin genes expressed in the macular epithelium. Among these genes were two members of the myosin I class, four members of the conventional myosin II class (a cardiac muscle subtype and three non-muscle subtypes), a myosin V, a myosin VI and a myosin VII. Assignment to known families indicated that one of the myosins (clone 10) represents a new myosin family, the tenth described, which is thus termed myosin X.

Identification of a candidate for the adaptation motor

Three myosins were chosen for antibody production. We picked myosin X for its novelty. Myosin VI was chosen because we speculated that the alternative splice forms would offer functional variability among cells expressing this gene and there is apparently some variability in adaptation among organs (Baird, 1992). Most attention was focused on the myosin I isoforms, for several reasons. First, the smaller physical dimensions of myosins I are more compatible with the size of tip-link insertion plaque, thought to move with adaptation (Shepherd *et al.*, 1991; Hudspeth and Gillespie, 1994). Also, a myosin I occurs in the microvilli of intestinal brush border, which share structural features with stereocilia. The 120 kDa size of the myosin-like protein identified in stereocilia suggested a myosin I, as did cross-reactivity with antibodies raised against a myosin I from bovine adrenal gland (Gillespie *et al.*, 1993). Of the two myosins I, the Iα form was a less likely candidate, because specific antibodies to the Iα form (a gift of M. Mooseker) do not label hair cell stereocilia (B. Barres and D. P. Corey, unpublished results). The myosin Iα form is also less abundant: 1% of all clones, compared to 16% for myosin Iβ. Thus myosin Iβ was chosen for antibody production.

Antibody production and labelling

For each of these three myosins, a segment of 15–17 amino acids was identified that was unrelated to other myosins. In addition, a search of the protein database showed no other known proteins with significant similarity to these segments. Peptides with these three sequences were synthesized and antipeptide antisera were generated in rabbits.

Frog saccular hair cells, prepared as whole mounts, dissociated cells or bundle blots, were not labelled with antisera to myosin X. Antibodies to myosin VI labelled hair cell bodies, but not stereocilia. It remains possible that these myosins are present in stereocilia, but the antisera are not strongly reactive or the epitope is hidden in the native protein. Both cell bodies and hair bundles were labelled with antisera to frog myosin Iβ. Cell body labelling tended to occur in the perinuclear region and just below the cuticular plate, but there was little label within the plate. Label within the stereocilia was variable, but when present appeared preferentially at and near the tips. Based on these preliminary results, we focused attention on myosin Iβ.

Full-length sequence of frog myosin Iβ

To determine the full-length sequence of bullfrog myosin Iβ, we screened a cDNA library with a probe derived from the head sequence. It did not seem practical to generate a cDNA library from bullfrog sacculae because of the small amount of tissue in each macula. Instead, we determined the distribution of myosin Iβ mRNA in other frog tissues with northern blots and found that it was expressed in brain, heart, intestine, kidney, lung, skeletal muscle, liver and eye (data not shown). A cDNA library was constructed from bullfrog kidney RNA with a mixture of both random and oligo (dT) primers. The library was screened with a probe generated by PCR from the sequence of clone 03. Fifteen positive library clones were isolated and two were found that contained the entire coding region. Both were sequenced and the sequences were identical.

The deduced amimo acid sequence of frog myosin Iβ is shown in Fig. 3. It contains 1028 amino acids, with a predicted molecular weight of 118,844 daltons. Similar myosins from rat, cow and *Drosophila* have recently been described (Ruppert *et al.*, 1993; Sherr *et al.*, 1993; Zhu and Ikebe, 1993; Morgan *et al.*, 1994; Reizes *et al.*, 1994). The frog myosin Iβ is 78% identical to the mammalian forms at the amino acid level, with no insertions or deletions.

The 80 kDa head region of myosin Iβ contains the consensus ATP- and actin-binding sites (underlined). Immediately following is a highly charged neck region (~9 kDa, 12.2 pI). This contains three repeats, at amino acids 697–766, of the 'IQ motif', which is thought to be a binding

```
                              gtagtcccggtggagtcgcatagtgggcagcgttcaggctccagcaatcgacgggatccgaataacc
MESALTARDRVGVGQDFVLLENYTSEAAFIENLRKRFKENLIYTYIGSVLVSVNPYKELEIYSKQHMERYRGVSFYEVSPHIYAIADNSYRSLRTERKDQC    100
ILISGESGAGKTEASKKILQYYAVTCPVSDQVETVKDRLLQSNPVIEAFGNAKTLRNDNSSRFGKYMDVQFDYKGAPVGGHILNYLLEKSRVVHQNHGER    200
     ATP
NFHIFYQLLEGGEEDLLRRLGLDKNAQNYQYLIKGQCARVSSINDKNDWKVVRRALSIINFNDDDIEELLSIVASVLHLGNVQFATDEHGAQVTTENQI    300
KYLARLLSVDSTVLRESLIHKKIIAKGEELNSPLNLEQAAYARDALAKAIYGRTFSWLVSKINKSLAYKGTDMHKLGSASVIGLLDIYGFEVFQHNSFEQ    400
FCINFCNEKLQQLFIELTLKSEQDEYESEGIAWEPVQYFNNKIICDLVEEKFKGIISILDEECLRPGEATDMTFLEKLEDTVKNHPHFVTHKLGDQKTRK    500
                                                                      peptide
VLGRDEFRLLHYAGEVNYSVAGFLDKNNDLLFRNLKEVMCDSGNPIAHQCFNRSELTDKKRPETAATQFKNSLSKLMEILMSKQPSYVRCIKPNDAKQPA    600
                                                                          actin
RFDEVLIRHQVKYLGLIENVRVRRAGFAYRRKYEIFLQRYKSLCPDTWPNWDGRAMDGVAVLVKSLGYKPEEYKMGRTKIFIRFPKTLFATEDALEVRKH    700
                                                                                            <
SIATFLQARWRGYHQRQKFLHMKHSAVEIQSWWRGTIGRRKAAKRKWAVDVVRRFIKGFIYRNQPRCTENEYFLDYIRYSFLMTLYRNQPKSVLDKSWPV    800
   IQ1          ><        IQ2       ><       IQ3        >
PPPSLREASELLREMCMNNMVWKYCRRINPEWKQQLEQKVVASEIFKDKKDNYPQSVPRLFINTRLGNDEINTKILQQLESQTLTYAVPVVKYDRKGYKP    900
RRRQLLLTQNAAYLVEEAMKQRIDYANLTGISVSSLSDNLFVLHVKCEDNKQKGDAILQSDHVIETLTKVAITAEKINNININQGSIKFTVGPGKEGII    1000
DFTAGSELLIAKAKNGHLSVVAPRLNSRtgaccactgcagtctgttaggaatctgatatacaaccaaactgacagcttcacatacatccatttcttccac
ttgttggacaatacctccttttaatacttctcgactctgcttttatattttatagttaccaa
```

Fig. 3. Full-length amino acid sequence of frog myosin Iβ. 5′ untranslated cDNA sequence is shown preceding the methionine intiation site, and 3′ untranslated cDNA is shown beginning at the stop codon. The coding region contains 1028 amino acids, with a predicted molecular weight of 118,844 daltons and an isoelectric point of 9.1. The conserved ATP (GESGAGKT) and actin (RCIKPN) binding sites are underlined, as is the region from which the synthetic peptide sequence was taken. The three IQ motifs, amino acids 698–766, are indicated. The complete cDNA sequence has been submitted to GenBank.

site for calmodulin or related regulatory subunits (Cheney and Mooseker, 1992). The 30 kDa tail of myosin Iβ has no conspicuous domains that might indicate function, but it is clearly different from the tail of myosin Iα which is thought to interact with membranes.

PCR of myosin Iβ from sacculus

Because this myosin was cloned in full from a frog kidney library, it is possible that the full-length sequence obtained represents a kidney isoform of a myosin Iβ that is different from a saccular macula isoform. The sequence of the head domain cloned from kidney is identical to the head obtained from macula, but the tails expressed in the two tissues could be different splice forms that serve different functions. Consequently, we used the myosin tail sequence from kidney to PCR the myosin tail from saccule cDNA. With primers that spanned 726 bases of the kidney tail, we generated a single PCR product of the same size from saccule. This product was sequenced and found to be identical to the kidney isoform. Thus the frog myosin Iβ shown in Fig. 3 is expressed in saccular macula.

Discussion

We have used a strategy of broadly screening a hair cell epithelium for myosins, in order to identify candidates for the adaptation motor. While this strategy was more time-consuming than testing candidate myosins, it was less likely to miss a novel myosin that might mediate the unique function of adaptation. Indeed, the myosin Iβ type we identified with this screen was only recently described in other species. Moreover, this method revealed a novel myosin family (myosin X), whose function and location remain to be determined.

The cloning and sequencing of myosin Iβ from the frog sacculus complement the biochemical and immunohistochemical studies of Gillespie *et al.* (1993). They found that monoclonal antibodies raised against a cow adrenal myosin I label a 120 kDa protein band in immunoblots of frog stereocilia and that these antibodies also label the tips of stereocilia. Yet it was unclear whether this simply represented cross-reactivity to a related saccular myosin. Our sequence of myosin Iβ cloned from frog saccular macula is nearly identical to the cow adrenal form (Reizes *et al.*, 1994) and to a rat brain form (Sherr *et al.*, 1993), indicating that these three are the same myosin isoform. We expect that the mammalian forms will be expressed in the saccule.

We still lack conclusive evidence that the myosin Iβ in stereocilia mediates adaptation. It will be important, therefore, to localize this myosin within the tips of stereocilia at the ultrastructural level. It will also be necessary to determine whether molecular probes derived from the myosin sequence can affect adaptation physiologically.

The sequence thus far yields few clues to indicate the relation of the myosin to other proteins of the transduction apparatus. Three putative calmodulin-binding sites have been identified at the junction between head and tail regions. This junction is thought to bind regulatory light chains in most myosins. Adaptation depends on calcium inside the tips of stereocilia (Assad *et al.*, 1989), it is blocked by some calmodulin inhibitors (Corey *et al.*, 1987) and calmodulin is located at the tips of stereocilia (Shepherd *et al.*, 1989). Thus it seems likely that calcium modulates adaptation through calmodulin binding to myosin Iβ.

Acknowledgements

Supported by the Howard Hughes Medical Institute, by NIH grant DC00304 and by a grant from the Whitaker Foundation. DPC is an Associate Investigator of the Howard Hughes Medical Institute.

References

Assad, J. A. and Corey, D. P. (1992). An active motor model for adaptation by vertebrate hair cells. *J. Neurosci.* **12:** 3291–3309.

Assad, J. A., Hacohen, N. and Corey, D. P. (1989). Voltage dependence of adaptation and active bundle movement in bullfrog saccular hair cells. *Proc. Natl. Acad. Sci. USA* **86:** 2918–2922.

Baird, R. A. (1992). Morphological and electrophysiological properties of hair cells in the bullfrog utriculus. *Ann. NY Acad. Sci.* **656:** 12–26.

Cheney, R. E. and Mooseker, M. S. (1992). Unconventional myosins. *Curr. Opin. Cell Biol.* **4:** 27–35.

Cheney, R. E., Riley, M. A. and Mooseker, M. S. (1993). Phylogenetic analysis of the myosin superfamily. *Cell Motil. Cytoskel.* **24:** 215–223.

Corey, D. P. and Hudspeth, A. J. (1983). Analysis of the microphonic potential of the bullfrog's sacculus. *J. Neurosci.* **3:** 942–961.

Corey, D. P., Smith, W. J., Barres, B. A. and Koroshetz, W. J. (1987). Calmodulin inhibitors block adaptation in vestibular hair cells. *Soc. Neurosci. Abstr.* Vol. 13, p. 538.

Crawford, A. C., Evans, M. G. and Fettiplace, R. (1989). Activation and adaptation of transducer currents in turtle hair cells. *J. Physiol.* **419:** 405–434.

Eatock, R. A., Corey, D. P. and Hudspeth, A. J. (1987). Adaptation of mechanoelectrical transduction in hair cells of the bullfrog's sacculus. *J. Neurosci.* **7:** 2821–2836.

Flock, A., Cheung, H. C., Flock, B. and Utter, G. (1981). Three sets of actin filaments in sensory cells of the inner ear. Identification and functional orientation determined by gel electrophoresis, immunofluorescence and electron microscopy. *J. Neurocytol.* **10:** 133–147.

Gillespie, P. G. and Hudspeth, A. J. (1991). High-purity isolation of bullfrog hair bundles and subcellular and topological localization of constituent proteins. *J. Cell Biol.* **112:** 625–640.

Gillespie, P. G. and Hudspeth, A. J. (1993). Adenine nucleoside diphosphates block adaptation of mechanoelectrical transduction in hair cells. *Proc. Natl. Acad. Sci. USA* **90:** 2710–2714.

Gillespie, P. G., Wagner, M. C. and Hudspeth, A. J. (1993). Identification of a 120 kd hair-bundle myosin located near stereociliary tips. *Neuron* **11:** 581–594.

Howard, J. and Hudspeth, A. J. (1987). Mechanical relaxation of the hair bundle mediates adaptation in mechanoelectrical transduction by the bullfrog's saccular hair cell. *Proc. Natl. Acad. Sci. USA* **84:** 3064–3068.

Hudspeth, A. J. and Gillespie, P. G. (1994). Pulling springs to tune transduction: adaptation by hair cells. *Neuron* **12:** 1–9.

Huxley, A. F. and Simmons, R. M. (1971). Proposed mechanism of force generation in striated muscle. *Nature* **233:** 533–538.

Morgan, N. S., Skovronsky, D. M., Artavanis-Tsakonas, S. and Mooseker, M. S. (1994). The molecular cloning and characterization of *Drosophila melanogaster* myosin-IA and myosin-IB. *J. Mol. Biol.* **239:** 347–356.

Reizes, O., Barylko, B., Li, C., Sudhof, T. C. and Albanesi, J. P. (1994). Domain structure of a mammalian myosin Iβ. *Proc Natl Acad Sci USA* **91:** 6349–6353.

Ruppert, C., Gogel, J., Reinhard, J. and Bahler, M. (1993). MYR-2 a novel class-I myosin identified in rat brain. GenBank submission #X74800.

Shepherd, G. M. G. and Corey, D. P. (1994). The extent of adaptation in bullfrog saccular hair cells. *J. Neurosci.* (in press).

Shepherd, G. M. G., Barres, B. A. and Corey, D. P. (1989). 'Bundle-blot' purification and initial purification and initial protein characterization of hair cell stereocilia. *Proc. Natl. Acad. Sci. USA* **86:** 4873–4977.

Shepherd, G. M. G., Corey, D. P. and Block, S. M. (1990). Actin cores of hair-cell stereocilia support myosin motility. *Proc. Natl. Acad. Sci. USA* **87:** 8627–8631.

Shepherd, G. M. G., Assad, J. A., Parakkel, M., Kachar, B. and Corey, D. P. (1991). Movement of the tip-link attachment is correlated with adaptation in bullfrog saccular hair cells. *J. Gen. Physiol.* 98, 25a.

Sherr, E. H., Joyce, M. P. and Greene, L. A. (1993). Mammalian myosin Iα, Iβ and Iτ: new widely expressed genes of the myosin I family. *J. Cell Biol.* **120:** 1405–1416.

Wolenski, J. S., Hayden, S. M., Forscher, P. and Mooseker, M. (1993). Calcium-calmodulin and regulation of brush border myosin-I MgATPase and mechanochemistry. *J. Cell. Biol.* **122:** 613–621.

Zhu, T. and Ikebe, M. (1993). A novel myosin I from bovine adrenal gland. *FEBS Lett.* **339:** 31–36.

Monitoring Cochlear Homeostasis with Automatic Analysis of the Low-Frequency Cochlear Microphonic

ROBERT PATUZZI

The Auditory Laboratory, Department of Physiology, University of Western Australia, Nedlands, Australia 6009

The sensitivity, frequency-selectivity and dynamic range of mammalian auditory nerve fibres are largely determined by the vibration of the organ of Corti within the cochlea (Patuzzi and Robertson, 1988). It is now clear that this vibration requires the application of pulsatile or oscillatory forces that act in synchrony with the vibration of the organ of Corti, and partially or wholly cancel the viscous forces that would otherwise limit mechanical sensitivity. The physiological process that generates these forces is known as the active process, while its action in reducing inherent friction is termed negative damping. Many experimental observations indicate that the generation or synchronization of these active forces requires the receptor currents through the outer hair cells (OHCs), while others suggest that the OHCs also act as the force generation elements *per se*. Whatever the cellular or molecular nature of the force generation process, it is clear that the mechanical sensitivity of the cochlea is increased by 55–60 dB when the OHCs are operating normally, and that sensitivity is reduced in a graded fashion as the ability of OHCs to generate their receptor current is impaired (Patuzzi *et al.*, 1989b,c). If the magnitude of this receptor current is reduced to the fraction, n, of its normal value, the loss of cochlear sensitivity or hearing loss in decibels (HL) is given by the simple empirical relationship $HL(dB) = 101\ dB \times (1 - n)/[1 + 0.85(1 - n)]$. In other words, sensitivity is initially lost at about 1 dB per 1% reduction in OHC receptor current.

The loss of OHC receptor current can be due to many cochlear disruptions, including a drop in endocochlear potential, depolarization of OHCs, blockage of transduction channels by ototoxic drugs or acoustic trauma, or by saturation of the current generation process at the apex of the OHCs by sufficiently large movements of their hair bundles (Patuzzi *et al.*, 1989b). These saturating movements can be caused by applied sounds, or slow mechanical biases originating within the cochlea itself. These slow movements could be due to small hydrostatic pressure fluctuations caused by fistulas or fluctuations in cerebrospinal fluid pressure, by osmotic imbalances due to loss of salt regulation within the cochlea, or possibly due to the motile activity of the OHCs themselves. It is worth noting here that the movement necessary to saturate the current generation process and produce a significant hearing loss is less than 10 nm, or a fraction of the width of a cell membrane.

Since auditory thresholds are normally stable within a few decibels over great periods of time, it is clear that the cochlea must possess a remarkable set of regulatory mechanisms that stabilize OHC receptor currents and/or the active forces *per se*. It should also be clear that an understanding of these regulatory processes and the deafness that results from loss of regulation requires the study of OHCs *in vivo*, where they are subject to the ongoing influences of their normal mechanical, electrical and biochemical environment. One problem that has hampered research into the regulatory or homeostatic processes within the cochlea is that it is not simple to measure small movements of the cochlear structures over long periods of time. This is largely because the surgery required to gain access to the cochlear structures is difficult, often causing deafness. The measurement techniques required can also be difficult, requiring drainage of cochlear fluids or placement of radioactive sources or optical reflectors. They are also not suited to the stable measurement of molecular displacements of the organ of Corti over the course of many minutes or hours, which would be required in the investigation of cochlear homeostasis.

Fortunately there is a way around these problems which allows us simultaneously to monitor OHC receptor current and small movements of the organ of Corti, rapidly and easily. The OHCs are themselves displacement transducers, capable of detecting displacements of atomic dimensions. They are also ideally situated within the organ of Corti to detect the micromechanical movements that may indicate loss of mechanical, electrical or biochemical regulation. For intense low-frequency stimulation, the receptor currents through these cells produce a small, distorted electrical signal within the fluids sur-rounding them (the cochlear microphonic or CM) and, in the basal region of the cochlea, this signal is easily measured using a wire electrode placed on the round window (Patuzzi *et al.*, 1989a; Fig. 1). The instantaneous magnitude of the fluctuations in current through the OHCs, and therefore the microphonic potential around the cells, is related to the displacement of the organ of Corti by a sigmoidal 'transfer curve'. By recording the low-frequency microphonic waveform and assuming that a pure-tone sound stimulus produces a sinusoidal displacement of the organ of Corti and the hair bundles of the OHCs, it is possible to reconstruct the non-linear transfer function. This is illustrated in Fig. 1. Here, the instantaneous microphonic potential at the round window of a guinea-pig, produced by a low-frequency tone presented at 112 dB SPL, is plotted against the assumed sinusoidal displacement to produce a Lissajous figure. This Lissajous figure can be taken as an approximation to the OHC transfer curve.

The non-linear transfer function is itself determined by the thermodynamics of the molecular gating process that causes the opening and closing of the stretch-activated channels at the apex of the hair cells. Work by Holton and Hudspeth (1986) has shown that the transduction channels of hair cells exist in discrete conduction states, and that the opening and closing of the stretch-activated channels of the

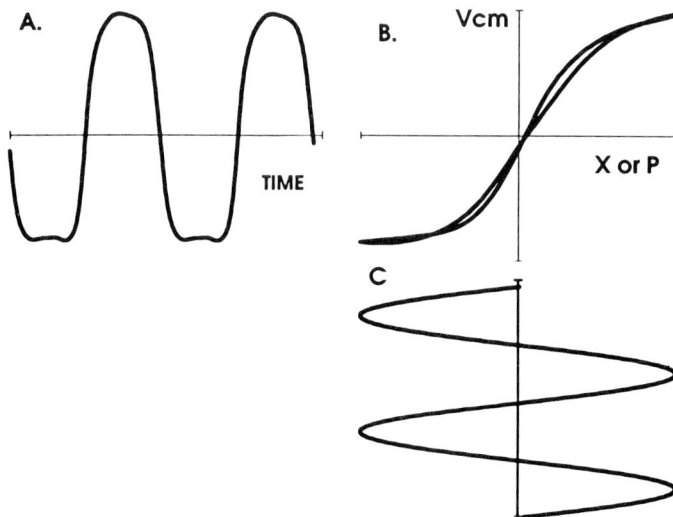

Fig. 1. Low-frequency (200 Hz) microphonic waveforms at the round window can be used to estimate the transfer curve relating ear canal sound pressure to changes in receptor current through local OHCs by plotting the instantaneous microphonic potential (upper left) against the assumed sinusoidal pressure stimulus in the cochlea (lower right). This creates a Lissajous figure (upper right) which is an approximation to the Boltzmann function describing the opening of transduction channels at the apex of OHCs (see text).

hair cells is well approximated by the Boltzmann function. That is, the probability of channel opening (p_o) is determined by the difference between the energy of the channel in its open state (E_{open}) and its energy when it is in the closed state (E_{closed}) such that $p_o = 1/\{1 + \exp[(E_{open} - E_{closed})/kT]\}$, where k is the Boltzmann constant of thermodynamics and T is the absolute temperature of the molecular population. (At mammalian temperatures the product $kT = 0.026$ eV). It is often assumed that the energy difference between the open and closed states of the transduction channels is proportional to the displacement of the hair bundles, x, such that $(E_{open} - E_{closed}) = G_z(x - x_o)$. The constant G_z defines the energy difference between the open and closed states in terms of the hair bundle displacement, and the constant x_o defines the hair bundle displacement required to give a probability of channel opening of 50%. In essence, the parameter x_o sets the operating or quiescent point on the hair cell transfer curve, while the parameter G_z defines the sensitivity of the transduction process to hair bundle displacement.

It is possible to make three assumptions that allow the estimation of the movement of the organ of Corti using the saturation of the microphonic potential, as shown in Fig. 1. We can assume that (i) the hair bundle displacement, x, is simply proportional to the instantaneous pressure fluctuation in the external ear canal caused by the low-frequency tone, with appropriate corrections for the phase rolls inherent in the acoustic and hydrodynamic coupling processes; (ii) the OHC receptor current is simply proportional to the probability of trans-duction channel opening (p_o); and (iii) the instantaneous microphonic potential at the round window is simply proportional to the instant-aneous receptor current through the hair cells of the basal cochlear turn. With these assumptions, the instantaneous microphonic potential, V_{CM}, is related to the instantaneous pressure, P, in the external ear canal by the Boltzmann relationship $V_{CM} = V_{sat}/\{1 + \exp[Z(P - P_o)/kT]\}$, where V_{sat} is the maximum peak-to-peak amplitude of the cochlear micro-phonic when the waveform is clipped and approaches a square wave, Z is a parameter relating the ear canal sound pressure, P, to the modulation of the energy difference between the open and closed states of the transduction channels. The additional parameter, P_o, represents the operating point on the microphonic transfer curve as an equivalent static ear canal pressure. (It should be stressed that the parameter P_o is only an *equivalent* sound pressure, because static pressures in the ear canal cannot produce static displacements of the organ of Corti.) When considering the receptor currrent, I_{ohc}, through the OHCs rather than the microphonic potential, V_{CM}, the saturation parameter, V_{sat} would be replaced by the equivalent current saturation parameter, I_{sat}.

In the whole cochlea, it is clear that changes in these parameters could indicate a great many things. A change in the saturation voltage of the cochlear microphonic, V_{sat}, could be due to: (i) a change in the endocochlear potential, (ii) a change in the membrane potential of the OHCs, (iii) a change in the electrical contact between the recording electrode and the fluids of the hair cells, or (iv) blockage or inactivation of transduction channels at the apex of hair cells. Similarly, a change in the sensitivity parameter, Z, could be due to a change in: (i) the amplitude of the pressure stimulus in the external ear canal; (ii) an acoustic shunt in parallel with the organ of Corti; (iii) conduction through the middle ear; (iv) the effective compliance of the organ of Corti; (v) the coupling between the displacement of the organ of Corti and the deflection of hair bundles or (vi) the relationship between deflection of hair bundles and modulation of the energy difference between the open and closed states of the transduction channels. And finally, a change in the operating point parameter, P_o, could be due to: (i) a mechanical displacement of the organ of Corti and/or hair bundle or (ii) an electrostatic influence at the transduction channel itself, not associated with an actual physical displacement.

Whichever is the case, the three parameters, V_{sat}, Z and P_o, can give an insight into the mechanical and electrical changes within the cochlea which influence the receptor currents through the OHCs, and ultimately the action of the active process. The last two parameters, Z and P_o, should be useful for *in vivo* monitoring of adaptation processes in hair cell current generation (Eatcock *et al.*, 1987; Howard and Hudspeth, 1987; Crawford *et al.*, 1989, 1991).

The low-frequency microphonic has already been used in this way by others (Neider and Neider, 1971). Most recently, analysis of the transfer characteristic has been used to determine small movements of the organ of Corti which would not be easily measured otherwise (Patuzzi and Rajan, 1990). In this case the clipped microphonic waveforms were stored on magnetic tape, averaged off-line by computer, and the Boltzmann parameters that gave the best approximation to the OHC transfer function were determined using statistical techniques to determine the least-squares fit to the data. The disadvantage of this method is that the off-line analysis takes as long as the experiment itself, and the data is not immediately available to allow effective decision making during experimentation. Initial attempts were made to use a digital signal processing system to speed up data acquisition and analysis, but the limitations on computation speed did not allow satisfactory real-time performance. To remedy this situation, an analogue computer was constructed to analyze automatically the distorted microphonic waveform in real time, so that the Boltzmann parameters V_{sat}, Z and E_o (see p. 150)

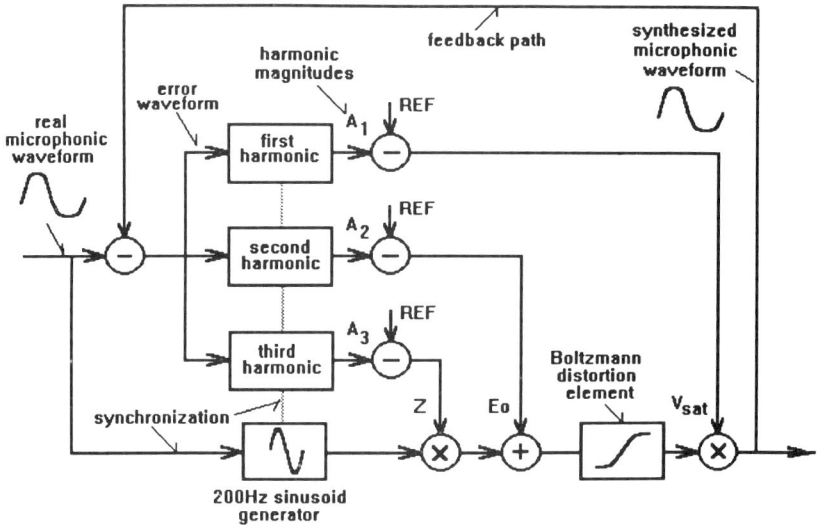

Fig. 2. The functional block diagram for the electronics which analyzes microphonic waveforms. A raw microphonic waveform is presented to the circuit, which matches a synthesized waveform in frequency, phase and distortion characteristics by adjusting its Boltzmann parameters Z, E_0 and V_{sat} to give minimum error. The synthesizer parameters are plotted continuously to a chart recorder and taken to be those of the OHC transduction process producing the real microphonic waveform.

could be determined within seconds of any changes within the cochlea. Essentially the system works by synthesizing a distorted microphonic waveform electronically, using a non-linear element with a Boltzmann characteristic, and matching the synthesized and real microphonic waveforms. The automatic waveform matching is accomplished by performing a Fourier analysis of the error waveform, itself produced by subtracting the real and synthesized microphonic waveforms. The Boltzmann parameters used by the synthesizer are set so that this error waveform contains no first, second or third harmonic components. By Fourier's theorem, the synthesized waveform is then the least-squares fit to the real waveform. The setting of the Boltzmann parameters is done automatically, using negative feedback techniques to minimize the error (see Fig. 2). The magnitude of the first harmonic in the error waveform is used to correct the final scaling of the synthetic waveform (V_{sat} of the Boltzmann curve), the second harmonic is used to adjust the operating point on the nonlinear element in the waveform synthesiser (E_0), and the third harmonic is used to adjust the amount of symmetrical clipping of the synthesized waveform (Z).

The waveform measured at the round window reflects the receptor

CHANGES IN MICROPHONIC WAVEFORM WITH TETRODOTOXIN PERFUSION

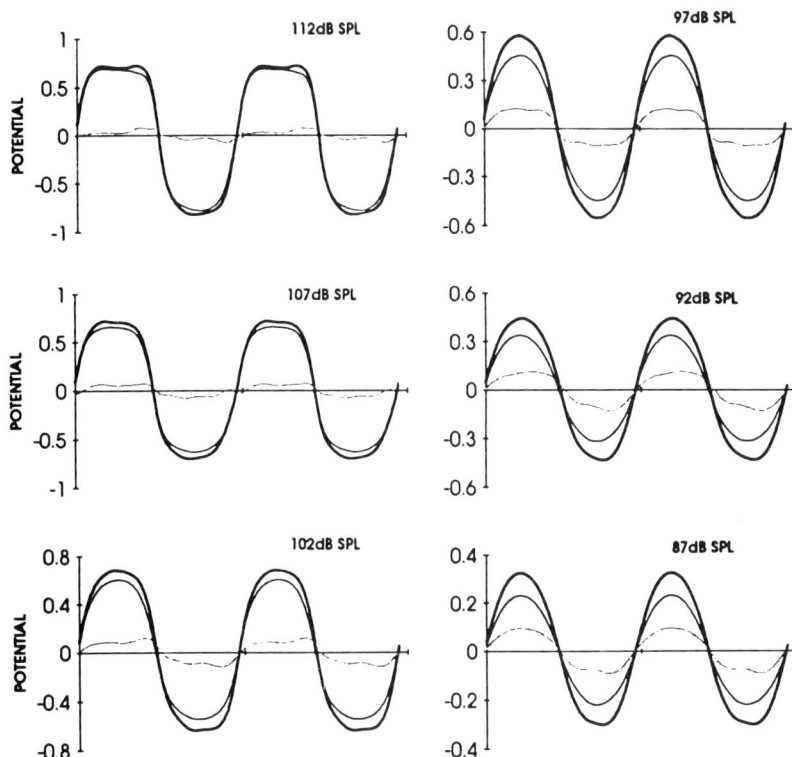

Fig. 3. Apart from a small drop in amplitude at all intensities, blocking all neural activity in the cochlea with tetrodotoxin (TTX) does not greatly affect the 200 Hz microphonic waveforms at the round window. The level of the 200 Hz pure-tone stimulus is shown next to each waveform. Heavy solid curves are waveforms obtained before application of TTX, lighter waveforms were obtained after TTX and light dotted waveforms show the difference between the before and after waveforms.

currents in the high-frequency region of the cochlea, and at high stimulus levels it is not contaminated significantly by neural components, as shown in Fig. 3. Here, the averaged microphonic waveforms recorded at the round window of a guinea-pig are shown for different levels of the low-frequency acoustic stimulus, before and after abolition of all neural responses with cochlear application of the neurotoxin, tetrodotoxin (TTX). While there were small changes in the microphonic waveforms, mainly at low stimulus intensities, the Boltzmann parameters were affected little by neural blockade. This can be seen in Fig. 4

148 Active hearing

in which the variations in the Boltzmann parameters with stimulus level before and after application of TTX are presented. Clearly there were similar changes in the Boltzmann parameters with stimulus intensity before and after TTX, in particular a rise in estimated saturation parameter, I_{sat}, and a fall in estimated sensitivity parameter, Z, with stimulus level. The operating point, E_o, also changed as the level of the 200 Hz stimulus tone was increased, with the operating point moving at the highest intensities in a direction equivalent to a movement of the organ of Corti towards scala tympani. The drop in the saturation parameter, I_{sat}, at all stimulus intensities is likely to represent a change in the tonic release of acetylcholine from the efferent terminals of the cochlea and a consequent change in the electrical impedance of the basolateral wall of OHCs, rather than a loss of neural contamination: similar changes were seen with the perfusion of strychnine, a post-synaptic blocker of the efferent neurotransmitter. In summary, these results indicate that the changes in equivalent Boltzmann parameters with level of the low-frequency tone are not neurally mediated, and that the cochlear microphonic potentials are not significantly contaminated by neural potentials.

The analogue computer method of analyzing the cochlear microphonic presented here is very rapid. Figure 5 shows the automatically derived Boltzmann parameters before, during and after a perfusion of scala tympani of the cochlea with isotonic artificial perilymph. The perfusion began at approximately 50 sec and ended at 150 sec. Note the abrupt movement of operating point, P_o, in a direction consistent with a movement of the organ of Corti towards scala vestibuli. This was

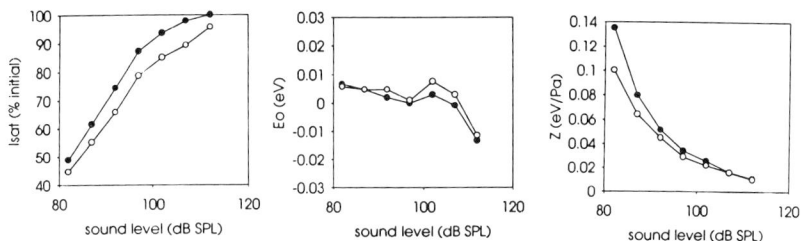

Fig. 4. Changes in Boltzmann parameters describing the waveforms of Fig. 3 with sound level. Before TTX (filled circles), the saturation parameter, I_{sat}, increased with stimulus level, while the sensitivity parameter, Z, fell. A significant change in operating point (E_o) occurred only at the highest levels, in a direction equivalent to movement of the organ of Corti towards scala tympani. After TTX (open circles), although microphonic amplitude fell at all sound levels (the saturation parameter, I_{sat}, decreased), the variations in all parameters with sound level were similar to those before TTX. Clearly, the changes were not neurally mediated, and were not due to contamination of the microphonic by neural responses.

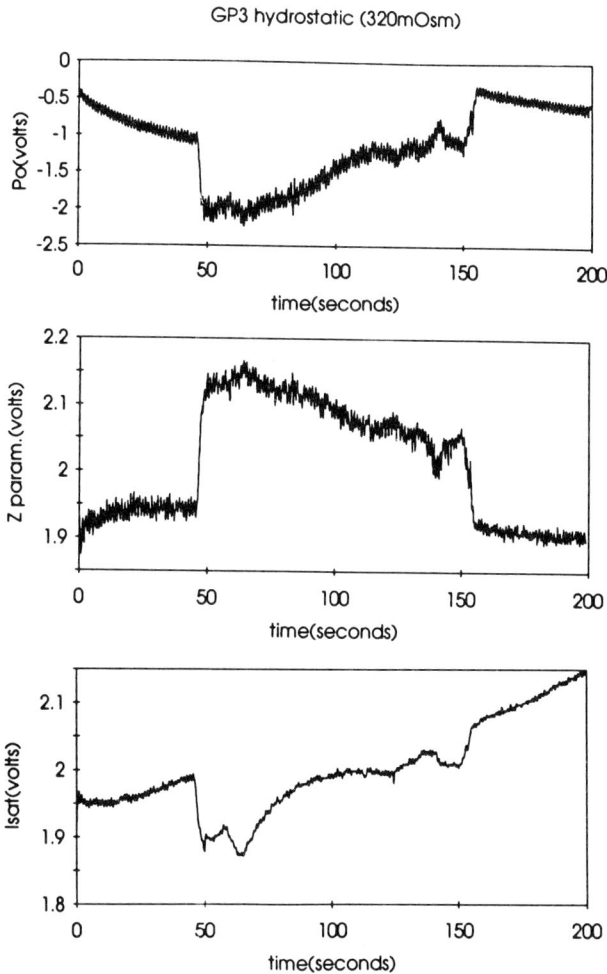

Fig. 5. Analysis of the microphonic can be rapid. Here, the automatically derived Boltzmann parameters are shown before, during and after perfusion of scala tympani with isotonic artificial perilymph. The perfusion began at approximately 50 sec and ended at 150 sec. Note the abrupt movement of operating point, P_o, in a direction consistent with a movement of the organ of Corti towards scala vestibuli. This was associated with a change in the sensitivity parameter, Z, and a small change in the saturation parameter, Isat. In this case, all parameters are expressed in arbitrary voltage units from the analysis electronics.

associated with a significant change in the sensitivity parameter, Z, and a small change in the saturation parameter, I_{sat}. In this case, all parameters are expressed in arbitrary voltage units from the analysis electronics. The example of Fig. 5 has been chosen to emphasize the point that the analysis technique is rapid and sensitive. The small cyclic fluctuations evident in the traces of the E_o and Z parameters were caused by a breathing artefact evident in the raw microphonic waveform, and abolished on stopping the ventilator. While not normal, the movement of the operating point with ventilation was presumably due to cyclic fluctuations in the pressure in the cochlear fluids (possibly due to analogous changes in CSF pressure). Preparations in which such rapid changes in operating point with isotonic perfusion or ventilation were present, suggesting a pressure-induced shift of the organ of Corti, would not normally be considered successful. Note also that the P_o and Z parameters of Fig. 5 did not show the instability evident in the I_{sat} parameter, which rose with time, even before perfusion. This demonstrates that the cross-talk between the three derived parameters produced by the analysis electronics was minimal, i.e. the three parameters are analyzed independently, and associated changes between parameters are physiological in origin.

Finally, the use of this technique has one other advantage: it possesses an inherent molecular calibration of hair bundle displacements, or at least of transduction molecule movement. The opening and closing of the OHC transduction channels would appear to be governed by the modulation of the relative energies of the open and closed conditions of the transduction channels. The Boltzmann parameters can therefore be expressed in equivalent molecular energy units. While the V_{sat} value can simply be expressed as a percentage of its normal or premanipulation value, the E_o parameter can be expressed in equivalent eV/molecule units, and the Z parameter can be expressed as an equivalent eV/molecule excursion across the Boltzmann curve (see Fig. 6). As seen from Figs 4 and 6, the sensitivity parameter varies almost 10-fold with stimulus level but, at high intensities, lies in the range 0.01–0.02 eV/Pa, which is in agreement with values obtained previously from hair cells *in vivo* and *in vitro* (see Patuzzi and Rajan, 1990).

The agreement between the values for the Boltzmann parameters obtained with off-line analysis or analogue computer analysis is very good. Figure 6 shows the variation in the three parameters recorded from one animal during a transient hypoxia. The open symbols represent the off-line determinations using waveform averaging and statistical fitting of the Boltzmann curves to the microphonic Lissajous figures. The solid line represents the continuous output from the analysis electronics. The calibration of each of the parameters in equivalent molecular energy units for the off-line and real-time estimates was done independently. Clearly the agreement between the two methods is extremely good. The data show that during transient hypoxia the magnitude of the microphonic waveforms falls, as

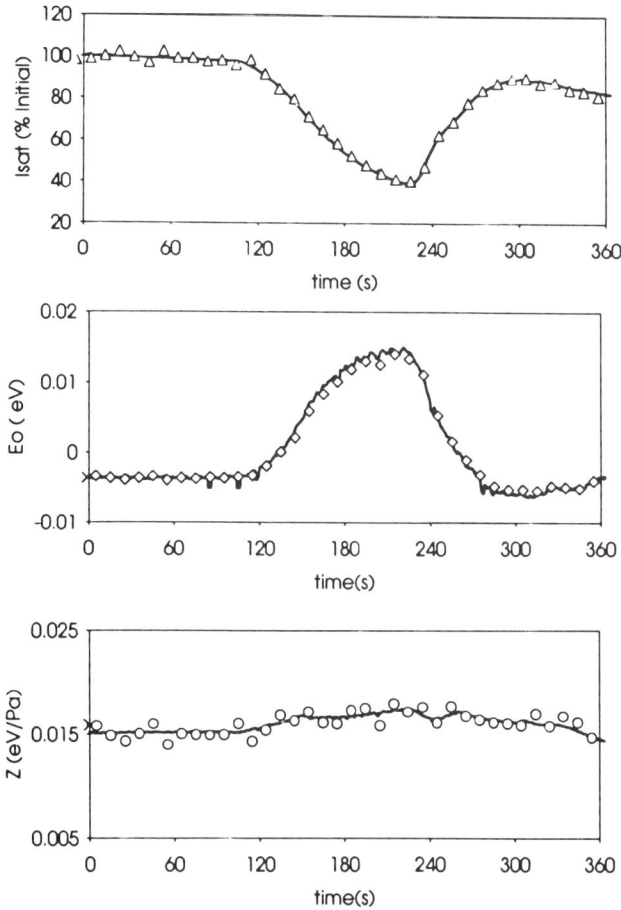

Fig. 6. Changes in Boltzmann parameters during transient hypoxia. At about 120 sec endocochlear potential fell, which was reflected in a drop in the saturation parameter, I_{sat}. Concurrently, the operating point parameter, E_o, changed in a direction consistent with movement of the organ of Corti towards scala vestibuli. There was little change in the sensitivity parameter, Z. Shortly after restoration of the oxygen supply at about 200 sec, the endocochlear potential rose abruptly, producing recovery of the saturation parameter, I_{sat}, and operating point, E_o. These results suggest that the sensitivity parameter, Z, which must be proportional to the low-frequency compliance of the organ of Corti, does not necessarily depend on the standing current through the OHCs.

indicated by the transient drop in V_{sat}, and that this fall is associated with a transient move of the organ of Corti towards scala vestibuli. This is evident in the transient changes in the E_o parameter. The drop in V_{sat} is almost entirely due to the transient drop in endocochlear potential during the hypoxia, and the transient move towards scala vestibuli is probably due to the transient osmotic imbalance caused by the dysfunction of the stria vascularis. It is likely that the salt concentration of scala media undergoes a small but significant decrease during hypoxia, causing a transient outward flow of water from scala media, and an upward lurching of the organ of Corti towards scala media as a consequence. The OHCs themselves are also likely to undergo some mechanical shape changes, either due to some motile process (see Dallos *et al.*, 1991) or due to simple osmotic imbalances. When the strial pumps recover with resupply of oxygen, this transient hypotonicity would be corrected, and the organ of Corti returns to its normal rest position. It is also interesting to note that the Z parameter varies insignificantly during this hypoxic transient. Why this should be the case is not clear, but the lack of change questions previous suggestions that the active process generates forces that act as a stiffness at low stimulus frequencies (McMullen and Mountain, 1985).

Whatever the explanation for these changes during hypoxia, the analysis method presented in this paper is one of the few techniques capable of demonstrating the fine changes that occur. It is also capable of giving estimates of changes in cochlear micromechanics in a stable way over many hours. It therefore holds great promise in the investigation of a wide range of cochlear phenomena, including basic cochlear function, ototoxicity, acoustic trauma, and homeostasis, including cochlear salt balance and neural control.

Acknowledgements

The development of this analysis technique has been carried out in collaboration with Mr Alex Moleirinho. The work was supported by a Program Grant from the National Health and Medical Research Council of Australia.

References

Crawford, A. C., Evans, M. G. and Fettiplace, R. (1989). Activation and adaptation of transducer currents in turtle hair cells. *J. Physiol.* **419:** 405–434.

Crawford, A. C., Evans, M. G. and Fettiplace, R. (1991). The actions of calcium on the mechano-electrical transducer current of turtle hair cells. *J. Physiol.* **434:** 369–398.

Dallos, P., Evans, B. N. and Hallworth, R. (1991). Nature of the motor element in electrokinetic shape changes of cochlear outer hair cells. *Nature* **350**: 155–157.

Eatcock, R. A., Corey, D. P. and Hudspeth, A. J. (1987). Adaptation of mechanoelectrical transduction in hair cells of the bulfrog's sacculus. *J. Neurosci.* **7**: 2821–2836.

Holton, T. and Hudspeth, A. J. (1986). The transduction channel of hair cells from the bull-frog characterized by noise analysis. *J. Physiol.* **375**: 195–227.

Howard, J. and Hudspeth, A. J. (1987). Mechanical relaxation of the hair bundle mediates adaptation in mechanoelectrical transduction by the bullfrog's saccular hair cell. *Proc. Natl. Acad. Sci. USA* **84**: 3064–3068.

Patuzzi, R. and Rajan, R. (1990). Does electrical stimulation of the crossed olivo-cochlear bundle produce movement of the organ of Corti. *Hearing Res.* **45**: 15–32.

Patuzzi, R. and Robertson, D. (1988). Tuning in the mammalian cochlea. *Physiological Rev.* **68**: 1009–1082.

Patuzzi, R. B., Yates, G. K. and Johnstone, B. M. (1989a). The origin of the low-frequency microphonic in the first cochlear turn of guinea-pig. *Hearing Res.* **39**: 177–188.

Patuzzi, R. B., Yates, G. K. and Johnstone, B. M. (1989b). Changes in cochlear microphonic and neural sensitivity produced by acoustic trauma. *Hearing Res.* **39**: 189–202.

Patuzzi, R. B., Yates, G. K. and Johnstone, B. M. (1989c). Outer hair cell receptor current and sensorineural hearing loss. *Hearing Res.* **42**: 47–72.

Neider, P. and Neider, I. (1971). Determination of microphonic generator transfer characteristics from modulation data. *J. Acoust. Soc. Am.* **49**: 478–493.

McMullen, T. A. and Mountain, D. C. (1985). Model of d.c. potentials in the cochlea: effects of voltage-dependent cilia stiffness. *Hearing Res.* **17**: **127–142.**

The Biomechanics of Outer Hair Cell Motility

RICHARD HALLWORTH

Department of Otolaryngology – Head and Neck Surgery, University of Texas Health Science Center, San Antonio, TX 78284, USA

Introduction

There is now a general consensus on the mechanism of electrically evoked fast outer hair cell (OHC) motility. It is believed that length changes are produced by the conformation changes of a membrane motor protein (Holley and Ashmore, 1988b; Dallos *et al.*, 1991; Kalinec *et al.*, 1992; Santos-Sacchi, 1993). The conformation changes occur in direct response to membrane potential change without intermediate biochemical steps. The motors are located in the area above the nucleus (Kalinec *et al.*, 1992; Hallworth *et al.*, 1993). Length changes are accompanied by a coordinated antiphasic active diameter change (Hallworth *et al.*, 1993). The unusual cylindrical shape of OHCs is maintained by a cortical meshwork of actin and spectrin filaments wound in such a way as to confer significant structural rigidity (Holley and Ashmore, 1988a).

Despite this progress, there is much less agreement on the role of OHC motility in cochlear function. *In vivo*, OHC motility would be driven by the cell's receptor potential and would exert such force on the organ of Corti. Many models require the OHC either to exert force (Neely and Kim, 1986; Geisler, 1993; Hubbard, 1993; Mammano and Nobili, 1993) or to modulate the partition stiffness (Kolston *et al.*, 1989). All prior data have been gathered using isolated OHCs that are operating essentially unloaded, i.e. against no resistance from surrounding tissue. Results from both *in vivo* (Mountain and Hubbard, 1989) and *in vitro* (Mammano and Ashmore, 1993) model systems suggest that OHCs do indeed exert significant force on the organ of Corti. While it has been demonstrated that cilia-evoked receptor potentials can evoke length changes in isolated cells (Evans and Dallos, 1993), it is not clear that the

155

OHC would be capable of doing so when appropriately loaded. In addition, models must take into account any internal loads specific to the cell itself. The experiments described in this paper were designed to determine the force generation capability of the isolated OHC. In addition, some aspects of the passive mechanical properties of the OHC have also been examined.

Methods

Guinea-pig OHCs were prepared by enzymatic dissociation. Guinea-pigs were anesthetized by Nembutal, decapitated and their cochleas removed into ice-cold L-15 medium. The organ of Corti from all four turns was dissected into fragments using fine needles. Organ fragments were transferred to a dissociation medium (L-15 plus 5 mM EGTA, 5 mM L-cysteine and 1.5 mg/ml papain [Sigma] for 15–30 min. Fragments were then gently triturated into the experimental medium, examined for defects and transferred individually to the experimental chamber using a suction pipette. The experimental medium contained L-15 plus 5 mM Cs, 10 mM TEA (Aldrich) and 2 mM Co to block active potassium and calcium currents (Housely and Ashmore, 1992). Cells were selected for measurement only if they met established criteria (Evans *et al.*, 1991). Cells from 20 to 80 μm in length were studied.

Glass fibers, tapered to 1–2 μm diameter, were pulled from molten glass and trimmed to 1.0–1.5 mm length. Fiber stiffness was measured using glass balls, 30–50 μm diameter, density 2.5 (Polysciences) (Howard and Hudspeth, 1988). One or several balls were attached to the fiber tip and the tip deflection measured when the balls were removed by a gentle air jet. Fibers with stiffnesses in the range 50–1000 μN/m were used in this study. To improve their optical density, fibers were sputter-coated with gold/10% palladium prior to use.

The experimental arrangement is illustrated in Fig. 1. Cells were held at their basal (synaptic) poles using a suction pipette. The pipette was made from a piece of soft glass tubing, 2.0 mm diameter, pulled to a microelectrode point and cut off at 30–60 μm diameter using a diamond knife. The pipette tip was then heat-polished down to an internal diameter of 4–6 μm, smaller than the diameter of an isolated OHC (8 μm). The pipette was mounted in an electrode holder on a triaxial hydraulic manipulator. Cells were drawn to the pipette using a micrometer-controlled suction device. Voltage commands were applied to the micropipette using a patch-clamp amplifier (Dagan).

The glass fiber was attached to a bimorph actuator and brought into close apposition to the OHC's apex using a triaxial mechanical

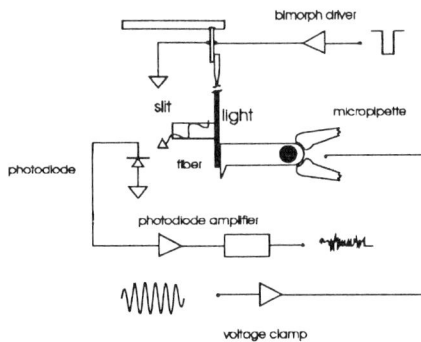

Fig. 1. Diagram of the experimental apparatus.

manipulator. The fiber was placed transversely to the OHC's long axis in such a way that the fiber's lateral motions would compress or relax the cell. Care was taken to avoid contact with cilia bundles that might be mistaken for contact with the cell apex proper. Fiber motions were measured using a photodiode-based measurement system similar to that described in previous publications (Evans *et al.*, 1991). By the use of a small-area photodiode, the bandwidth of the present system has been extended to 17.34 kHz. Calibrations used the optical lever method (Clark *et al.*, 1990).

Data acquisition was performed using an IBM-PC clone and data acquisition software provided by the Dallos laboratory. Input wave-forms were acquired at 400 samples/sec after prefiltering at 200 Hz low-pass and averaged 2–16 times. Stimulus waveforms were also generated by computer. Probe stimulus waveforms were shaped by a 120 Hz low-pass filter to minimize bimorph resonance.

Results

Cells are linear in compression for brief small amplitudes

When step displacements of the glass fiber probe in the direction of cell compression were applied, the resulting displacements of the probe tip resembled those shown in Fig. 2. The initial displacements were rapid, being complete in only a few milliseconds. The observed displacements were a linear function of the applied voltage. This applied for displacements of up to 400 nm. Some evidence of creep in

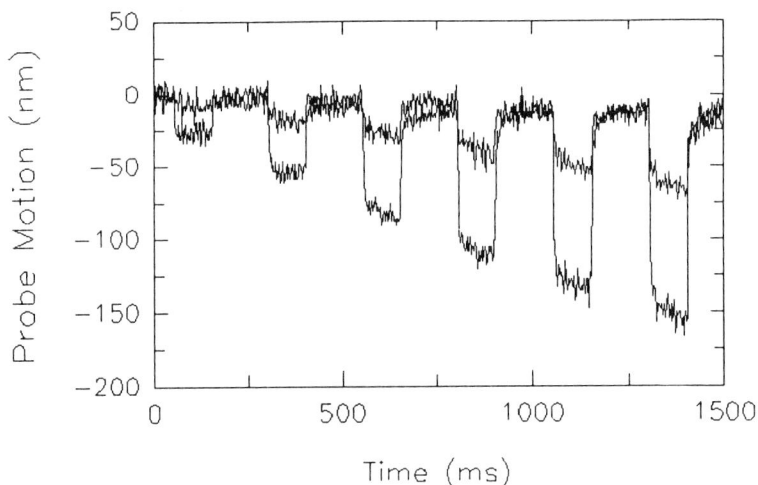

Fig. 2. Displacement waveforms of glass fiber probes apposed to the apex of OHCs when the fiber base was displaced in the direction of cell compression. The applied displacements were 100 msec steps increasing in 50 nm steps to a maximum of 300 nm. Large displacements: cell length 59 μm, compliance 960 m/N. Small displacements: cell length 38 μm, compliance 180 m/N.

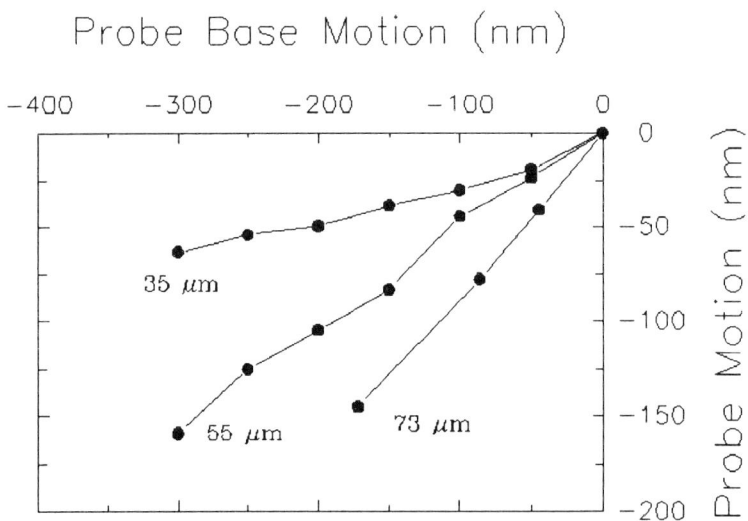

Fig. 3. Amplitude of initial motion of a probe apposed to an OHC as a function of the displacement applied to the probe base. Cell 1, length 35 μm, compliance 330 m/N. Cell 2, length 55 μm, compliance 1280 m/N. Cell 3, length 73 μm, compliance 10.0 km/N.

the compressive direction, evident in this figure, was frequently but not consistently observed, indicating that viscoelastic properties of the cell can become evident even for the brief static displacements used in these studies. No such creep was observed when the fiber was not in contact with a cell. Figure 3 shows examples of the linearity of the displacement amplitudes with increasing probe displacement, for cells of different lengths. Displacement measures were taken at the end of the initial rise portion of the response.

Cell compliance is a function of cell length

Cell compliance was calculated from the deflections of the fiber tip against the cell, knowing the fiber stiffness. The data in Fig. 3 exemplify the general observation that short cells (from the more basal portion of the cochlea) were less compliant than longer, more apical cells. Measurements of cell compliance were obtained for 38 cells. The results are plotted as a function of cell length in Fig. 4. There was considerable variability in measured compliance, but, to the level of precision possible with this technique, a clear trend was evident. Short cells were consistently less compliant (more stiff) than long cells, varying by approximately an order of magnitude from 30 to 80 μm.

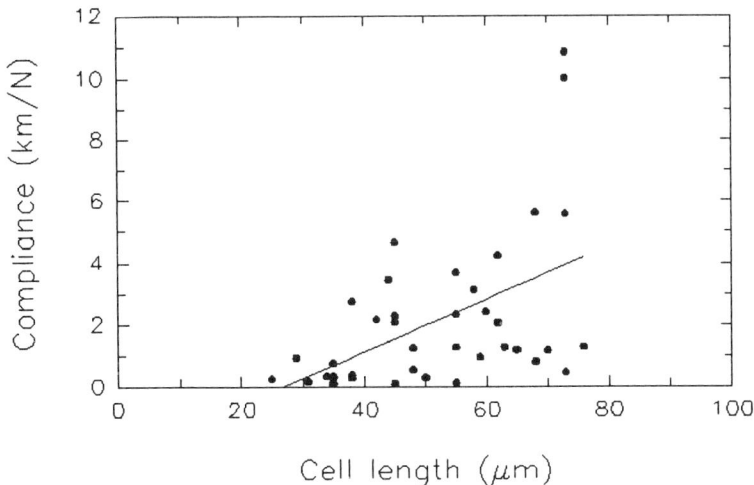

Fig. 4. Compliances of 38 cells plotted with respect to cell length. The regression line is given by $C = 0.0805L - 2.27$, where C is the compliance in km/N and L is the cell length in μm.

Motile force is of the order of 1–20 pN/mV

   The force generation capability of an isolated OHC was determined
by measuring the displacement it exerted, when electrically stimulated,
on a fiber of known stiffness. The fiber was applied to the cell with a
small (less than 0.5 μm) static compression so that fiber deflections
could be measured in response to both contraction and extension cycles
of the response. Figure 5 (top part) shows the displacements of a fiber
when driven by an OHC stimulated by a 240 mV 20 Hz sinusoidal
voltage command. Figure 5 (bottom part) shows the length change

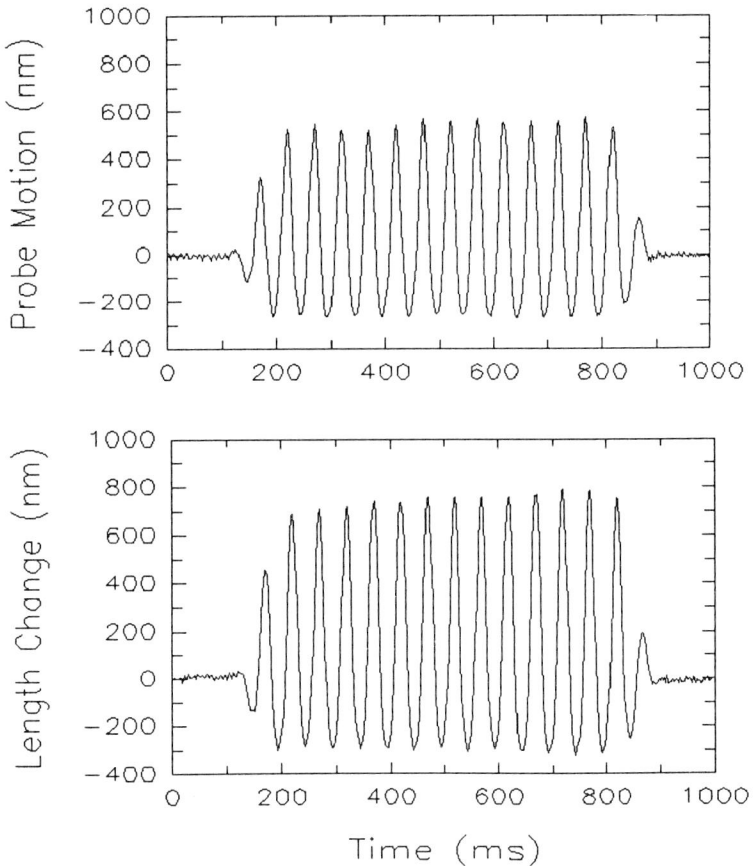

Fig. 5. (Top) Motion of a glass fiber probe apposed to an OHC driven by a 240 mV peak
sinusoidal voltage command. Positive values indicate motion resulting from cell extension,
negative values from cell contraction. Cell length 73 μm, fiber stiffness 270 μM/m. (Bottom)
Displacement response of the apex of the same cell to the same estimulus when unloaded.

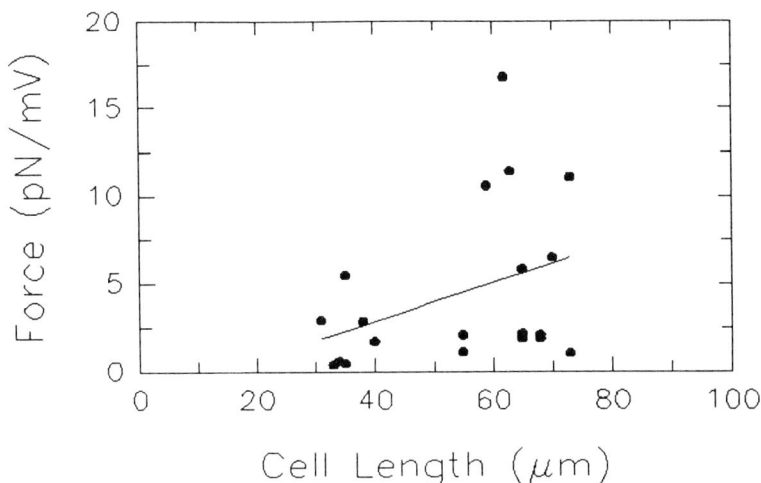

Fig. 6. Peak motile force per millivolt of applied voltage command generated by 20 cells in response to sinusoidal voltage commands of 240 or 120 mV peak, as determined by the deflections of a calibrated glass fiber probe apposed to the cell apex. The values of force per mV have been corrected for cell membrane area as described in the text and plotted against cell length. The equation of the regression line is given by $F = 0.11L - 1.54$, where F is the motile force in pN/mV and L is the cell length in μm.

responses of the same cell to the same voltage command with no fiber loading it down.

Cell force is a function of cell length

The force exerted by the cell was calculated from the known stiffness of the fiber. In order to relate the motile force to the membrane potential change, the effective command voltage delivered to the cell was calculated as follows. The cell membrane was assumed to form a resistive voltage divider such that the applied voltage command drops first across the membrane in the pipette orifice, then across the remaining cell membrane. The effective command potential applied was estimated by assuming the cell had a uniform membrane specific resistance, a constant diameter of 8 μm, a flat apical end with no cilia and a hemispherical basal pole. The hemispherical basal pole was assumed to constitute the membrane segment within the orifice. The calculated cell peak force per millivolt is plotted as a function of cell length for 20 cells (Fig. 6). The effective cell force per millivolt was found

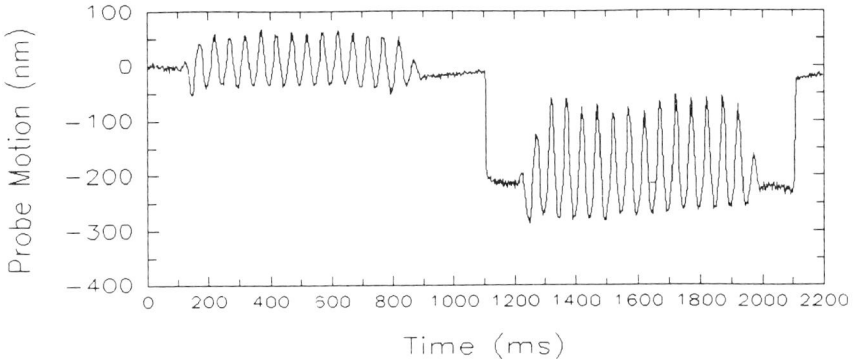

Fig. 7. Motion of a glass fiber probe (stiffness 550 μN/m) driven by the length changes of an OHC (cell length 38 μm). The voltage command waveform, applied twice, was a 20 Hz sinusoidal burst, 800 msec, 100 msec rise and fall. At 1100 msec, a compressive force was applied to the cell by displacing the probe base by 600 nm. The peak force generated by the cell, which was 1.16 pN/mV prior to compression, was enhanced to 2.25 pN/mV during the compression. The values have been corrected for cell membrane area as described in the text.

to increase with cell length by about a factor of 3 between the shortest and longest cells.

Motile force is increased by cell compression

The ability of the OHC to respond to electrical stimulation is significantly affected by the cell's internal pressure. It is known that reduction of internal pressure by aspiration of cell contents reduces or abolishes responses to voltage stimulation (Santos-Sacchi, 1991). It is reasonable, therefore, to suggest that some of the variability observed in the ability of OHCs to generate force may also be due to loss of internal pressure, perhaps due to damage during isolation.

Increasing internal pressure by applying compressive force was regularly observed to increase force generation by OHCs. In a substantial fraction of cells studied (9/17), compression of the cell by amounts of up to several hundred nanometers increased the amplitude of the deflection of a force measuring probe. An example of an experiment of this type is shown in Fig. 7. No concomitant changes in current evoked by the voltage command were observed, suggesting that the increased force was not secondary to an increase in seal resistance at the pipette orifice. Increases in motile force due to compression were up to a factor of 2. A constant probe base displacement was used, thus the amount of cell compression attained varied with both fiber and cell compliance. The detectability of force increases due to compression was not a function of cell length, since it was observed for cells of all lengths studied.

Discussion

Cell compliance

The data on cell compliance presented here may be compared with previous direct and indirect attempts to measure the same quantity. Holley and Ashmore (1988a) measured compliance with glass fibers using somewhat larger deflections than in this study. They reported cell compliances in the range 0.92–6.8 km/N for cells between 53 and 75 μm in length, a range closely comparable to the data reported here (see Fig. 4). They did not report data for cells shorter than 53 m. Gitter *et al.* (1993) used suction-induced distortion of cell shape to estimate a cell compliance of 5 km/N, also comparable to the measurements reported here. In contrast, Mountain and Hubbard (1994) estimated the stiffness of a typical hair cell to be 63 mN/m, equivalent to a compliance of 15.9 m/N, more than an order of magnitude lower than those in this study.

The trend reported here of decreasing compliance with cell length is compatible with a number of possible models of cell structure. Anatomically, OHCs are similar in diameter and internal anatomy, differing mainly in the length of the section between the nucleus and the cuticular plate. Using the regression line in Fig. 4, the relationship between cell length and compliance reported here may be modeled by assuming that each cell consists of a stiff section at the nuclear pole, 28.2 μm in length, plus a section of variable length and a compliance of 0.0805 km/N/μm. However, models based on the theory of plates and shells (e.g. Iwasa and Chadwick, 1992) also give linear relationships between applied force and compression.

The measurements of compliance presented are probably the least reliable for the longest cells. All attempts were made to select straight cells. However, many longer cells are to some extent bowed. While the small deflections used in this study probably preclude any contribution from buckling, it is possible that deflection modes other than simple compression could have been involved when long cells were compressed.

Force generation

The data reported here represent the first direct measurements of the force generation ability of OHCs. As such, the data may help to constrain possible models of OHC function in cochlear transduction. Other estimates have been made by less direct methods. For example, Gitter *et al.* (1993) proposed a force in the isolated cell of 200 pN for 1 μm contraction, from which a motile force of 2 pN/mV may be estimated, comparable to the data reported here. Iwasa and Chadwick (1992) estimated a force of 500 pN/mV, rather larger than the present observations. Hubbard (1993) used a force of 500 pN/mV in his model, a value which now seems optimistic in the light of the present data. Mountain and Hubbard (1994) estimated a force of 1.25 nN/mV, also a high value compared to the present results.

From the data, the ability of long cells to generate force appears to be greater than that of short cells by a factor of about 3. This is surprising since the basilar membrane in the basal turn is much stiffer than in the apex. However, some reservations need to be expressed about the data gathered so far. The relationship between cell compression and motile force (see next section) suggests that the small prior compression used in these studies is a probable source of variability. Moreover, since the number of cells reported is still small and the variability large, it is possible that the apparent trend may not be evident when more data are gathered.

The relationship of force generation to internal pressure

The finding that cell compression increases the motile force emphasizes the important role that internal pressure must play in the maintenance of force generation. The changes in cell force reported here, when observed at all, are surprisingly large. Only small (0.5%) compression were required to evoke them, which is unlikely to be enough to cause substantial changes in internal pressure. That cell compression was not always exhibited may possibly be related to the amount of compression that could be delivered given a fixed fiber step but a variable fiber stiffness in relation to the cell's compliance.

Iwasa (1993) has shown that OHC membrane capacitance can be manipulated by changes in internal pressure. This capacitance change has been related to a form of charge transfer, termed gating current, that is associated with OHC motility (Santos-Sacchi, 1991). It is possible that the change in internal pressure caused by cell compression could shift the transfer function of motility and thereby change its gain. However, the effect of internal pressure was shown to be equivalent to a hyperpolarizing shift in the function relating charge transfer to membrane potential. If the pressure-sensitive capacitance is directly related to motility, then any increased internal pressure would reduce the motile force generated, since the slope gain of motility is smaller at more hyperpolarized potentials (Evans *et al.*, 1991).

Acknowledgements

I thank Peter Dallos for the generous gift of data acquisition software, Brian Clark for assistance in the maintenance of same, Nancy K. R. Smith for coating the fibers and Kristin Scott for technical assistance.

References

Clark, B. A., Hallworth, R. and Evans, B. N. (1990). Calibration of photo-diode measurements of cell motion by a transmission optical lever method. *Pflüger's Arch.* **415**: 490–493.

Dallos, P., Evans, B. N. and Hallworth, R. (1991). Nature of the motor element in electrokinetic shape changes of cochlear outer hair cells. *Nature* **350**: 155–157.

Evans, B. N. and Dallos, P. (1993). Stereocilia displacement induced somatic motility of cochlear outer hair cells. *Proc. Natl. Acad. Sci. USA* **90**: 8347–8351.

Evans, B. N., Hallworth, R. and Dallos, P. (1991). Outer hair cell motility: the sensitivity and vulnerability of the DC component. *Hearing Res.* **52**: 288–304.

Geisler, C. D. (1993). A realizable cochlear model using feedback from motile outer hair cells. *Hearing Res.* **68**: 253–262.

Gitter, A. H., Rudert, M. and Zenner, H.-P. (1993). Forces involved in length changes of cochlear outer hair cells. *Pflüger's Arch.* **424**: 9–14.

Hallworth, R., Evans, B. N. and Dallos, P. (1993). The location and mechanism of electromotility in guinea pig outer hair cells. *J. Neurophysiol.* **70**: 549–558.

Holley, M. C. and Ashmore, J. F. (1988a). A cytoskeletal spring in cochlear outer hair cells. *Nature* **335**: 635–637.

Holley, M. C. and Ashmore, J. F. (1988b). On the mechanism of a high frequency force generator in outer hair cells isolated from guinea pig cochlea. *Proc. R. Soc. Lond. B* **232:** 413–429.

Housley, G. D. and Ashmore, J. F. (1992). Ionic currents of outer hair cells isolated from the guinea-pig cochlea. *J. Physiol.* **448:** 73–98.

Howard, J. and Hudspeth, A. J. (1988). Compliance of the hair bundle associated with gating of mechanical transduction channels in the bull-frog's saccular hair cell. *Neuron* **1:** 189–199.

Hubbard, A. (1993). A traveling-wave amplifier model of the cochlea. *Science* **259:** 68–71.

Iwasa, K. H. (1993). Effect of stress on the membrane capacitance of the auditory outer hair cell. *Biophys. J.* **65:** 492–498.

Iwasa, K. H. and Chadwick, R. S. (1992). Elasticity and active force generation of cochlear outer hair cells. *J. Acoust. Soc. Am.* **92:** 3169–3173.

Kalinec, F., Holley, M., Iwasa, K., Lim, D. and Kachar, B. (1992). A membrane based force generation mechanism in auditory sensory cells. *Proc. Natl. Acad. Sci. USA* **89:** 8671–8675.

Kolston, P. J., Viergiver, M. A., de Boer, E. and Diependaal, R. J. (1989). Realistic mechanical tuning in a micromechanical cochlear model. *J. Acoust. Soc. Am.* **86:** 133–140.

Mammano, F. and Ashmore, J. F. (1993). Reverse transduction measured in the isolated cochlea by laser Michelson interferometry. *Nature* **365:** 838–841.

Mammano, F. and Nobili, R. (1993). Biophysics of the cochlea: Linear approximation. *J. Acoust. Soc. Am.* **93:** 3320–3332.

Mountain, D. C. and Hubbard, A. E. (1989). Rapid force production in the cochlea. *Hearing Res.* **42:** 195–202.

Mountain, D. C. and Hubbard, A. E. (1994). A piezoelectric model of outer hair cell function. *J. Acoust. Soc. Am.* **95:** 350–354.

Neely, S. T. and Kim, D. O. (1986). A model for active elements in cochlear biomechanics. *J. Acoust. Soc. Am.* **79:** 1472–1480.

Santos-Sacchi. J. (1991). Reversible inhibition of voltage-dependent outer hair cell motility and capacitance. *J. Neurosci.* **11:** 3096–3110.

Santos-Sacchi, J. (1993). Harmonics of outer hair cell motility. *Biophys. J.* **65:** 2217–2227.

Labyrinthine Lateral Walls: Cochlear Outer Hair Cell Permeability and Mechanics

W. E. BROWNELL[1*], J. T. RATNANATHER[1], A. S. POPEL[2], M. ZHI[1] AND P. S. SIT[2]

[1]*Center for Hearing Sciences and* [2]*Department of Biomedical Engineering, The Johns Hopkins University School of Medicine, Baltimore, MD 21205, USA*

The outer hair cell (OHC) is a cylindrical epithelial cell that is partitioned into three regions with different functional roles in hearing. The mechanoelectrical transduction that is characteristic of all hair cells occurs at the apical pole where stereocilia are located. Electrochemical transduction takes place at the afferent and efferent synapses associated with the cell's basal pole. Electromechanical transduction occurs along the lateral wall and is responsible for electromotility at acoustic frequencies. OHC electromotility is a unique form of cell motility that is thought to contribute to the sensitivity and frequency selectivity of mammalian hearing. The trilaminate lateral wall is nearly cylindrical and extends from the tight junctional complexes at its apical end to the infranuclear region at its basal end. The structural features of the OHC's apical and basal poles are common to all hair cells but the lateral wall is highly specialized both structurally and functionally.

The lateral wall is laminated with three distinct coaxial components, all found within 100 nm of the cell's external surface. The outermost component is the cell's cytoplasmic membrane. The innermost component of the lateral wall is a concentric, membrane-bound organelle known as the subsurface cisterna (SSC). Sandwiched between

*Current address: Department of Otorhinolaryngology and Communicative Sciences, Baylor College of Medicine, Houston, TX, 77030, USA

the cytoplasmic membrane and the SSC is a cytoskeletal matrix known as the cortical lattice (for review see Brownell, 1990). The *in vitro* ultrastructural appearance of the SSC has been described (Dieler *et al.*, 1991; Evans, 1990; Slepecky and Ligotti, 1992). The SSC consists of a nearly continuous pair of crisp, double-layer membranes lying within 50 nm of the cytoplasmic membrane extending the entire length of the lateral wall. The inner and outer cisternal membranes (referenced to the cell's axis) are separated by 20–30 nm. The content of the intracisternal space is unknown. Very few fenestrae are observed in the SSC of quiescent cells (Dieler *et al.*, 1991; Slepecky and Ligotti, 1992) but they appear after prolonged exposure to salicylate or after electrical stimulation resulting in tonic OHC length changes (Evans, 1990). Disruption of the SSC results in an elevation in the threshold for electromotility (Dieler *et al.*, 1991; Evans, 1990). Further evidence that the SSC is required for the full expression of OHC electromotility comes from developmental studies in guinea-pigs (Pujol *et al.*, 1991) and gerbils (He *et al.*, 1994; Weaver and Schweitzer, 1994). The onset of electromotiliy in developing OHCs coincides with the morphological expression of a single layer of the SSC.

We have recently labeled the SSC membranes with C_6-NBD-ceramide and $DiOC_6$ (Pollice and Brownell, 1993). Both probes demonstrate distinct fluorescence along the lateral wall of the OHC. The lateral wall of inner hair cells is not labeled. Salicylate insult results in broadening of the lateral wall fluorescence, which is consistent with ultrastructural vesiculation and dilation of the SSC. The incorporation of the fatty acid ceramide or its metabolic by-products (e.g. sphingomyelin), coupled with the paucity of fenestrations in the lateral most cisternal layer, means that the SSC could function as an insulator, electrically dividing the cell into an axial compartment and an extracisternal compartment (i.e. the cytoplasmic space between the SSC and the lateral cytoplasmic membrane). The SSC partitions the cytoplasm into two compartments: an axial central core and the extracisternal space. The fatty acid composition of the SSC contributes to its elastic properties and those of the lateral wall. In addition to a nearly continuous layer, the SSC of guinea-pig OHCs has multiple layers (or stacks) of membranes over approximately 15% of the OHC's basal-lateral surface (Dieler *et al.*, 1991). The cisternal stacks may be transient structures involved in an ongoing membrane recycling process that maintains the integrity of the outermost layer. The rate of membrane turnover is not known, but remodeling of the SSC is complete within 30 min of the end of salicylate insult (Dieler *et al.*, 1991).

The OHC is a cellular hydrostat comprised of an elastic shell and a fluid-filled central core that is maintained at a positive pressure

(Brownell, 1990; Chertoff and Brownell, 1994). The cell's hydraulic skeleton permits rapid cell movements and provides the compressive strength required to maintain the structure of the organ of Corti. It is advantageous for a cellular hydrostat to have a low water permeability in order to maintain its cytoplasmic turgor pressure. We have recently determined the OHC's effective water permeability by measuring the OHC's rate of volume change in response to a hypotonic challenge (Ratnanather *et al.*, 1993). This permeability is as low as that of some plant cells which are also cellular hydrostats. This report provides further characterization of the cell's permeability. It also provides experimental data on the elastic and plastic properties of the lateral wall.

The Impact of Aspirin on Water Permeability

We have previously demonstrated that exposing the OHC to salicylate while electrically stimulating it causes the cell to lose volume and collapse (Shehata *et al.*, 1991). The possibility that salicylate may increase water permeability seemed to be reinforced by the observation that longer exposures resulted in disruption of the SSC (Deiler *et al.*, 1991). These observations led us to speculate that the SSC membranes contribute to the cell's low water permeability (Ratnanather *et al.*, 1993). We have tested this hypothesis by disrupting the SSC membranes with salicylate and then measuring the cell's effective water permeability. Our proceedures were a variation of the constant flow chamber experiment of Chertoff and Brownell (1994). Isolated OHCs were subjected to hypotonic challenges after preincubation in a standard bathing media or in media containing 10 mM sodium salicylate. A morphometric analysis of the cells was made from video images. The time-course of the OHC volume changes were determined from the morphometric analysis.

Figure 1 shows the time-course for OHC volume changes in response to perfusion by a hypotonic solution. The dashed lines show the volume response of OHCs that were preperfused with a standard bathing media. There is no change in the volume of these cells during the preperfusion period (Chertoff and Brownell, 1994). The solid line shows the response of cells that are preperfused with 10 mM sodium salicylate added to the bathing media. The left panel shows the response to −45 mOsm challenge, and the right panel shows the response to a −30 mOsm challenge.

Volume changes were normalized to the OHC's volume at the begining of the hypotonic challenge (1800 sec). The mean initial volume

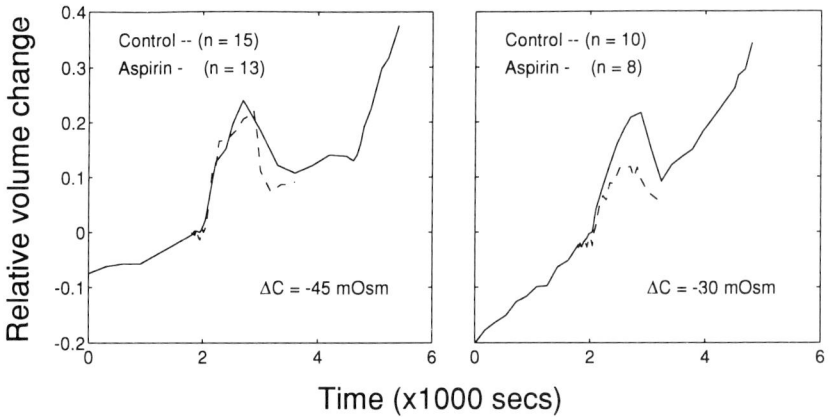

Fig. 1. Timecourse for OHC volume change in response to perfusion by hypotonic solution. Left side shows response to hypotonic challenge of –45 mOsm. Right side shows response to hypotonic challenge of –30 mOsm. Test perfusion begins at 1800 sec. Control cells, dashed line; salicylate intoxicated cells, solid line.

for the control cells in the left panel was 4.2 ± 0.3 (SEM) × 10^3 µm³; the mean initial value for the salicylate-intoxicated cells was 5.5 ± 0.3 (SEM) × 10^3 µm³. The mean initial volume for the control cells in the right panel was 4.0 ± 0.4 (SEM) × 10^3 µm³, while the mean initial value for the salicylate-intoxicated cells was 5.2 ± 0.6 (SEM) ×10^3 µm³. Note that the salicylate-exposed cells show a volume increase during preperfusion (0–1800 sec). The response to the hypotonic challenge at 1800 sec is a volume increase for all cells. The data in the left panel of Fig. 1 are based on 21 salicylate-intoxicated cells. Eight of these cells burst when exposed to the hypotonic challenge and could not be analyzed further. None of the control cells burst. The control cells undergo a volume increase of about 20% and then return to about 9% above their initial value when reperfused with the standard PBS solution (reperfusion begins at 2700 sec). The salicylate cells undergo a total change in volume of about 25% and are then allowed to 'recover' in PBS solution for another 1800 sec at which time they are subjected to a second hypotonic challenge.

Inspection of the four traces in Fig. 1 reveals that the slope of the volume increase in response to hypotonic challenge is essentially the same whether or not the cell has been preperfused with salicylate. The slope allows a calculation of the OHC's effective water permeability (Ratnanather *et al.*, 1993): the fact that the slopes are essentially the same means that aspirin does not increase the cell's effective water permeability. An exponential fit was used to calculate the effective water permeability of all 46 cells in Fig. 1 and was found to be = 3.3 ±

0.4 (SEM) $\times$ 10^{-14} m/s/Pa. The absence of a water permeability increase suggests that: (i) the water permeability of the cytoplasmic membrane is not affected by salicylate; and (ii) the extracisternal space provides a low-resistance pathway for water so that disruption of the SSC has little effect on water movement. These results are consistent with our previous observation that changes in whole-cell conductance observed in salicylate-intoxicated OHCs were not correlated with changes in cell volume (Shehata *et al.,* 1991).

The increase in volume of the cell during preperfusion with salicylate may result from one of two potential mechanisms. One possibility is that the OHC can regulate the osmolarity of its cytoplasm and the drug interferes with this regulation. The second is that the drug induces a mechanical change in the lateral wall, making it more compliant and allowing an osmotically driven water influx that is normally prevented by the elevated intracellular hydrostatic pressure. The fact that both normal and intoxicated cells did not return to their original volume on return to the standard bathing media suggests plastic changes occur with time.

Mechanically Induced Permeability Changes?

The osmotically induced volume changes reported in Fig. 1 required 3 min to produce a 10% volume increase. We have previously reported experiments in which only 3 sec were required to produce a 10% volume increase (Brownell *et al.,* 1989). The earlier experiments presented a sugar solution by pressure injection from a pipette. The tip of the pipette was greater than 10 μm in diameter and located within 20 μm of the cell. The flow chamber used in our more recent studies provided control and test solutions at constant flow, and the test solution arrived as a step function with a time-course of less than 30 sec (measured optically). We challenged OHCs in the constant-flow chamber with the same sugar solutions that produced the dramatic volume increase we had observed earlier and were surprised to observe that the cells collapse rather than increase in volume. We then repeated the original experiments and observed rapid volume increase. This observation caused us to compare a variety of different solutions in terms of their response to pressure injection vs. a solution change in a constant flow chamber. The results of these comparisons are presented in Table 1.

The difference in the rate of volume increase for the hypotonic challenge (column 2) suggests the cell's effective water permeability is greater when it is subjected to the pressure injection. The stresses

Table 1. Rate of normalized volume change

	PBS (control)	Hypotonic 250 mOsm	Raffinose 300 mOsm	Sucrose 300 mOsm	Potassium 300 mOsm
Constant flow	0	0.001 s^{-1}	<0	<<0	<0
Injected	0	0.020 s^{-1}	0	0.014 s^{-1}	<0

applied to the cell remain the same in the constant-flow chamber, while pressure injection could result in spatially different and temporally changing stress. The collapse of the cell in response to raffinose and sucrose solutions may reflect the absence of monovalent cations in the test solution. The resulting efflux of potassium would be accompanied by a loss of water from the cell. We have previously presented results that suggest OHC is impermeable to raffinose. If pressure injection renders the cytoplasmic membrane permeable to sucrose, then it would enter the cell, down its concentration gradient, and water would follow. The collapse in 300 mOsm potassium solution in the constant flow chamber may reflect the micromolar Ca^{2+} concentration in the test solution. The collapse with pressure injection is at odds with the increase reported by Brownell *et al.* (1990) and may be the result of a smaller pipette tip (5 µm diameter) used in the present experiments.

Elastic Properties of the Lateral Wall

We have examined the mechanical properties of the OHC lateral wall using an experimental paradigm that is essentially the same as that developed to study the mechanical properties of the red blood cell membrane (Evans and Skalak, 1980) and the endothelial cell (Sato *et al.*, 1987). A fire-polished suction pipette with an inside diameter of 3.8–4.2 µm is brought into contact with the lateral wall and negative pressure is applied. The resulting deformation of the lateral wall is recorded on video tape and the images are subjected to morphometric analysis. The pipette is attached via a 12 in. length of #90 PE tubing to a water column housed in 3/4 in. ID (1 in. OD) PE tube. A ruler is zeroed to the level of the microscope stage. All pressures are recorded as the distance (in cm) of the top of the water column away from zero. Once a cell is located, the microscope stage is rotated to orient the OHC's major axis orthogonal to the suction pipette. The pipette tip is brought into contact with the

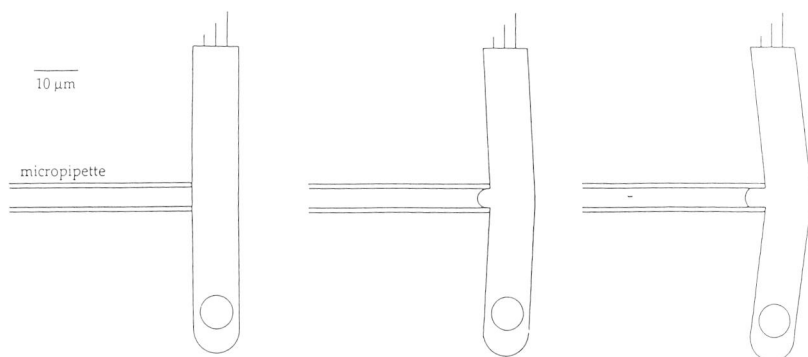

Fig. 2. Scheme showing elastic deformation of OHC lateral wall in response to aspiration. (Left panel) The micropipette is attached to the OHC cytoplasmic membrane. No negative pressure is applied and the wall is not deformed. (Middle panel) After a modest negative pressure is applied a portion of the lateral wall ('tongue') is aspirated into the micropipet. The length of the tongue increases when the magnitude of the pressure is increased (right panel). If the pressure is reduced (and less than 5 min have passed) the tongue will retract to the original value for the lower pressure.

OHC's lateral wall. The water column is dropped to produce a constant negative pressure. Figure 2 presents a scheme of the resulting cell deformation, and Fig. 3 presents analysis of the resulting deformations.

The left panel of Fig. 3 reveals that there is a linear relation between the length of the deformation and the pressure used; the slope of this relation is commonly referred to as the 'stiffness parameter' ($\delta PR_p^2/\delta L_t$, where R_p is the micropipette inner radius and L_t is the length of the tongue aspirated into the pipette). For comparison, the dashed lines in the left panel of Fig. 3 show the average pressure vs. length relationship for aortic endothelial cells (Sato *et al.*, 1987). The stiffness parameter of both the OHCs and endothelial cells is two orders of magnitude higher than that of red blood cells. The values of the stiffness parameter are $\delta PR_p^2/\delta L_t$ 1–2 dyn/cm for the four cells in Fig. 3. An analysis has been developed (Dr A. Spector, personal communication) for deformations of the OHC lateral wall that is analogous to the analysis Evans and Skalak (1980) developed for the red blood cell. This analysis relates the stiffness parameter to the shear modulus of the lateral wall. Preliminary results yield values for the shear modulus compatible with those of Iwasa and Chadwick (1992).

The right panel of Fig. 3 suggests the possibility of an inverse relation between the stiffness parameter and the length of the cell. Some, but not all, data points based on cells obtained from different cochleas fell on

Active hearing

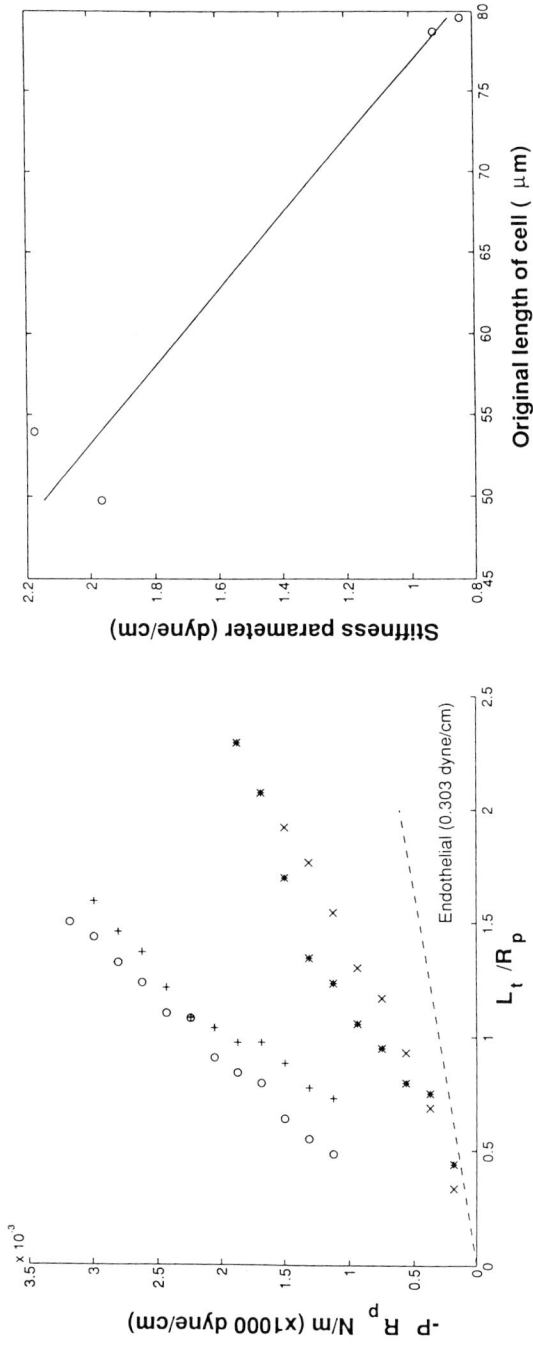

Fig. 3. (Left panel) Plot of deformation as a function of pressure for four different cells from the same cochlea (Lt = length of the tongue; P = negative pressure; and Rp = radius of the pipet). (Right panel) Stiffness parameter obtained in micropipette aspiration experiments for the same four cells plotted against the length of the cell.

the line connecting the four points in the right panel. When there was a difference between the stiffness parameters of cells from the same cochlea, the shorter cells were stiffer. If this relation proves to be generally true it could contribute to the production of the OHC length-dependent differences in mechanical resonance that have been reported (Brundin and Russell, 1994).

Plastic Change in the Lateral Wall

When a cell is subjected to a prolonged exposure at pressures greater than -10 cm H_2O, a plastic deformation occurs. A scheme for this plastic deformation is presented in Fig. 4.

Several observations support our interpretation that the cytoplasmic membrane pulls away from the SSC and the cortical lattice. The SSC can be seen near the pipette tip when large pressures are used. We conclude that the cortical lattice remains with the SSC because of the presence, in some cells, of a dimple on the opposite side of cell from the tongue. The cortical lattice is the only structure with a circumferential organization

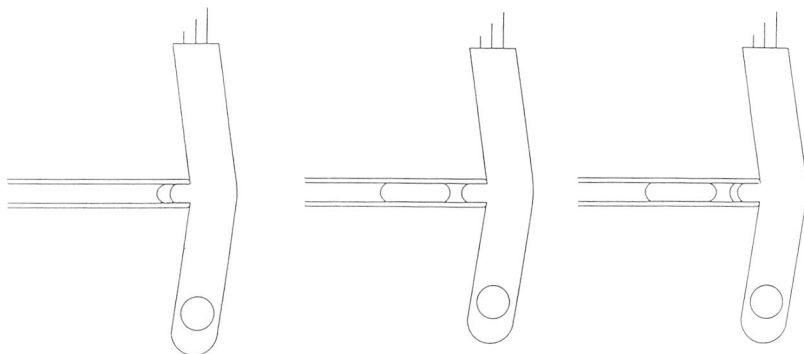

Fig. 4. A scheme showing the plastic phase of a micropipette aspiration experiment. After going through the elastic phase described in Figs 2 and 3, the cell begins to undergo a plastic deformation. (Left panel) When a critical pressure (the vesiculation pressure) is reached, typically around -10 cmH$_2$O, the cytoplasmic membrane is separated from the underlying structures and the length of the aspirated cytoplasmic membrane starts to grow. The bending of the cell associated with the entry of the lateral wall into the micropipette becomes conspicuous at this stage. (Middle panel) The leading edge of the cytoplasmic membrane continues its penetration into the micropipette at the constant vesiculation pressure and the aspirated portion of the cytoplasmic membrane pinches off and a vesicle is formed. The cytoplasmic membrane reseals. (Right panel) The vesicle moves slowly towards the widening end of the micropipet. Another portion of the cytoplasmic membrane is separated from the underlying structures. This leads to the formation of a second vesicle and theprocess can be repeated many times.

that might produce a dimple of comparable dimensions to the SSC deformation in the pipette. The cytoplasmic membrane begins to vesiculate when pressures exceed a value that varies from cell to cell (the pressure is typically greater than –10 cm H_2O). The cytoplasmic membrane reseals as each vesicle is released into the pipette. Vesiculation continues until a cell-specific limit is reached and no more vesicles are produced. This limit seems greatest for the longest cells, for which as many as 22 vesicles have left the cell.

The cytoplasmic membrane adjacent to the SSC is diaphanous and undulating (Dieler *et al.*, 1991; Saito, 1983). We interpret the vesicles as representing membrane that originates in the wave-like crenulations of the cytoplasmic membrane. It is possible that only the lipid bilayer flows into the pipette, leaving behind some integral membrane proteins connected with the cortical lattice. The crenulations have been measured to contain up to 10% more membrane than would be present if the cytoplasmic membrane were straight (Ulfendahl and Slepecky, 1988). The function of these crenulations may be to permit cell length changes without stressing the cytoplasmic membrane. They would work in much the same way that the air-impermeable pleated walls of an accordion permit significant changes in the volume and length of the instrument.

The ease with which the cytoplasmic membrane is detached from the cortical lattice has strong implications for current molecular models for electromotility based on transmembrane voltage sensors. The first reports of OHC electromotility noted that the active, force-generating region of the cell appeared to reside in the cell's lateral wall (Brownell, 1984; Brownell *et al.*, 1985). These qualitative observations have been confirmed with experiments that electrically partitioned the cell and demonstrated that no movement occurred when regions other than the lateral wall were stimulated. The same experiments demonstrated differences in movement magnitude depending on the amount of the lateral wall that is stimulated (Dallos *et al.*, 1991). Several authors have speculated that molecular motors residing in the cytoplasmic membrane change their conformation in response to changes in the transmembrane potential. The force generated by these unit motors is then summed across the active region and transmitted either by the cytoplasmic membrane or the cortical lattice to the ends of the cell. If the force generated by these unit motors exceeds a critical value, the cytoplasmic membrane could buckle rather than transmit force to the ends of the cell. The fact that the cytoplasmic membrane is loosely tethered to the cortical lattice limits the force that might be transmitted along the length of the cell via the lattice.

Discussion

The biological phenomenon of voltage-driven cell motility at frequencies in excess of 1 kHz has only been observed in the OHC, and the force generator for this electromotility is associated with the lateral wall. No other vertebrate cell (including the inner hair cell) displays electromotility, is a cellular hydrostat, or possesses as extensive a continuous subsurface cisternal system along its lateral wall. Our experiments indicate that the SSC does not contribute to the cell's effective water permeability. Micropipette aspiration experiments suggest that the SSC plays an important role in determining the mechanical properties of the OHC lateral wall, and that the overlying cytoplasmic membrane is a compliant component that is loosely tethered to the cortical lattice. The cytoplasmic membrane could easily buckle in response to shear forces in the plane of the membrane and reduce the effectiveness of postulated unit motor molecules in generating electrically evoked shape changes. Further characterization of the cell's permeability and the viscoelastic properties of its lateral wall is required so that OHC voltage–displacement functions may be converted into voltage–force functions. A precise knowledge of the OHC force generation is required to assess the cell's role as cochlear amplifier.

Acknowledgements

Research supported by NIDCD grant R01-DC00354, the W. M. Keck Foundation, a DRF research grant to J. T. R., and instrumentation grants from the Office of Naval Research and the Hasselblad Foundation. We thank Ms Clara Lee for invaluable assistance in the performing the morphometric analysis.

References

Brownell, W. E. (1984). Microscopic observation of cochlear hair motility. *Scan. Electr. Microsc.* 1401–1406.

Brownell, W. E. (1990). Outer hair cell electromotility and otoacoustic emissions. *Ear & Hearing* **11**: 82–92.

Brownell, W. E., Bader, C. R., Bertrand, D. and de Ribaupierre, Y. (1985). Evoked mechanical responses of isolated cochlear outer hair cells. *Science* **227**: 194–196.

Brownell, W. E., Shehata, W. E. and Imredy, J. P. (1990). Slow electrically and chemically evoked volume changes in guinea pig outer hair cells. In *Biomechanics of Active Movement and Deformation of Cells* (N. Akkas, eds), pp. 493–498. Springer-Verlag, New York.

Brundin, L. and Russell, I. (1994). Tuned phasic and tonic responses of isolated outer hair cells to direct mechanical stimulation of the cell body. *Hearing Res.* **73:** 35–45.

Chertoff, M. E. and Brownell, W. E. (1993). Characterization of cochlear outer hair cell turgor. *Am. J. Physiol.* **266:** C467–C479.

Dallos, P., Evans, B. N. and Hallworth, R. (1991). Nature of the motor element in electrokinetic shape changes of cochlear outer hair cells. *Nature* **350:** 155–157.

Dieler, R., Shehata-Dieler, W. E. and Brownell, W. E. (1991). Concomitant salicylate-induced alterations of outer hair cell subsurface cisternae and electromotility. *J. Neurocytol.* **20:** 637–653.

Evans, B. N. (1990). Fatal contractions: ultrastructural and electromechanical changes in outer hair cells following transmembraneous electrical stimulation. *Hearing Res.* **45:** 265–282.

Evans, E. A. and Skalak, R. (1980). *Mechanics and Thermodynamics of Biomembranes*. CRC Press, Boca Raton, FL.

He, D. Z. Z., Evans, B. and Dallos, P. (1994). First appearance and development in neonatal gerbil outer hair cells. *Hearing Res.* **78:** 77–90.

Iwasa, K. H. and Chadwick, R. S. (1992). Elasticity and active force generation of cochlear outer hair cells. *J. Acoust. Soc. Am.* **92:** 3169–3173.

Pollice, P. A. and Brownell, W. E. (1993). Characterization of the outer hair cell's lateral wall membranes. *Hearing Res.* **70:** 187–196.

Pujol, R., Zajic, G., Dulon, D., Raphael, Y., Altschuler, R. A. and Schacht, J. (1991). First appearance and development of motile properties in outer hair cells isolated from guinea-pig cochlea. *Hearing Res.* **57:** 129–141.

Ratnanather, J. T., Brownell, W. E. and Popel, A. S. (1993). Mechanical properties of the outer hair cell. In *Biophysics of Hair Cell Sensory Systems* (H. Duifhuis, J. W. Horst, P. van Dijk and S. van Netten, eds), pp. 199–206. World Scientific, Groningen.

Saito, K. (1983). Fine structure of the sensory epithelium of guinea pig organ of Corti: subsurface cisternae and lamellar bodies in the outer hair cells. *Cell Tissue Res.* **220:** 467–481.

Sato, M., Levesque, M. J. and Nerem, R. M. (1987). Micropipette aspiration of cultured bovine aortic endothelial cells exposed to shear stress. *Arteriosclerosis* **7**: 276–286.

Shehata, W. E., Brownell, W. E. and Dieler, R. (1991). Effects of salicylate on shape, electromotility and membrane characteristics of isolated outer hair cells from guinea pig cochlea, *Acta Otolaryngol.* **111**: 707–718.

Slepecky, N. and Ligotti, P. (1992). Characterization of inner ear sensory hair cells after rapid-freezing and freeze-substituion. *J. Neurocytol.* **21**: 374–381.

Ulfendahl, M. and Slepecky, N. (1988). Ultrastructural correlates of inner ear sensory cell shortening. *J. Submicrosc. Cytol. Pathol.* **20**: 47–51.

Weaver, S. P. and Schweitzer, L. (1994). Development of gerbil outer hair cells after the onset of cochlear function: an ultrastructural study. *Hearing Res.* **72**: 44–52.

Structure of the Electromechanical Transduction Mechanism in Mammalian Outer Hair Cells

FEDERICO KALINEC AND BECHARA KACHAR

Section on Structural Cell Biology, Laboratory of Cellular Biology, NIDCD, NIH, MD, USA

Outer Hair Cell Motility

The unique characteristic of the mammalian outer hair cells (OHCs), not shared by other cochlear cells, is their motor function. They are able to generate mechanical forces by at least two separate mechanisms. Very large changes over a timecourse of seconds or minutes (slow motility) can be elicited by chemical (Dulon *et al.*, 1988, 1990; Flock *et al.*, 1986; Zenner, 1986;) and physical (Brundin *et al.*, 1989; Canlon *et al.*, 1988) stimuli. Findings by Zenner (1986) that contractions induced by ATP and calcium were blocked by cytochalasin suggested the involvement of actin in the process. Studying the effects of calcium, Dulon *et al.* (1990) concluded that contraction was related to circumferential forces in the cell cortex. Later, Holley and Kachar (1992) suggested that the slower mechanism lies in the structurally dynamic inner layer which includes the cortical cytoskeleton and the lateral cisternae. An efferent control of OHC slow motility has been suggested, but the finding that neither acetylcholine, carbachol nor GABA induce any detectable change in the length of isolated OHCs raised doubts on this interpretation (Bobbin *et al.*, 1990).

In addition to slow motility, OHCs can convert high-frequency changes of plasma membrane potential into high-frequency changes of cell length (Ashmore, 1987; Brownell *et al.*, 1985; Kachar *et al.*, 1986). The mechanism of electromotility is distributed along the length of the OHC basolateral membrane and it is not directly dependent upon ATP or calcium (Ashmore, 1987; Kachar *et al.*, 1986). Depolarization decreases and hyperpolarization increases the cell length; simultaneously an antiphasic radial component is present, with depolarization increasing and hyperpolarization decreasing the cell diameter (Hallworth *et al.*, 1993). Different experiments have suggested that OHC electromotility is produced by the concerted direct action of a large number of independent molecular motors (Dallos *et al.*, 1991; Holley and Ashmore, 1988; Kalinec *et al.*, 1992), and the correlated behavior of motile response and non-linear capacitive currents suggests that it involves the movement of charges in membrane-bound molecules (Santos-Sacchi, 1991). We have recently demonstrated that the OHC lateral plasma membrane is the holder of the electromechanical transduction mechanism, and postulated a model involving a highly ordered array of transmembrane proteins present in the region of electromechanical transduction as the core of this novel motor (Kalinec *et al.*, 1992). In our model, the forces generated in the membrane are transmitted to the cortical cytoskeleton through the pillars, and conversely through these connections the cytoskeleton could be able to modulate the motor functioning. The nature of the pillars is unknown. However, the presence of AE and 4.1 proteins (which are involved in the plasma membrane–cytoskeleton connection in erythrocytes) in the OHC lateral cortex suggest the association of these proteins with the pillars (Kalinec *et al.*, 1993).

The two motile responses of the OHCs raise interesting structural problems. How can OHCs accommodate the large shape changes associated with slow motility without compromising the performance of the electromotile (fast) mechanism? Part of the answer may be in the organization of the cortical cytoskeleton.

The OHC Cortical Cytoskeleton

The structure of the cortical cytoskeleton was examined in our laboratory by electron microscopy in thin-sectioned, freeze-etched and negatively stained preparations (Holley *et al.*, 1992) (Fig. 1). Skeletons extracted in low salt buffers and stabilized with phalloidin show a lattice with relatively long, thick and circumferentially oriented filaments cross-linked by thinner filaments. Results from our lab and others suggest that longer filaments are F-actin, and that non-erythroid spectrin (fodrin) is the constituent of the cross-links (Holley *et al.*, 1992; Nishida *et al.*, 1993). Analysis of the orientation of circumferential filaments shows that the cortical cytoskeleton is composed of discrete domains with different orientations. The mean angle of the circumferential filaments in relation to the transverse axis of the cell was 9 deg,

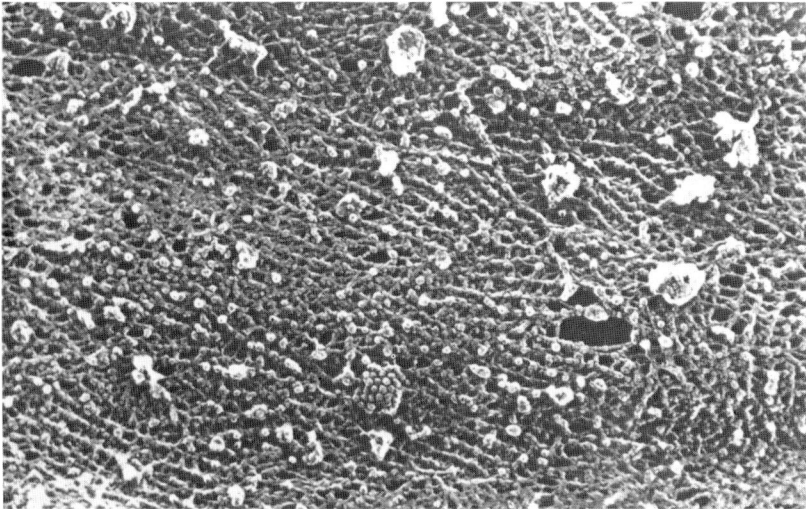

Fig. 1. Freeze-etched preparation of the OHC cortical cytoskeleton. The discrete domains of parallel, cross-linked actin filaments are clearly visible.

with values ranging from –55 to 74 deg. The domains varied in size, including only two or three filaments 200 nm long to at least 10 filaments 1 μm long (Holley *et al.*, 1992). The irregular contours and variable lengths of the cross-links suggest that they are more elastic and more easily deformable than the circumferential filaments. This idea is consistent with the known properties of actin and spectrin. Since the pillars link the plasma membrane only to actin (Holley *et al.*, 1992), any forces generated in the plane of the plasma membrane should encounter more resistance circumferentially than longitudinally along the axes of the cross-links. The cortical lattice may thus provide a vectorial component to the high frequency motor.

Mechanical deformations of the cortical cytoskeleton during high frequency length changes are about 5%. Spectrin would probably accommodate such length changes without disrupting the organization of the lattice. In contrast, the ability of the cortical cytoskeleton to accommodate the larger shape changes that are associated with slow motility may lie in its 'modular' organization. If our model of domains is correct, then we would expect the links between domains to have a lower resistance to sustained forces than those within domains. Sustained forces applied to the lattice could lead to movement between domains without compromising the unit (intradomain) structure.

If the structural domains within the cortical cytoskeleton can move relative to each other, and they are connected to the plasma membrane through the pillars, then there should be movements between corresponding domains within the plasma membrane. Since we have postulated that the electromechanical transduction mechanism is in the plasma membrane associated with the lattice of transmembrane proteins, the previous statement suggests that the membrane motors are also organized in small domains with different orientations.

Orientation of the Force vectors in the OHC Lateral Plasma Membrane

We have used video microscopy to measure the displacement of microspheres attached to the surface of electrically stimulated OHCs (Fig. 2) (Zajic and Schacht, 1991). Previous descriptions of OHC shape changes during fast motility were performed by measuring displacement of discrete cellular organelles (Holley and Ashmore, 1988; Kalinec et al., 1992). However, since the motor is in the plasma membrane, this method is intrinsically unable to distinguish if the cell contractions and elongations are uniform along the cell axis. The microspheres method is more suitable for this purpose (Zajic and Schacht, 1991).

As expected from a mechanism composed by many elementary motors working in a cumulative manner, the microspheres' displacement increases with the distance to the pipette attachment site (Fig. 3a). However, the relation is only roughly linear (correlation coefficient ~0.68), and the dispersion of the values cannot be explained by the uncertainty associated to the measuring procedure.

Figure 3(b) shows the frequency distribution in 10 deg angle classes of the orientation of the displacement vector for each microsphere (i.e. the angle between the vector and the cell's longitudinal axis). The full range was around 170 deg, with 30% of the points with angles bigger than 40 deg. The estimated mean orientation to the longitudinal axis of the cell, calculated from 148 measurements, was –7 deg, but the mode was in the –10 to –20 deg class.

The wide dispersion in the values of the moduli and orientation of the microspheres' displacement vectors can be explained assuming that the direction of the forces generated locally in the two-dimensional plane of the plasma membrane is not uniform. A close inspection of the displacement of microspheres attached to the OHC lateral plasma membrane very close one to another, as shown in Fig. 4, demonstrate that this is the case. This result suggests that the motor elements themselves are not homogeneously distributed.

In agreement with previous reports (Kalinec et al., 1992), no changes

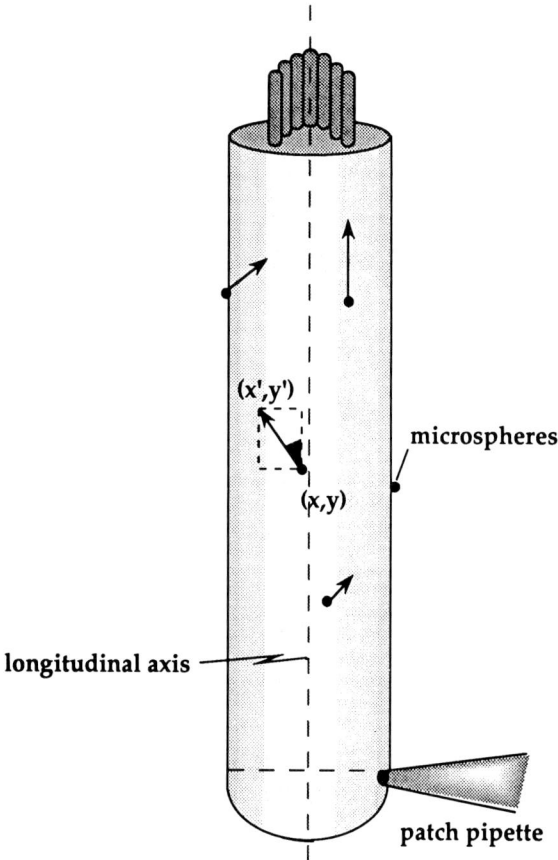

Fig. 2. Polybead-carboxylate monodisperse microspheres were used as markers. The microspheres' displacement was estimated, in digitized video images, by determining the coordinates of each bead on depolarized and hyperpolarized OHCs. The distance between each bead and the pipette's attachment site was recorded. A longitudinal axis was selected. The angle between this axis and the displacement vector was measured for each microsphere.

in length were detected in the basal (infranuclear) or apical (cuticular) regions of the lateral plasma membrane.

In short, our results, using this approach, support the hypothesis that the electromechanical transduction mechanism is circumscribed to the region between the nucleus and the apical tight junction belt (Kalinec *et al.*, 1992), and organized in small domains with different orientations. In spite of the dispersion of the motor's orientation angles, the mean angle (–7 deg) and the mode (–10 to –20 deg) are consistent with a mechanism

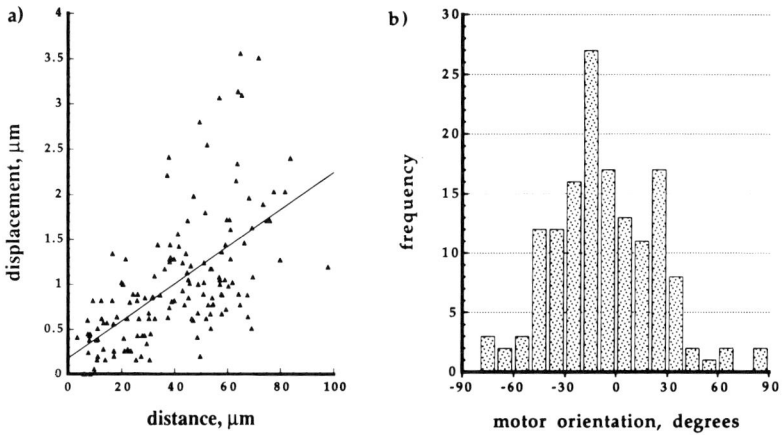

Fig. 3. (a) Microspheres' displacement over their distance to the pipette attachment site (n = 148). Displacement increases with the distance, but the correlation coefficient is only ~0.68. (b) Frequency distribution in 10 deg angle classes of the direction of the displacement vector for each microsphere in respect to the cell longitudinal axis. The estimated mean was –7 deg, and the mode was in the –10 to –20 deg class.

designed to produce longitudinal shape changes in OHCs and with experimental results about the motor orientation in guinea-pig OHCs (Dallos *et al.*, 1993).

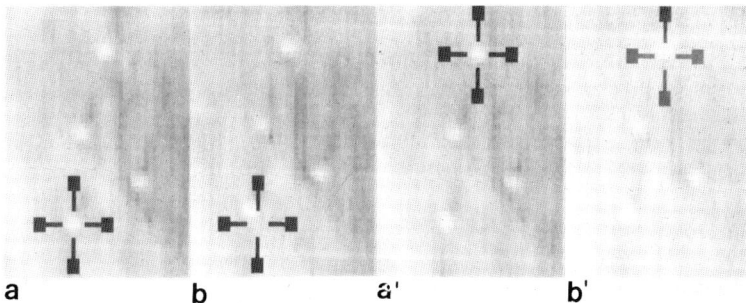

Fig. 4. Four microspheres attached to the lateral plasma membrane of an OHC in a depolarized (a, a') and in a hyperpolarized (b, b') condition. The displacement vector of the most apical of the microspheres is smaller in modulus and possesses a different direction than the vector of the microsphere closer to the pipette attachment site.

Structural Domains in the OHC Lateral Plasma Membrane

We examined freeze-etched replicas of the lateral plasma membrane of OHCs partially extracted with Triton X-100 in a low salt buffer. The images show clearly that the array of transmembrane proteins in the OHC lateral plasma membrane is organized in microdomains with different orientations (Fig. 5). The microdomains are not at random, but they look oriented around a preferred direction. This segregation of the regular array of transmembrane proteins in small domains resembles the pattern described in the cortical lattice (Holley *et al.*, 1992).

Motor Reorientation

We have used diamide to test if changes in the orientation of the force-generating domains would result in a reorientation of the electromotile forces.

Diamide is a bifunctional thiol reagent that specifically oxidizes the sulfhydryl group of cysteine to the disulfide (Becker *et al.*, 1986).

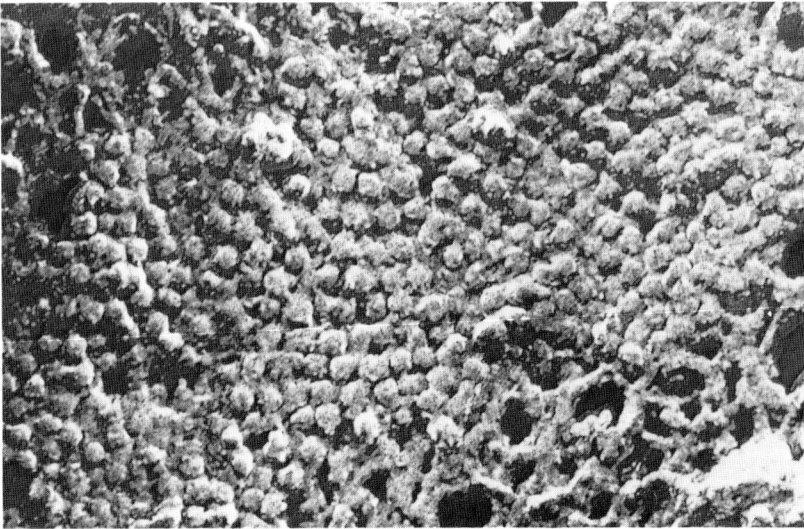

Fig. 5. Freeze-etched replica of the lateral plasma membrane of an OHC partially extracted with Triton X-100 in a low salt buffer. The transmembrane proteins are organized in small domains with different orientations. The proteins are still attached to the cortical cytoskeleton, which is visible as a filamentous structure beneath the plasma membrane.

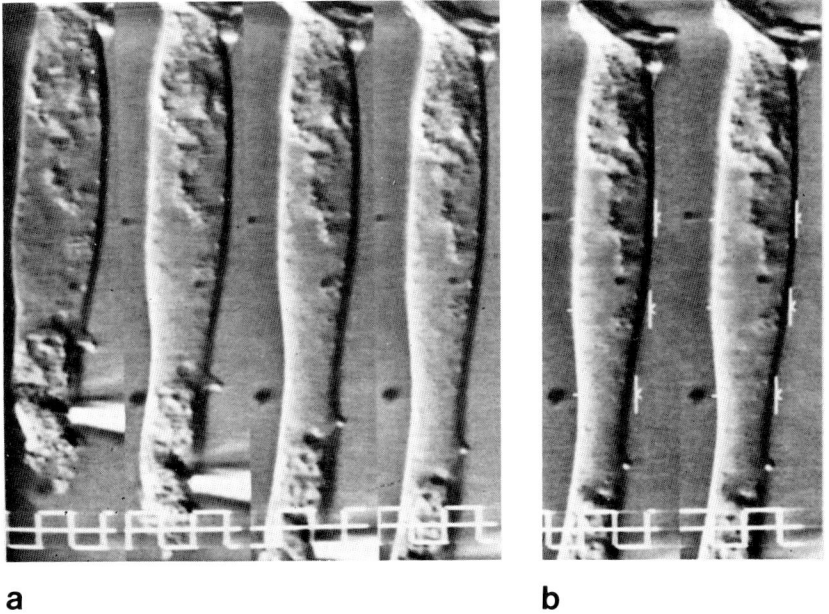

a   b

Fig. 6. (a) OHCs incubated with diamide (1 mM, Calbiochem) attached to the glass by the apical region and stretched using a patch pipette attached to the basal end. The OHCs yield easily to stretch. Following moderate stretching, the biggest cell deformations were observed in the supranuclear region. (b) When stretched, diamide-treated OHCs show a different pattern of electromotility. Hyperpolarization increases and depolarization decreases the cell diameter. The radial changes are more visible in the supranuclear region.

Spectrin, protein 4.1 and the cytoplasmic domain of AE proteins, but not actin, have been reported as sensitive to diamide (Becker *et al.*, 1986).

OHCs incubated with diamide became less rigid, yielding easily to stretching (Fig. 6a). Deformations around 25–30% were recoverable, but beyond this point the cylindrical shape was permanently lost and the cytoskeleton did not recover its original shape when the cell was demembranated with Triton X-100. In contrast, non-treated cells stretched in the same way showed very slight elastic deformations with total recovery.

As was previously reported (Kalinec and Kachar, 1993), diamide-treated cells respond to electrical stimulation like control cells. This result is completely consistent with the postulated independence of the OHC motor from the structural integrity of the cortical cytoskeleton (Kalinec *et al.*, 1992). However, the pattern of motility changes in mechanically stretched cells. Membrane hyperpolarization now results in a radial expansion, whereas depolarization diminishes cell diameter (Fig. 6b). When we measured the diameter along the cell body, we did not observe electrically stimulated radial movements at the cuticular plate level or in the infranuclear region, but they were evident in the middle region. As shown in Fig. 6(b), the change in diameter was maximal (around 10%) in the supranuclear region.

In erythrocytes, spectrin cross-link by diamide results in reduction of erythrocyte deformability as measured with a high-shear rheoscope (Maeda *et al.*, 1983). The different response of OHCs could be related to structural differences between the membrane skeleton of both cell populations. In erythrocytes, short actin filaments are interconnected with five or six spectrin tetramers in a spider-web fashion (Luna and Hitt, 1992). The cytoskeleton–membrane interaction is mediated by protein 4.1 and ankyrin, with this second protein linking the integral membrane protein AE1 (Band 3) to the β strand of spectrin at a site located in the midregion of spectrin tetramers. Diamide action in this cell population probably results in the loss of the 4.1-mediated actin–spectrin linkage and the cross-linking of spectrin filaments in a new, extensive, near all-spectrin cytoskeleton. The reported increment in erythrocyte rigidity would be a direct consequence of the mechanical properties of this new spectrin meshwork still attached to the plasma membrane through ankyrin. In contrast, the action of diamide on OHCs must result, presumably, in the formation of small clusters of the shorter spectrin filaments detached from the actin ones. Although the membrane–cytoskeleton coupling through the pillars would remain probably intact, the 'pitch' between actin filaments could be easily changed by stretching, resulting in the reported increment in cell deformability.

The alteration in the force direction when OHCs were stretched can be interpreted as a reorientation of the domains of membrane proteins. As were discussed in relation to the organization of the cortical cytoskeleton (Holley *et al.*, 1992), sustained forces applied to the cell could lead to movements between domains relative to each other to accommodate large cell shape changes without compromising its unit structure. In this way, cell stretching as performed in our experiments represents an extreme situation. The work performed by the motor in the longitudinal (parallel to the main axis of the cell) direction is no more energetically favorable, and the domains turn to a more convenient orientation. This experiment also suggests that the motor would be intrinsically anisotropic, an idea also supported for the observation that following cortical cytoskeleton disruption by trypsin, the cell movement remains anisotropic (Kalinec *et al.*, 1992). However, we cannot rule out the possibility of an anisotropic cortical lattice still providing the vectorial component to an isotropic motor.

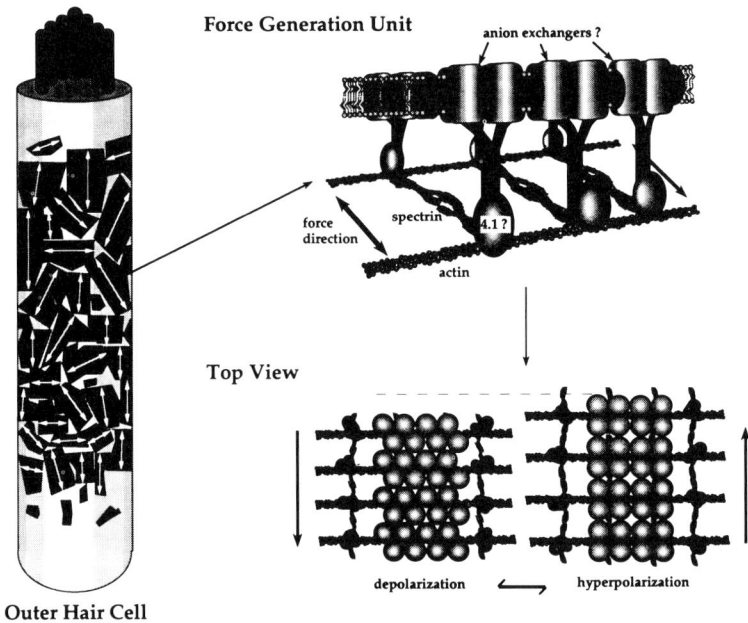

Fig. 7. We suggest that the lateral wall of the OHCs is covered with many three-layered 'force-generation units' composed of small domains of membrane-bound motor proteins and actin–spectrin cytoskeleton, with the pillars coupling both structures. (Top view) The rearrangement of tetrameric aggregates of motor proteins is the core of the force-generation mechanism. For simplicity the lipid bilayer and other membrane components have not been represented.

Structure of the Electromechanical Transduction Mechanism

We postulate that the OHC electromechanical transduction mechanism possesses a modular organization. Each module would be structured as a mosaic of three-layered 'force-generation units' composed of small domains of membrane-bound motor proteins and actin–spectrin cytoskeleton, with the pillars coupling both structures (Fig. 7). The entire lateral wall of the OHC would be covered with many of these force-generating units with slightly different orientations. Motor proteins would be organized as multimeric (tetrameric?) aggregates. Changes in membrane potential would produce conformational changes in the motor proteins, leading to the rearrangement of the multimeric aggregates in a way that increases the plasma membrane area when the cell is hyperpolarized and decreases it when it is depolarized (Fig. 7, top view). We envisioned the rearrangement of the multimeric aggregates as the basic process behind the force generation mechanism. Since this rearrangement is constrained for the proteins anchored to the actin filaments in the cortical lattice, the final result would be a displacement directed in the plane of the membrane in a direction perpendicular to the actin filaments, with a consequent contraction (expansion) in the orthogonal direction.

Conclusions

Structural and functional requirements of the OHCs as well as previous results about the organization of the cortical cytoskeleton led to us to think in terms of an electromechanical transduction mechanism structured in small domains. To test this hypothesis we have performed three different experiments:

- By attaching microspheres to OHCs we determined that the direction of the force generated is not uniform along the lateral plasma membrane.
- Using freeze-etching electron microscopy we found that the closely packed transmembrane proteins associated with the site of electromechanical transduction are segregated in clusters, forming arrays with different orientations.
- By stretching diamide-treated OHCs, we found that hyper-polarization increases and depolarization decreases the cell diameter, suggesting that the force vectors can be reoriented in the plane of the membrane.

Based on these results, we suggest that the OHC force generation mechanism is structured as a mosaic of three-layered 'force-generation units' composed by small domains of membrane-bound motor proteins and actin–spectrin cytoskeleton, with the pillars coupling both structures. Relative movements between domains could allow substantial changes of cell shape (as those associated to slow motility) without disrupting the unit structure of the motor, thus allowing the cell cortex to retain its elastic responses to high frequency deformations.

References

Ashmore, J. F. (1987). A fast motile response in guinea pig outer hair cells: the cellular basis for the cochlear amplifier. *J. Physiol.* **388**: 323–347.

Becker, P. S., Cohen, C. M. and Lux, S. E. (1986). The effect of mild diamide oxidation on the structure and function of human erythrocyte spectrin. *J. Biol. Chem.* **261:** 4620–4628.

Bobbin, R. P., Fallon, M., Puel, J.-L., Bryant, G., Bledsoe, S. C., Zajic, G. and Schacht, J. (1990). Acetylcholine, carbachol, and GABA induce no detectable change in the length of isolated outer hair cells. *Hearing Res.* **47:** 39–52.

Brownell, W. E., Bader, C. R., Bertrand, D. and de Ribaupierre, Y. (1985). Evoked mechanical responses of isolated cochlear outer hair cells. *Science* **227:** 194–196.

Brundin, L., Flock, Å. and Canlon, B. (1989). Sound-induced motility of isolated cochlear outer hair cells is frequency-specific. *Nature* **342:** 814–816.

Canlon, B., Brundin, L. and Flock, Å. (1988). Acoustic stimulation causes tonotopic alterations in the length of isolated outer hair cells from guinea pig hearing organ. *Proc. Natl. Acad. Sci. USA* **85:** 7033–7035.

Dallos, P., Evans, B. N. and Hallworth, R. (1991). Nature of the motor element in electrokinetic shape changes of cochlear outer hair cells. *Nature* **350:** 155–157.

Dallos, P., Hallworth, R. and Evans, B. N. (1993). Theory of electrically driven shape changes of cochlear outer hair cells. *J. Neurophysiol.* **70:** 299–323.

Dulon, D., Aran, J. M. and Schacht, J. (1988). Potassium-depolarization induces motility in isolated outer hair cells by an osmotic mechanism. *Hearing Res.* **32:** 123–130.

Dulon, D., Zajic, G. and Schacht, J. (1990). Increasing intracellular free calcium induces circumferential contractions in isolated cochlear outer hair cells. *J. Neurosci.* **10:** 1388–1397.

Flock, A., Flock, B. and Ulfendahl, M. (1986). Mechanisms of movement in outer hair cells and a possible structural basis. *Arch. Otorhinolaryngol.* **243:** 83–90.

Hallworth, R., Evans, B. N. and Dallos, P. (1993). The location and mechanism of electromotility in guinea pig outer hair cells. *J. Neurophysiol.* **70:** 549–558.

Holley, M. C. and Ashmore, J. F. (1988). On the mechanism of a high-frequency force generator in outer hair cells isolated from the guinea pig cochlea. *Proc. R. Soc. Lond. B* **232:** 413–429.

Holley, M. C. and Kachar, B. (1992). The cortical cytoskeleton and cell shape changes in mammalian outer hair cells. *Adv. Biosci.* **83**: 27–33.

Holley, M. C., Kalinec, F. and Kachar, B. (1992). Structure of the cortical cytoskeleton in mammalian outer hair cells. *J. Cell Sci.* **102**: 569–580.

Kachar, B., Brownell, W. E., Altschuler, R. and Fex, J. (1986). Electrokinetic shape changes of cochlear outer hair cells. *Nature* **322**: 365–367.

Kalinec, F. and Kachar, B. (1993). Inhibition of outer hair cell electromotility by sulfhydryl specific reagents. *Neurosci. Lett.* **157**: 231–234.

Kalinec, F., Holley, M. C., Iwasa, K., Lim, D. J. and Kachar, B. (1992). A membrane-based force generation mechanism in auditory sensory cells. *Proc. Natl. Acad. Sci. USA* **89**: 8671–8675.

Kalinec, F., Jaeger, R. and Kachar, B. (1993). Mechanical coupling of the outer hair cell plasma membrane to the cortical cytoskeleton by anion exchanger and 4.1 proteins. In *Biophysics of Hair Cell Sensory Systems* (H. Duifhuis, J. W. Horst, P. van Dijk and S. M. van Netten, eds), pp. 175–181. World Scientific, Singapore.

Luna, E. J. and Hitt, A. L. (1992). Cytoskeleton–plasma membrane interactions. *Science* **258**: 955–963.

Maeda, N., Kon, K., Imaizumi, K., Sekiya, M. and Shiga, T. (1983). Alteration of rheological properties of human erythrocytes by crosslinking of membrane proteins. *Biochim. Biophys. Acta* **735**: 104–112.

Nishida, Y., Fujimoto, T., Takagi, A., Honjo, I. and Ogawa, K. (1993). Fodrin is a constituent of the cortical lattice in outer hair cells of the guinea pig cochlea: Immunocytochemical evidence. *Hearing Res.* **65**: 274–280.

Santos-Sacchi, J. (1991). Reversible inhibition of voltage-dependent outer hair cell motility and capacitance. *J. Neurosci.* **11**: 3096–3110.

Zajic, G. and Schacht, J. (1991). Shape changes in isolated outer hair cells: measurements with attached microspheres. *Hearing Res.* **52**: 407–410.

Zenner, H.-P. (1986). Motile responses in outer hair cells. *Hearing Res.* **22**: 83–90.

Ca²⁺ Signaling in Deiters Cells of the Guinea-Pig Cochlea: Active Process in Supporting Cells?

DIDIER DULON

Laboratoire d'Audiologie Expérimentale, INSERM et Université de Bordeaux II, Hôpital Pellegrin, 33076 Bordeaux, France

Introduction

The organ of Corti is a marvel of technology that allows the inner hair cells (IHCs) to perceive displacement of the basilar membrane at sub-nanometric levels for specific frequencies along the cochlear partition (for review see Dallos, 1992; Ruggero, 1992). It is now generally accepted that this amazing sensory performance is based on an active mechanical process which originates from the outer hair cells (OHCs). These OHCs are believed to perform an essential duty, namely to increase the incoming sound within the cochlea via a reverse transduction, resulting in a tremendous sharpening of the mechanical resonance of the basilar membrane (Brownell *et al.*, 1985; Mammano and Ashmore, 1993).

It is obvious, however, that the unique mechanical performance of the organ of Corti is in large part conferred by the supporting cells, which position the OHCs in an ideal configuration. Among the supporting cells in the organ of Corti, the Deiters cells certainly play an essential role because they provide a direct mechanical support for the electromotile OHCs (Fig. 1).

While the OHCs have received considerable interest in the last few years, we know very little about the physiology of the Deiters cells. These cells have been generally considered as passive support cells uninvolved in synaptic transmission or active micromechanics. Our recent studies demonstrating Ca²⁺ signals upon K⁺ depolarization (Moataz *et al.*, 1992) and via ATP purinergic receptors (Dulon *et al.*, 1993)

195

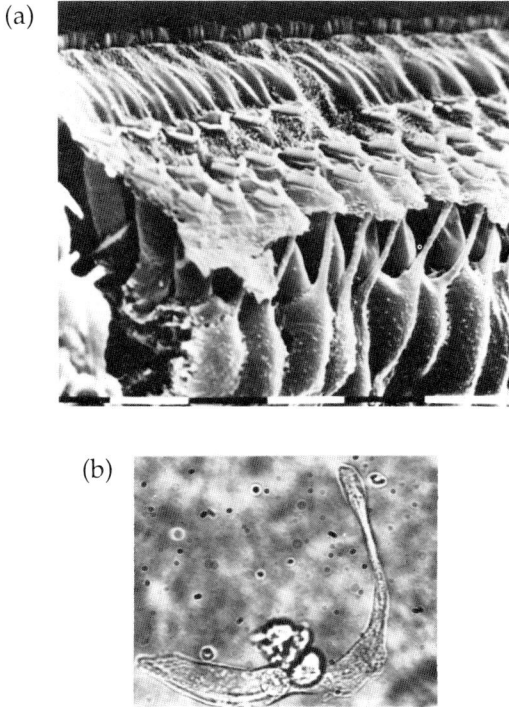

Fig. 1. (a) Lateral view under scanning electron microscopy of the outermost row of Deiters cells supporting the OHCs in the basal turn of the guinea-pig cochlea. The phalangeal processes of the Deiters cells, rich in endocellular filaments, participate strongly in the morphological organization and stability of the outer part of the organ of Corti. (b) Photomicrograph of a Deiters cell isolated from the guinea-pig cochlea.

raise the possibility that the Deiters cells may be more than passive bystanders undergoing the mechanical constraint of the OHCs.

Extracellular K^+: A Possible Signaling Agent

Because of the large electrochemical gradient of K^+ between the endolymph, the hair cell cytoplasm and the perilymph, there are good reasons to believe that there is a continuous flow of K^+ into the space of Nuel around the supporting cells, and that this flow of K^+ could increase with hair cell activity (Johnstone *et al.*, 1989). Therefore, the physiology of the supporting cells of the organ of Corti might be under the influence of local and transient variations of the concentration of K^+ ions, in similar fashion to the supporting glial cells in the CNS or in the retina. The neuroglia and the retinal glial cells have long been recognized to react to and regulate changes in extracellular K^+ that result from neuronal and photoreceptor activity (for review see Pentreath, 1982; Newman 1985). It is interesting to note in this context that receptor potentials have been recorded in the supporting cells of the cochlea during sound stimulation *in situ*, and that a slow component of these electrical responses has been suggested to be reminiscent of slow potentials triggered by extracellular K^+ in the CNS neuroglia (Oesterle and Dallos, 1990).

We have recently studied the effects of rising extracellular K^+ ions on the cytosolic concentration of free Ca^{2+} ($[Ca^{2+}]_i$) of isolated Deiters cells *in vitro*, using indo-1 ratiometric microspectrofluorometry. The isolated Deiters cells have generally, in our experimental conditions *in vitro*, a resting $[Ca^{2+}]_i$ of 100 ± 32 nM ($n = 16$), against a Ca^{2+} concentration of 1.25 mM in the external medium. This indicates, as we have previously reported in isolated hair cells (Dulon *et al.*, 1991), that isolated Deiters cells are able to maintain a low concentration of $[Ca^{2+}]_i$ against a large electrochemical potential, giving a large driving force for Ca^{2+} entry. The application of an isotonic high K^+-solution (20–150 mM KCl) to these cells produced a rapid and transient increase in $[Ca^{2+}]_i$ (Fig. 2). The Ca^{2+} sigr .l, increasing with the concentration of K^+, started to be detectable

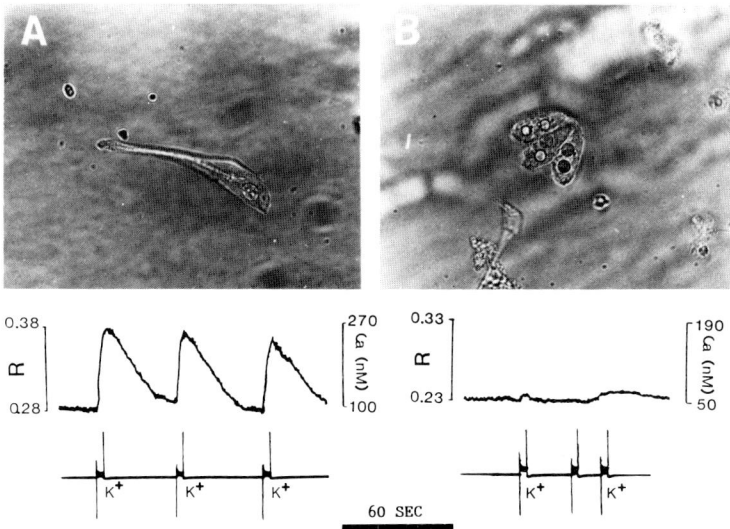

Fig. 2. Isotonic high K^+ solution (75 mM) produced reversible increases of intracellular Ca^{2+} in isolated Deiters cells (A), but not in Hensen cells (B). Calcium was measured with indo-1 microspectrofluorometry (R is the emission fluorescence ratio F_{405}/F_{480} nm).

at a K^+ concentration of 20 mM and reached a maximum around 100 mM of external K^+. The observed increment in $[Ca^{2+}]_i$ during K^+ stimulation was likely coming from a Ca^{2+} influx because removing extracellular free Ca^{2+} entirely suppressed the response. These results suggested the presence of voltage-gated Ca^{2+} channels in Deiters cells.

In order to characterize further the K^+-induced Ca^{2+} signals in isolated Deiters cells, indo-1 microspectrofluorometric and whole-cell patch-clamp recordings were combined. Surprisingly, depolarization of membrane potential under current clamp experiments did not generate significant increment in $[Ca^{2+}]_i$ within the whole cell's interior. However, this does not preclude the existence of voltage-gated Ca^{2+} channels in Deiters cells. Activation of voltage-gated Ca^{2+} channels can solely induce localized increase in intracellular Ca^{2+}, beneath the plasma membrane for example and only at particular active zones as suggested in neurons or hair cells (Roberts *et al.*, 1991). These localized changes in $[Ca^{2+}]_i$ would be difficult to measure using our whole-cell indo-1 microspectrofluorometric technique.

Interestingly, as mentioned above while depolarization steps do not increase whole-cell $[Ca^{2+}]_i$, a consecutive application of high K^+ solutions depolarized the cells and generated a rise in whole-cell $[Ca^{2+}]_i$ (Fig. 3). These K^+-induced Ca^{2+} signals were not inhibited by the dihydropyridine antagonists nifedipine or isradipine which are known to block L-type Ca^{2+} channels. Another intriguing observation was that the K^+-induced Ca^{2+} signals can still be observed when the cells are

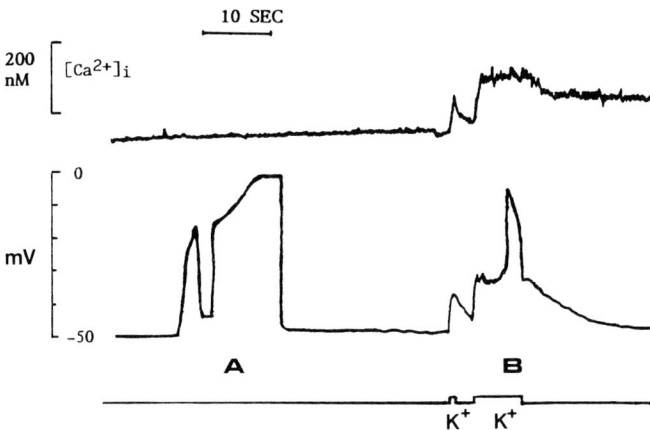

Fig. 3. Calcium (indo-1 fluorescence) and voltage responses of a current-clamped Deiters cell (perforated patch). Note that prolonged depolarization steps of the cells (A) do not increase intracellular Ca^{2+}, while subsequent application of isotonic high K^+ solution (70 mM) depolarizes the cells and increases intracellular Ca^{2+} (B).

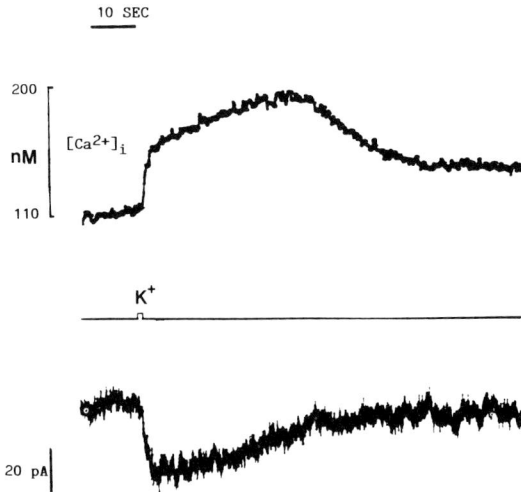

Fig. 4. Calcium (indo-1 fluorescence) and current responses of a voltage-clamped Deiters cell (voltage clamped at –80 mV under perforated patch). Extracellular puff application of an isotonic high K^+ solution (100 mM) increases intracellular Ca^{2+} concurrently with inward current.

voltage clamped at –80 mV (Fig. 4). These observations suggested that the influx of Ca^{2+} observed here, which induces an increment in whole-cell $[Ca^{2+}]_i$, was essentially triggered by a rise of extracellular K^+ ions rather than a drop in membrane potential.

Further investigation are now needed in order to characterize this K^+-induced Ca^{2+} influx observed in Deiters cells in order to know whether it originates from the activation of a new type of Ca^{2+} channel sensitive to external K^+, or whether high external K^+ increases the permeability of voltage-gated Ca^{2+} channels as observed in neuro-secretory GH_3 cells (Suzuki *et al.*, 1989).

The effect of K^+ stimulation was also tested on another type of supporting cell of the organ of Corti, the Hensen cells. The Hensen cells are also able to maintain a low level of resting $[Ca^{2+}]_i$ *in vitro* below 100 nM, but unlike the Deiters cells, the Hensen cells do not increase their $[Ca^{2+}]_i$ upon the application of high K^+ solutions (Fig. 2). The present results indicate a differential K^+ sensitivity in terms of Ca^{2+} response between the two types of supporting cells.

Extracellular ATP: Another Possible Intercellular Signaling Agent

There is recent evidence that ATP receptors, characterized as P_2-

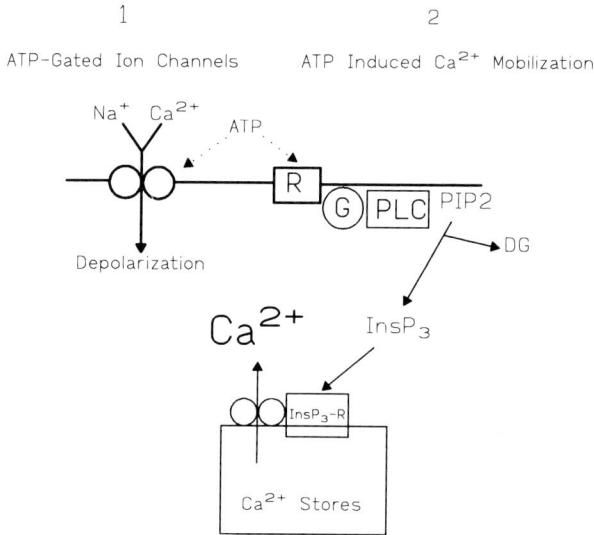

Fig. 5. Hypothetical transduction pathways via P$_2$-purinergic receptors for extracellular ATP in Deiters cells: (1) ATP-gated non-specific cation channels, and (2) ATP-induced Ca^{2+} mobilization from intracellular Ca^{2+} stores via metabotropic receptors.

purinergic receptors, are expressed in supporting cells of the organ of Corti (Ashmore and Ohmori, 1990; Dulon *et al.*, 1993). We have observed that extracellular ATP at sub-micromolar concentrations evokes transient increases in [Ca^{2+}]$_i$ in single Deiters cells. The results suggested that the response originates from 'metabotropic' receptors that release Ca^{2+} from intracellular organelles and from ATP-gated ion channels (Fig. 5).

Experiments using the caged precursor of InsP$_3$, loaded into the cell via the patch-pipette, demonstrated the presence of InsP$_3$-sensitive Ca^{2+} stores in Deiters cells. These results suggested that the Ca^{2+} mobilization observed upon activation of purinergic receptors by ATP in Deiters cells could use InsP$_3$ as an intracellular messenger (Dulon *et al.*, 1993).

Extracellular ATP has been also shown to increase [Ca^{2+}]$_i$ in OHCs (Ashmore and Ohmori, 1990), but the Ca^{2+} signals of OHCs have a much slower kinetic than Deiters cells and, more importantly, OHCs have an apparent K_d for ATP more than 10 times higher (Nilles *et al.*, 1994). These observations suggest that the Deiters cells would be much more sensitive to external ATP than the attached OHC *in vivo*.

What could be the possible physiological source of extracellular ATP in the fluid of the space of Nuel?

There are at least two obvious possibilities. First, since ATP is known to be above millimolar concentration in the cytosol of most cells, ATP could be released on sudden breakage of intact cells, such as OHCs during severe noise trauma for example. Second, ATP could be co-packaged with ACh in exocytosis vesicles at the medial efferent nerve endings at the base of the OHCs. It is interesting to note that the basal part of the OHCs sits in a cup-shaped structure formed by the Deiters cell. This cup of the Deiters cell constitutes a confined area in close contact with the OHCs. Housley *et al.* (1991) observed that the ATP-gated ion channels are mainly localized at the apical endo-lymphatic surface in OHCs. Our observations using a similar protocol suggested that the ATP gated-ion receptors are rather located at the perilymphatic surface in Deiters cells, and more precisely at the beginning of the phalangeal process, i.e. near the place where the OHC sits *in situ* (Fig. 6), while Ca²⁺ mobilization appeared to begin in the phaleangeal process (Dulon et al. 1993). Again, these observations suggest that the Deiters cells would be much more sensitive to extracellular ATP than OHCs in the space of Nuel.

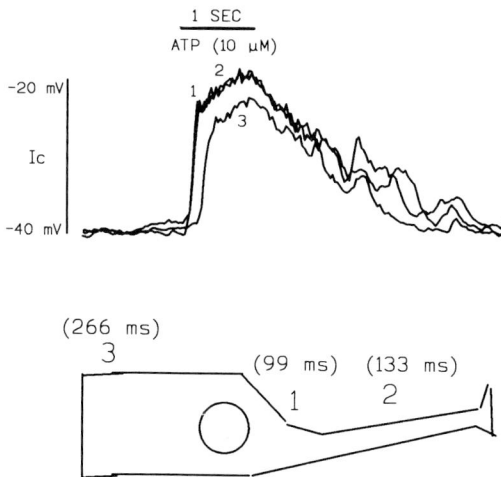

Fig. 6. Localization of purinergic receptors on an isolated Deiters cell. Voltage responses under whole-cell current-clamp are shown for extracellular applications of ATP (pressure puffs) at different locations of the cell. Note that the depolarization amplitudes and the onset latencies suggest that the purinergic receptors are essentially located in regions 1 and 2, the regions contacting the base of the OHC and the phaleangeal process, respectively.

Ca^{2+}: An Intracellular Messenger for Active Mechanical Responses?

It is clear that the Deiters cells, because of their unique structural organization, constitute the most direct damping device for the transversal electromechanical responses of the OHCs. The basal part of the Deiters cell, resting on the basilar membrane, forms a cup-shaped structure in which sits an OHC. From this cup structure, a slender phalangeal process extends apically and obliquely to contact another OHC at its apical circumfererence, participating in the formation of the reticular lamina (Fig. 1). The phaleangeal process of each Deiters cells constitutes a stiff arc containing longitudinal parallel arrays of microfilaments and microtubules, firmly encoring the OHC's apical end in the reticular lamina (Flock *et al.*, 1982; Slepecky and Chamberlain, 1987).

The aim of the following study was to determine whether intracellular Ca^{2+} may change the mechanical properties of the phalangeal process of the Deiters cells (Dulon *et al.*, 1994). Loaded via the patch-pipette in the whole-cell patch-clamp configuration, the caged Ca^{2+} molecule DM-nitrophen was used to increase $[Ca^{2+}]_i$ rapidly and reversibly in isolated Deiters cells of the organ of Corti. This technique is advantageous because it can rapidly elevate $[Ca^{2+}]_i$ on brief UV exposure (100–200 msec) without any mechanical perturbation of the cell, unlike protocols that involve perfusion or pressure puff of solutions around the cell. Immediately after the photoliberation of Ca^{2+}, a small movement of the head of the phalangeal process could be observed in 75% of cells. This mechanical movement, with an amplitude ranging between 0.5 and 1 μm within few hundred milliseconds, consisted of an extension of the phalanges away from the cell body.

Measurement of phalangeal stiffness in transversal flexion toward the cell body ranged from 15 to 440 pN/μm. The stiffness values of the phalangeal processes measured in our experimental conditions were about 1/5 to 1/10 of the longitudinal stiffness reported in isolated OHCs (Holley and Ashmore, 1988). An extrapolation to micromechanics *in situ* suggests that the Deiters cells participate significantly in the transverse stiffness of the organ of Corti. Stiffness of the process was observed to increase by 28 to 51% after $[Ca^{2+}]_i$ had increased.

We do not have any simple explanation to account at present for the precise molecular mechanism underlying the observed Ca^{2+}-induced mechanical response in Deiters cells. Calcium could affect the gelation stage or generate polymerization–depolymerization of cytoskeletal elements of the Deiters cells. Another alternative, which seems, however, more likely because of the rapid onset of the movement and its associated change in stiffness, is that Ca^{2+} activates cross-linking proteins, producing relative motion of the parallel arrays of microtubules and microfilaments in the cell's process. The results suggest, however, Ca^{2+} as a potential intracellular messenger for active mechanical responses in Deiters cells.

Active Electrical Responses: Electrical Resonance

Electrical resonance, manifested as damped membrane-potential oscillations elicited by the injection of current pulses, were observed in isolated Deiters cells. In whole cell current-clamp conditions, the membrane potential of the Deiters cells showed a damped sinusoidal oscillation (ranging from 20 to 45 Hz in 16 cells) when depolarized by steps of injected current (100–200 pA) from holding potential –70 mV to –40 mV (Fig. 7). This voltage ringing resembled that described in the hair cells of lower vertebrates (for review see Fuchs, 1992). The resonant frequency was voltage dependent and increased with depolarization as

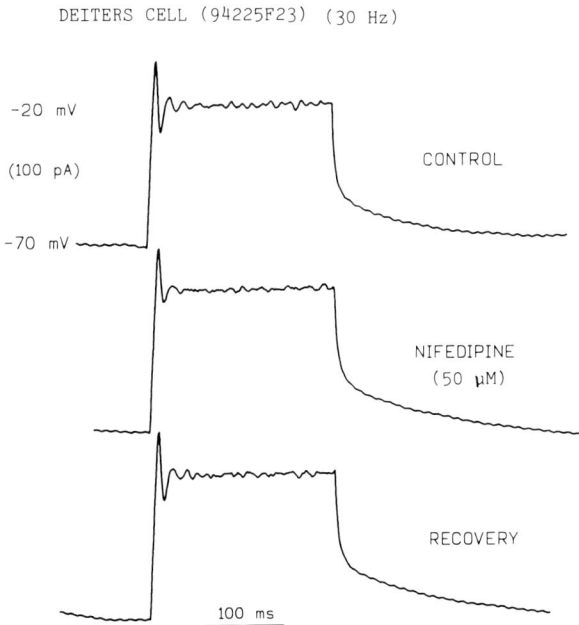

Fig. 7. Damped oscillations of the membrane potential at about 30 Hz (electrical resonance) produced by a 100 pA pulse of depolarizing current in a Deiters cell. The voltage oscillations were reversibly reduced by the L-type Ca^{2+} channels antagonist nifedipine, suggesting the implication of Ca^{2+} current followed by Ca^{2+}-activated K^+ current.

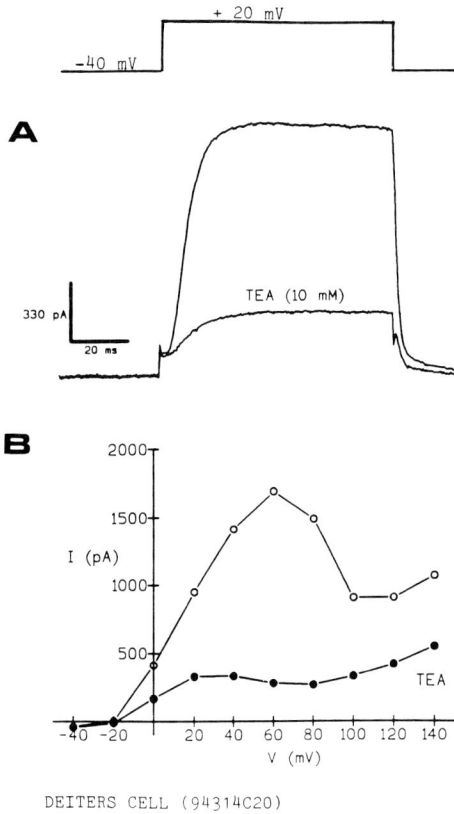

Fig. 8. Whole-cell ionic currents of a Deiters cells under voltage-clamp. (A) An example of outward current in response to a depolarizing voltage pulse from –40 to +20 mV. This current is largely reduced by 10 mM TEA. (B) Steady-state current–voltage relationships. Note the N-shaped IV characteristic of a Ca^{2+}-dependent K^+ current. The outward current could be reversibly blocked greater than 50% by 10 mM TEA.

previously shown in chick hair cells. The damped electrical oscillation in Deiters cells was generally observed to start around –40 to –20 mV. The damped electrical oscillations could reversibly be suppressed when cells were exposed to 10 mM tetraethylammonium ions (TEA) (n = 4), 50 μM nifedipine (n = 9), or to a Ca^{2+}-free solution containing 5 mM EGTA (n = 3) (Fig. 7), suggesting the implication of a voltage-dependent inward Ca^{2+} current (I_{Ca}) and Ca^{2+}-activated potassium current (K_{Ca}).

Under voltage-clamp, depolarizing voltage steps from a holding membrane potential of –80 to –40 mV typically generated a rapidly activating, non-inactivating outward current above –40 to –20 mV (Fig. 8). The current–voltage relation of this current was N-shaped at positive membrane potential, generally reaching a maximum around 60 mV. This behavior suggested that this outward current was activated by influx of Ca^{2+} and was therefore strongly dependent on the driving force for Ca^{2+} entry. The equilibrium or zero-current potential of Ca^{2+} in our experimental conditions was calculated to be around 180 mV. The Ca^{2+} dependence of this outward current was additionally suggested by the reversible reduction with the application of 50 μM nifedipine ($n = 11$), 5 mM Cd^{2+} ($n = 5$) or a Ca^{2+}-free solution containing 5 mM EGTA ($n = 5$). This current could also be largely inhibited by 10 mM TEA ($n = 4$) (Fig. 8). These findings strongly suggested that a large part of the outward current originates from a Ca^{2+}-activated K^+-current (K_{Ca}) similar to that described in hair cells.

The amplitude of the outward current was strongly dependent on the nature of the intracellular Ca^{2+} buffer as shown in saccular hair cells (Roberts, 1993). All the experiments related above (Figs 7 and 8) were carried out using the fast Ca^{2+} buffer BAPTA (1 mM) in the internal pippette solution. When the slow Ca^{2+} buffer EGTA (1–5 mM) was used instead of BAPTA, the outward K^+ currents had much higher conductance and the N-shape of the I/V curve was no longer observed. This suggested that the K_{Ca} current, in the presence of a slow internal Ca^{2+} buffer, was already fully Ca^{2+}-activated and only demonstrated its voltage dependence. However, even with BAPTA (1–5 mM) as Ca^{2+} buffer, the amplitude of the outward current progressively increased after 5–10 min of whole-cell recording.

Conclusions

It is quite reasonable to assume that the phalangeal processes of the Deiters cells represent a significant source of mechanical damping for the transversal electromotile responses of the OHCs in the organ of Corti. The important question now is to know whether this mechanical damping is simply passive or whether it can be actively modulated in particular physiological circumstances.

The observed Ca^{2+} signaling upon extracellular ATP or K^+ and the electrical resonant properties of their plasma membrane reveal the Deiters cells as potentially excitable cells in the organ of Corti. The observation of Ca^{2+}-evoked mechanical responses suggests that these cells may have an active mechanical role in the organ of Corti. Future investigations will tell whether these supporting cells participate in a fast, active damping and/or in a long-term regulation of cochlear micromechanics during, for example, intense sound stimulation or other trauma.

Acknowledgements

I thank Jean-Marie Aran, Carlos Erostegui and Masashi Sugasawa for critical reading of the manuscript, and Anne Guilhaume for the SEM micrograph of the organ of Corti.

References

Ashmore, J. F. and Ohmori, H. (1990). Control of intracellular calcium by ATP in isolated outer hair cells of the guinea-pig cochlea. *J. Physiol.* **42:** 109–131.

Brownell, W. E., Bader, C. R., Bertrand, D. and de Ribeaupierre, Y. (1985). Evoked mechanical responses of isolated outer hair cells. *Science* **227:** 194–196.

Dallos, P. (1992). The active cochlea. *J. Neurosci.* **12:** 4575–4585.

Dulon, D., Mollard, P. and Aran J.-M. (1991). Extracellular ATP elevates cytosolic Ca^{2+} in cochlear inner hair cells. *Neuroreport* **2:** 69–72.

Dulon, D., Moataz, R. and Mollard, P. (1993a). Characterization of Ca^{2+} signals generated by extracellular nucleotides in supporting cells of the organ of Corti. *Cell Calcium* **14:** 245–254.

Dulon, D., Zajic, G. and Schacht, J. (1993b). InsP3 releases intracellular stored calcium in Deiters cells of the organ of Corti. *Abstracts of the 16th Midwinter Research Meeting of ARO*, p. 117, no. 466.

Dulon, D., Blanchet, C. and Laffon, E. (1994). Photo-released intracellular Ca^{2+} evokes reversible mechanical responses in supporting cells of the guinea-pig organ of Corti. *Biochem Biophys. Res. Commun.* **201** (3): 1263–1269.

Flock, A., Bretscher, A. and Weber, K. (1982). Immunohistochemical localization of several cytoskeletal proteins in inner ear sensory and supporting cells. *Hearing Res.* **7:** 75–90.

Fuchs, P. A. (1992). Ionic current in cochlear hair cells. *Prog. Neurobiol.* **39:** 493–505.

Holley, M. C. and Ashmore, J. F. (1988). A cytoskeletal spring in cochlear outer hair cells. *Nature* **335:** 635–637.

Housley, G. D., Greenwood, D. and Ashmore, J. F. (1991). Localization of cholinergic and purinergic receptors on outer hair cells isolated from the guinea-pig cochlea. *Proc. R. Soc. Lond. B* **249:** 265–273.

Johnstone, B. M., Patuzzi, R., Syka, J. and Sykova, E. (1989). Stimulus-related potassium changes in the organ of Corti of guinea pig. *J. Physiol.* **408:** 77–92.

Mammano, F. and Ashmore, J. F. (1993). Reverse transduction measured in the isolated cochlea by laser Michelson interferometry. *Nature* **365:** 838–841.

Moataz, R., Saito, T. and Dulon, D. (1992). Evidence for voltage sensitive Ca^{2+} channels in supporting cells of the organ of Corti: characterization by indo-1 fluorescence. In *Advances in the Biosciences*, Vol. 63: *Auditory Physiology and Perception* (Y. Cazals, L. Demany and K. Horner, eds), pp. 53–59. Pergamon, Oxford.

Newman, E. A. (1985). Regulation of potassium levels by glial cells in the retina. *Trends Neurosci.* **8:** 156–159.

Nilles, R., Järlebark, L., Zenner, H. P. and Heilbronn, E. (1994). ATP-induced cytoplasmic $[Ca^{2+}]$ increases in isolated cochlear outer hair cells. Involved receptor and channel mechanisms. *Hearing Res.* **73:** 27–34.

Oesterle, E. and Dallos, P. (1990). Intracellular recordings from supporting cells in the guinea pig cochlea: DC potentials. *J. Neurophysiol.* **2:** 617–636.

Pentreath, V. W. (1982). Potassium signaling of metabolic interactions between neurons and glial cells. *Trends Neurosci.* **5:** 339–345.

Roberts, W. M. (1993). Spatial calcium buffering in saccular hair cells. *Nature* **363:** 74–76.

Roberts, W. M., Jacobs, R. A. and Hudspeth, A. J. (1991). The hair cell as a presynaptic terminal. *Ann. NY Acad. Sci.* **635:** 221–233.

Ruggero, M. A. (1992). Responses to sound of the basilar membrane of the mammalian cochlea. *Curr. Opin. Neurobiol.* **2:** 449–456.

Suzuki, N., Yoshioka T, Okano, Y., Nozawa, Y. and Kano, M. (1989). Anomalous increase of Ca^{2+} current by high concentration K^+ stimulation in whole cell clamped GH3 cells. *Biochem. Biophys. Res. Commun.* **158:** 534–540.

Slepecky, N. and Chamberlain, S. C. (1987). Tropomyosin co-localizes with actin microfilaments and microtubules within supporting cells of the inner ear. *Cell Tissue Res.* **248:** 63–66.

Slow Motility of Outer Hair Cells

JOCHEN SCHACHT, JAMES D. FESSENDEN AND GARY ZAJIC

Kresge Hearing Research Institute, University of Michigan, Ann Arbor, MI 48109-0506, USA

Introduction

Motility or the ability to change shape is a property of a variety of cell types. Muscle cells, for example, can perform mechanical work by contracting. Lymphocytes, neutrophils and macrophages can actively move to sites of injury. In the mammalian cochlea, the motility of outer hair cells (OHCs) is closely linked to transduction mechanisms as it serves to amplify and fine tune the response of the basilar membrane to auditory stimulation. Several distinctly different motile responses can be observed in OHCs *in vitro* and differentiated by their kinetics and their underlying mechanism. Based on their kinetics, two major forms of shape changes have been described as 'fast' and 'slow' motility.

Fast Motility

'Fast motility' refers to movements of OHCs occurring in the same timeframe as the acoustic vibrations traveling through the cochlea. These fast contractions and elongations can be electrically evoked *in vitro* ('electromotility,' Brownell *et al.*, 1985; Kachar *et al.*, 1986) or triggered by mechanical displacement of the stereocilia (Evans and Dallos, 1993). These fast movements are independent of calcium and the immediate availability of cellular energy. Rather, they seem to be driven by changes in membrane potential (Ashmore, 1986; Santos-Sacchi and Dilger, 1988). Fast motility is thought to act as an amplifier of the acoustically evoked vibrations of the basilar membrane (Mammano and Ashmore, 1993).

209

Slow Motilities

'Slow motility' is the term generally applied to OHC movements observed *in vitro* in response to chemical, osmotic or mechanical stimuli and which takes place over seconds (see review by Dulon and Schacht, 1992). This broad classification, however, is inadequate since the various slow responses differ in their biochemical mechanisms, particularly their dependence on calcium and their presumed physiological or pathophysiological roles. A subdivision of slow motile responses into 'passive shape changes' independent of calcium and an active 'calcium-dependent motility' seems best suited for a discussion of these phenomena. Motility induced by somatic stimulation of outer hair cells (Canlon *et al.*, 1988) cannot as yet be classified since the calcium-dependence of this process is unknown.

Passive shape changes

Passive shape changes are induced by alterations in the K^+ concentration (K^+-depolarization) or the osmolarity of the external medium (Goldstein and Mizukoshi, 1967; Zenner *et al.*, 1985; Dulon *et al.*, 1987). These shape changes do not require calcium but result from changes in the osmotic pressure of the cell (Dulon *et al.*, 1988). Upon K^+-depolarization or exposure to a hyposmotic medium, the cell swells, its diameter increases and the concomitant decrease in cell length suggests a 'contraction'. It is uncertain whether these responses contribute to the normal homeostasis of cell shape and turgor. Rather, they may be associated with pathophysiologies such as Menière's disease (Zenner *et al.*, 1985; Dulon *et al.*, 1987).

Calcium-dependent motility

Calcium-dependent motility has been observed in detergent-permeabilized and intact OHCs *in vitro* (Flock *et al.*, 1986; Zenner, 1986; Dulon *et al.*, 1990). It has been likened to muscle contraction because of the presence of actin and calmodulin in hair cells (Flock *et al.*, 1986) and the dependence of this motile response on ATP and actomyosin (Zenner, 1986; Zenner *et al.*, 1988). However, the concept of a muscle-like OHC contraction may not be strictly applicable since there are distinct morphological differences between the contractile apparatus of muscle and OHC (Slepecky and Savage, 1994). Nevertheless, an analogous mechanism may be operating.

This notion is supported by the fact that calmodulin antagonists block calcium-induced hair cell motility (Dulon *et al.*, 1990), suggesting the participation of calmodulin in the contractile process. Furthermore, calcium-dependent motility is accompanied by protein phosphorylation and inhibitors of calcium/calmodulin-dependent protein kinases block both motility and protein phosphorylation (Coling *et al.*, 1994).

Calcium is a potent bioactive molecule and a most important ion in cellular physiology. Small and rapid fluctuations of the intracellular concentration of free Ca^{2+} modulate a variety of cellular processes, including biochemical pathways, transduction mechanisms, cell-to-cell signaling and motility (for reviews see Tash *et al.*, 1983; Goldmann and Brenner, 1987; Kater *et al.*, 1991). It seems most likely that the physiologically relevant form of slow motility is the calcium-dependent motility of OHCs. In contrast to the emerging consensus that fast motility is a positive feedback to basilar membrane motion, the precise role of slow motility has yet to be defined. It can be hypothesized, however, that by altering the response of the basilar membrane to sound, slow motility can modulate the transduction process (Zenner *et al.*, 1988; Dulon and Schacht, 1992; Zenner, 1993).

Calcium-dependent contraction or elongation?

Calcium-dependent motility shows apparently contradictory forms in intact and in permeabilized OHCs. OHCs permeabilized by detergents or penetrated by a pippette contract longitudinally upon addition of calcium (Zenner, 1986; Flock *et al.*, 1986; Dulon *et al.*, 1988). In contrast, intact OHCs elongate when intracellular calcium levels are increased by the application of a calcium-ionophore (Dulon *et al.*, 1990). In addition, whereas external calcium is not required for K^+-induced cell shortening, it is required for a subsequent elongation (Zenner *et al.*, 1985; Zenner, 1986). Similarly, calcium-influx during K^+-depolarization opposes the K^+-induced cell shortening (Dulon *et al.*, 1989). Conversely, the removal of extracellular calcium, which should result in lowered intracellular calcium, decreases OHC length (Pou *et al.*, 1991).

Cortical Contractions—The Slow Motor

These results can be reconciled if we accept a contractile, actin-based cytoskeleton in OHCs as the anatomical substrate of slow motility (Flock *et al.*, 1986; Zenner, 1986; Dulon *et al.*, 1990), perhaps a structure similar to the cortical lattice described by Holley and his colleagues

(Holley and Ashmore, 1990a, b; Holley *et al.*, 1992). The cortical lattice consists of circumferential actin filaments that are pitched at a mean angle of 10 deg relative to the transverse axis of the cell. Cross-linking is provided by spectrin or its analog fodrin (Holley and Ashmore, 1990; Nishida *et al.*, 1993). It has been suggested that the cortical lattice provides the structural basis of hair cell shape and couples forces generated in the plasma membrane to changes in cell shape (Holley *et al.*, 1992).

In both intact and permeabilized cells, calcium-dependent motility is indeed characterized by a cortical (circumferential) contraction of the cell (Flock *et al.*, 1986; Schacht and Zenner, 1987; Dulon *et al.*, 1990). However, a cortical force exerted by a lattice-like structure would lead to different patterns of contraction in the two preparations. In an intact cell, a cortical contraction translates into a compensatory increase in cell length since the cell volume remains approximately constant. The longitudinal direction of this turgor-dependent deformation agrees well with the idea that a cortical lattice is more elastic longitudinally than circumferentially (Holley *et al.*, 1992). In a permeabilized cell, elongation cannot occur since maintenance of cell turgor and volume is disrupted. Instead, the oblique arrangement of the contractile apparatus will compress the cell both circumferentially and longitudinally. Thus, calcium-dependent motility operates by a single mechanism based upon a cortical contraction.

The Source of Calcium

The source of calcium is important in determining the stimulus, mechanism and regulation of slow motility. Intracellular free calcium levels can be rapidly elevated by calcium entry through voltage-gated calcium channels and by two neurotransmitter-mediated mechanisms. While all these mechanisms operate in OHCs, calcium entry via voltage-gated calcium channels may be the major or even the only trigger of slow motility. Mediation by neurotransmitters, although still controversial, seems less likely.

Motility and calcium entry via voltage-gated channels

The activation of voltage-gated calcium channels leads to cytoplasmic elevations of calcium as seen with electrical depolarization (Ashmore and Ohmori, 1990) or depolarization by external K^+ (Dulon *et al.*, 1989; Ashmore and Ohmori, 1990). A link between calcium entry via these channels and motility can indeed be established from the interactions between K^+-depolarization, calcium entry and cell elongation. As mentioned above, K^+-depolarization shortens the cell by osmotic forces independent of calcium (Dulon *et al.*, 1988). Calcium concomitantly entering through voltage-gated channels can activate the motile apparatus: it counteracts the shortening (Dulon *et al.*, 1989) or induces a subsequent elongation of the cell Zenner (1986).

Neurotransmitters and calcium pools in outer hair cells

Two other mechanisms that elevate intracellular calcium operate in OHCs, namely entry through ligand-gated channels and release from intracellular stores. Both pathways are controlled by the neurotransmitters acetylcholine and ATP via different receptor mechanisms. It seems unlikely, however, that calcium arising from these different sources will trigger the same physiological response, namely slow motility.

Effects of calcium are highly local. Limited diffusion and compartmentalization of calcium confines its effectiveness to specific cellular locations. The free diffusion range of calcium before being buffered by binding proteins may be as small as 0.1 µm (Allbritton *et al.*, 1992). Thus, it must be expected that Ca^{2+} originating from different sources acts on different targets. An example of such strict calcium compartmentalization occurs in cultured hippocampal neurons. Calcium influx through plasma membrane channels exerts completely different effects on the physiology of these cells, depending on whether influx is regulated through voltage-gated calcium channels or through neurotransmitters (Bading *et al.*, 1993).

Spatial calcium buffering is highly efficient in saccular hair cells where calcium-binding proteins capture entering calcium within microseconds and limit its effective range (Roberts, 1993). Similar spatial restrictions of different calcium pools apply to OHCs. In contrast to the cytoplasmic elevations after entry via voltage-gated channels (Dulon *et al.*, 1989; Ashmore and Ohmori, 1990), neurotransmitter-mediated increases in calcium appear limited and local. Entry via ATP-gated channels is apparently confined to the apical end of the cell (Ashmore and Ohmori, 1990). Mobilization from intracellular stores by ATP and acetylcholine-mediated increases originate at the base (Ashmore and Ohmori, 1990; Housley *et al.*, 1992; Doi and Ohmori, 1993) and may not lead to an overall increase in cytoplasmic calcium (Ashmore and Ohmori, 1990; Dulon *et al.*, 1990).

Another pool of calcium originates from intracellular binding sites.

Acetylcholine and ATP can release calcium from intracellular stores through the action of the lipid-derived second messenger, inositol trisphosphate (IP_3; Berridge, 1993). For ATP, this is documented by its ability to elevate intracellular calcium in isolated OHCs in the absence of extracellular calcium (Ashmore and Ohmori, 1990; Nilles *et al.*, 1994). For acetylcholine, there is ample pharmacological evidence for the presence of the appropriate muscarinic receptors in the cochlea (Guiramand *et al.*, 1990; Niedzielski *et al.*, 1992; Ogawa and Schacht, 1993) and their expression in outer hair cells (Safieddine and Eybalin, 1994).

Neurotransmitter-Mediated Outer Hair Cell Motility?

If acetylcholine or ATP trigger motility of OHCs, they cause cell elongation. Both the elevation of intracellular calcium and the hyperpolarization of the cell observed with the application of acetylcholine (Doi and Ohmori, 1993) would affect cell shape in an identical fashion. Hyperpolarization by deflection of the stereociliary bundle (towards the smaller stereocilia) indeed causes OHCs to elongate, a direction of movement which is in accordance with the effects of electrical hyperpolarization (Evans and Dallos, 1993). However, acetylcholine has been reported to 'contract' cells longitudinally (Brownell *et al.*, 1985; Slepecky *et al.*, 1988; Plinkert *et al.*, 1990). The contractile response, occasionally followed by relaxation, may require high doses of acetylcholine (0.5–5 mM) and develops over an extended period of time (2–4 min). Its kinetics thus differs significantly from the kinetics of acetylcholine-induced calcium elevations. Bobbin *et al.* (1990) have argued that motile responses induced by neurotransmitters in isolated OHCs are likely to be artifacts. Observations in over 500 hair cells exposed to cholinergic agents (Table 1) and over 50 hair cells exposed to purinergic agents (Table 2) support this view. Under strictly controlled conditions of osmolarity, temperature and pH, no shape changes were seen.

Thus, calcium released from intracellular stores may be a distinct pool not associated with motility. Contractions induced by IP_3 and ATP in permeabilized OHCs (Schacht and Zenner, 1987) do not argue against this notion. In detergent-treated cells, the liberated calcium is no longer compartmentalized and may act on intracellular targets otherwise inaccessible. While the question has yet to be unambiguously resolved, there is currently no clear evidence that neurotransmitter action may trigger slow motility.

Table 1. Cholinergic agents do not alter outer hair cell shape

No. of cells observed	>500
Conditions	1 µM–5 mM acetylcholine + eserine
	10 µM–10 mM carbachol
Typical group ($n = 80$)	contractions = 1
	elongations = 4
	no response = 76

In part from Bobbin *et al.* (1990).

Table 2. Purinergic agents do not alter outer hair cell shape

No. of cells observed	>50
Conditions	2–200 µM ATP or ATP-γ-S
Typical group ($n = 9$)	length change = $-0.1 \pm 0.5\%$

Role of Neurotransmitters and Second Messengers

If the efferent neurotransmitters and their second messengers do not trigger motility directly, then the question arises whether their actions may be associated with slow motility at all. While this cannot be answered with certainty, a variety of homeostatic functions documented in other tissues could operate in OHCs and impact on transduction mechanisms. For example, cholinergic or purinergic activation could influence motility by modulating intracellular calcium levels or the gating properties of channels. Neurotransmitters may indeed regulate the characteristics of voltage-dependent Ca^{2+} channels through an IP_3-dependent mechanism (de Waard *et al.*, 1992). This could alter the kinetics or amplitude of the motile response. Likewise, modulation of a plasma membrane-bound Ca^{2+}-ATPase (Fraser and Sarnacki, 1992) or the inhibition of a membrane-bound Na^+/Ca^{2+}

exchanger (Fraser and Sarnacki, 1990) would affect levels of internal free Ca^{2+} and thus influence a calcium-dependent motility.

Conclusion

The concept developed here is intriguing since it postulates that both fast and slow motility essentially originate from the same stimulus, namely depolarization induced by mechanotransduction. Calcium-dependent slow motility could interact (in the form of a negative feedback) with the positive feedback of fast motility. Such an interaction need not be strictly coupled but allow for both amplification and attenuation of the acoustic stimulus. If voltage-gated calcium channels in OHCs activate at high sound intensities, slow motility would be an effective protective system against noise trauma. At low stimulus intensities, however, slow motility would not be triggered, thus not interfering with the amplification mechanisms of fast motility. The role of neurotransmitters and second messengers would be one of modulators rather than effectors of slow motility (Dulon *et al.*, 1990; Dulon and Schacht, 1992). Their potential ability to alter the kinetics of the voltage-gated calcium channels could enhance the efficacy of this mechanism, a model consistent with the assumed protective role of efferent stimulation on noise trauma (Puel *et al.*, 1988). Crosstalk among various second messenger pathways, common in other systems, could further fine-tune OHC motility and basilar membrane mechanics.

Acknowledgements

The authors wish to thank their colleagues Don Coling, Matthew Holley and Kathryn Kimmel for stimulating discussions and input into this manuscript. The authors' research is supported by NIH grant DC-00078.

References

Allbritton, N. L., Meyer, T. and Stryer, L. (1992). Range of messenger action of calcium ion and inositol 1,4,5-trisphosphate. *Science* **258:** 1812–1815.

Ashmore, J. F. (1986). A fast motile response in guinea-pig outer hair cells: the cellular basis of the cochlear amplifier. *J. Physiol.* **388:** 323–347.

Ashmore, J. F. and Ohmori, H. (1990). Control of intracellular calcium by ATP in isolated outer hair cells of the guinea-pig cochlea. *J. Physiol.* **428:** 109–131.

Bading, H., Ginty, D. D. and Greenberg, M., E. (1993). Regulation of gene expression in hippocampal neurons by distinct calcium signaling pathways. *Science* **260:** 181–186.

Berridge, M. J. (1993). Inositol trisphosphate and calcium signalling. *Nature* **361:** 315–325.

Bobbin, R. P., Fallon, M., Puel, J.-L., Bryant, G., Bledsoe, S. C., Zajic, G. and Schacht, J. (1990). Acetylcholine, carbachol and GABA induce no detectable change in the length of isolated outer hair cells. *Hearing Res.* **47:** 39–52.

Brownell, W. E., Bader, C. R., Bertrand, D. and de Ribeaupierre, Y. (1985). Evoked mechanical responses of isolated outer hair cells. *Science* **227:** 194–196.

Canlon, B., Brundlin, L. and Flock, A. (1988). Acoustic stimulation causes tonotopic alterations on the length of isolated outer hair cells from guinea pig hearing organ. *Proc. Natl. Acad. Sci. USA* **85:** 7033–7035.

Coling, D., Bartolami, S. and Schacht, J. (1994). Involvement of protein phosphorylation in calcium-dependent outer hair cell motility. *Abstr. Assoc. Res. Otolaryngol.* Vol. 17, p. 74.

de Waard, M., Seagar, M., Feltz, A. and Couraud, F. (1992). Inositol phosphate regulation of voltage-dependent calcium channels in cerebellar granule neurons. *Neuron* **9:** 497–503.

Doi, T. and Ohmori, H. (1993). Acetylcholine increases intracellular Ca^{2+} concentration and hyperpolarizes the guinea-pig outer hair cell. *Hearing Res.* **67:** 179–188.

Dulon, D. and Schacht, J. (1992). Motility of cochlear outer hair cells. *Am. J. Otol.* **13:** 108–112.

Dulon, D., Aran, J.-M. and Schacht, J. (1987). Osmotically induced motility of outer hair cells: implication for Meniere's disease. *Arch. Otorhinolaryngol.* **244:** 104–107.

Dulon, D., Aran, J.-M. and Schacht, J. (1988). Potassium-depolarization induces motility in isolated outer hair cells by an osmotic mechanism. *Hearing Res.* **32:** 123–130.

Dulon, D., Zajic, G., Aran, J.-M. and Schacht, J. (1989). Aminoglycoside antibiotics impair [K$^+$]-depolarization-induced calcium entry but not motility in outer hair cells. *J. Neurosci. Res.* **24:** 338–346.

Dulon, D., Zajic, G. and Schacht, J. (1990). Increasing intracellular free calcium induces circumferential contractions in isolated cochlear outer hair cells. *J. Neurosci.* **10:** 1388–1397.

Evans, B. N. and Dallos, P. (1993). Stereocilia displacement induced somatic motility of cochlear outer hair cells. *Proc. Natl. Acad. Sci. USA* **90:** 8347–8351.

Flock, A., Flock, B. and Ulfendahl, M. (1986). Mechanisms of movement in outer hair cells and a possible structural basis. *Arch. Otorhinolaryngol.* **243:** 83–90.

Fraser, C. L. and Sarnacki, P. (1990). Inositol 1,4,5–trisphosphate may regulate Ca$_i^{2+}$ by inhibiting membrane bound Na$^+$–Ca^{2+} exchanger. *J. Clin. Invest.* **86:** 2169–2173.

Fraser, C. L. and Sarnacki, P. (1992). Regulation of plasma membrane-bound Ca^{2+}-ATPase pump by inositol phosphates in rat brain. *Am. J. Physiol.* **262:** F411–F416.

Goldmann, Y. E. and Brenner, B. (1987). Molecular mechanism of muscle contraction. *Ann. Rev. Physiol.* **49:** 629–636.

Goldstein, A. J. and Mizukoshi, O. (1967). Separation of the organ of Corti into its component cells. *Ann. Otol. Rhinol. Laryngol.* **76:** 414–426.

Guiramand, J., Mayat, E., Bartolami, S., Lenoir, M., Rumigny, J.-F., Pujol, R. and Récasens, M. (1990). A M3 muscarinic receptor coupled to inositol phosphate formation in the rat cochlea? *Biochem. Pharmacol.* **39:** 1913–1919.

Holley, M. C. and Ashmore, J. F. (1990a). Spectrin, actin and the structure of the cortical lattice in mammalian cochlear outer hair cells. *J. Cell Sci.* **96:** 283–291.

Holley, M. C. and Ashmore, J. F. (1990b). A cytoskeletal spring for the control of cell shape in outer hair cells isolated from the guinea pig cochlea. *Eur. Arch. Otorhinolaryngol.* **247:** 4–7.

Holley, M. C., Kalinec, F. and Kachar, B. (1992). Structure of the cortical cytoskeleton in mammalian outer hair cells. *J. Cell Sci.* **102:** 569–580.

Housley, G. D., Greenwood, D. and Ashmore, J., F. (1992). Localization of cholinergic and purinergic receptors in outer hair cells isolated from the guinea pig cochlea. *Proc. R. Soc. Lond. B* **249:** 265–273.

Kachar, B., Brownell, W. E., Altschuler, R. and Fex, J. (1986). Electrokinetic shape changes of cochlear outer hair cells. *Nature* **322:** 365–368.

Kater, S. B. and Mills, L. R. (1991). Regulation of growth cone behavior by calcium. *J. Neurosci.* **11:** 891–899.

Mammano, F. and Asmore, J. F. (1993). Reverse transduction measured in the isolated cochlea by laser Michelson interferometry. *Nature* **365:** 838–841.

Niedzielski, A. S., Ono, T. and Schacht, J. (1992). Cholinergic regulation of the phosphoinositide second messenger system in the organ of Corti. *Hearing Res.* **59:** 250–254.

Nilles, R., Järlebark, L., Zenner, H. P. and Heilbronn, E. (1994). ATP-induced cytoplasmic [Ca^{2+}] increases in isolated cochlear outer hair cells. Involved receptor and channel mechanisms. *Hearing Res.* **73:** 27–34.

Nishida, Y., Fujimoto, T., Tagaki, A., Honjo, I. and Ogawa, K. (1993). Fodrin is a constituent of the cortical lattice in outer hair cells of the guinea pig: immunocytochemical evidence. *Hearing Res.* **65:** 274–280.

Ogawa, K. and Schacht, J. (1993). Receptor-mediated release of inositol phosphates in the cochlear and vestibular sensory epithelia of the rat. *Hearing Res.* **69:** 207–214.

Plinkert, P. K., Gitter, A. H., Zimmermann, U., Kirchner, T., Tzartos, S. and Zenner, H. P. (1990). Visualization and functional testing of acetylcholine receptor-like molecules in cochlear outer hair cells. *Hearing Res.* **44:** 25–34.

Pou, A. M., Fallon, M., Winbery, S. and Bobbin, R. P. (1991). Lowering extracellular calcium decreases the length of isolated outer hair cells. *Hearing Res.* **52:** 305–311.

Puel, J. L., Bobbin, R. P. and Fallon, M. (1988). An ipsilateral cochlear efferent loop protects the cochlea during intense sound exposure. *Hearing Res.* **37:** 65–70.

Roberts, W. M. (1993) Spatial calcium buffering in saccular hair cells. *Nature* **363:** 74–76.

Santos-Sacchi, J. and Dilger, J. P. (1988). Whole cell currents and mechanical responses of isolated outer hair cells. *Hearing Res.* 35, 143–150.

Safieddine, S. and Eybalin, M. (1994). Expression of neurotransmitter receptors mRNA in rat and guinea pig auditory system. *Abstr. Assoc. Res. Otolaryngol.* **17:** 59.

Schacht, J. and Zenner, H. P. (1987). Evidence that phosphoinositides mediate motility in cochlear outer hair cells. *Hearing Res.* **31:** 155–160.

Slepecky, N. B. and Savage, J. E. (1994). Expression of actin isoforms in the guinea pig organ of Corti: muscle isoforms are not detected. *Hearing Res.* **73:** 16–26

Slepecky, N., Ulfendahl, M. and Flock, A. (1988). Shortening and elongation of isolated outer hair cells in response to application of potassium gluconate, acetylcholine and cationized ferritin. *Hearing Res.* **34:** 119–126.

Tash, J. S. and Means, A. R. (1983). Cyclic adenosine 3′,5′ monophosphate, calcium and protein phosphorylation in flagellar motility. *Biol. Reprod.* **28:** 75–104.

Zenner, H. P. (1986). Motile responses in outer hair cells. *Hearing Res.* **22:** 83–90.

Zenner, H. P. (1993). Possible roles of outer hair cell, d.c. movements in the cochlea. *Br. J. Audiol.* **27:** 73–77.

Zenner, H. P., Zimmermann, U. and Schmitt, U. (1985). Reversible contraction of isolated mammalian cochlear hair cells. *Hearing Res.* **18:** 127–133.

Zenner, H. P., Zimmermann, R. and Gitter, A. H. (1988). Active movements of the cuticular plate induce sensory hair motion in mammalian outer hair cells. *Hearing Res.* **34:** 233–240.

Purinergic Modulation of Outer Hair Cell Electromotility

G. D. HOUSLEY, B. J. CONNOR AND N. P. RAYBOULD

Department of Physiology, School of Medicine, University of Auckland, Private Bag 92019, Auckland, New Zealand

Introduction

While adenosine 5'-triphosphate (ATP) acts via P_2-purinoceptors as a co-transmitter or even the principal neurotransmitter in such diverse systems as smooth muscle of rat and guinea-pig urogenital tract and rabbit ear blood vessel; coeliac ganglion; nodose ganglion; spinal cord and brain (for reviews see Gordon, 1986; Dubyak, 1991; Barnard *et al.*, 1994), very little is presently known of the actions of purines, and ATP in particular, on the sensory transduction process in the cochlea. Interestingly, studies on isolated cochlear hair cells demonstrated definitively that P_2 receptors, activating a non-selective cation current, were present (Nakagawa *et al.*, 1990) and that this inward current produced both direct and indirect increase in intracellular Ca^{2+} (Ashmore and Ohmori, 1990; Nilles *et al.*, 1994). Recently the inner hair cells (IHCs) and supporting cells have also been shown to have P_2 receptors and to similarly show an increase in intracellular Ca^{2+} upon exposure to ATP (Dulon *et al.*, 1991, 1993; Housley *et al.*, 1993).

Classification of the P_2 receptors on cochlear hair cells

In considering the function of ATP on the sensory hair cells it is important to analyse the characteristics of the response to ATP. A number of pharmacological studies have attempted to classify the receptor into a particular P_2 subtype. This has led to the conclusion that the sensory hair cells possess either a P_{2y} or P_{2z} type of receptor. P_{2y} is generally classified as a G-protein linked system (Barnard *et al.*, 1994),

221

while P_{2x} and P_{2z} subtypes purinoceptors gate non-selective channels (Dubyak, 1991). In outer hair cells (OCHs), an onset latency of less than 10 msec for the ATP-activated inward current (Housley *et al.*, 1992), the sensitivity of the ATP-activated intracellular calcium rise to block by the ADP-ribosylating pertussis toxin (Nilles *et al.*, 1994), and the fact that Kakehata *et al.* (1993) failed to show an effect of PTX on the ATP-activated inward current itself, suggest that both direct and G-protein coupled P_2 receptor systems may be present on OHCs. There is also evidence for a heterogenous P_2-purinoceptor population in the supporting cells of the organ of Corti (Dulon *et al.*, 1993). Recently Illes and Nörenberg (1993) have proposed a classification of P_2 receptors in neural tissue which may be transportable to the cochlea. This includes a P_{2x}-type and additional subclassification of the P_{2y} class into both direct ligand gated ($P_{2y\alpha}$) and G-protein coupled ($P_{2y\beta}$) types. Consideration of purinergic modulation of cochlear function must therefore consider a number of different cell types and take both direct gating of non-selective cation channels, and delayed intracellular release of Ca^{2+} from intracellular stores into consideration.

Site of action of extracellular ATP

Given the well-established efferent regulation of cochlear sensitivity via the OHCs (for a review see Eybalin, 1993) and the evidence for ATP acting as a co-transmitter in noradrenergic and cholinergic systems (for a review see Housley *et al.*, 1992), it is not surprising that ATP has been proposed to act as co-transmitter in the olivocochlear innervation (Nakagawa et al., 1990; Dulon *et al.*, 1993; Nille *et al.*, 1994). However, when the existence of P_2 receptors on IHCs and supporting cells is considered (Kolston and Ashmore, 1992; Dulon *et al.*, 1993; Housley *et al.*, 1993) – systems not directly linked to the centrifugal control pathway, then a more global mechanism of action by ATP appears likely. This view is reinforced by the evidence that P_2 receptors on OHCs are not co-localized on the basolateral membrane in juxtaposition to the cholinergic receptors, as would be predicted if ATP were an efferent co-transmitter, but rather, from electrophysiological and imaging studies, the receptors for the ATP-gated ion channels appear to be associated with the stereocilial region of the cells (Housley *et al.*, 1992, 1993; Ashmore *et al.*, 1993; Mockett *et al.*, 1994). Based on this data, the endolymphatic compartment is implicated in the purinergic mechanism. Within this emerging complexity of purinoceptor subtypes in organ of Corti cells, the present data aimed to investigate the potential for purinergic modulation of OHC function and electromotility

(Brownell *et al.*, 1985; Ashmore, 1987; Santos-Sacchi, 1989; Dallos *et al.*, 1993) in particular.

Methodology

OHCs were isolated from guinea-pig cochlea following a procedure previously described (after Ashmore, 1987), as approved by the University of Auckland Animal Ethics Committee. Briefly, following rapid cervical dislocation, temporal bones were removed and the bulla exposed. The bony capsule of the cochlea was immersed in an artificial perilymph-like solution (external solution) of the following composition (in mM): NaCl 150; KCl 4; $CaCl_2$ 1.5; $MgCl_2$ 1; NaH_2PO_4 2; Na_2HPO_4 8; D-glucose 3. The pH was adjusted to 7.25 with 1 N NaOH and the osmolarity of the solution was approximately 315 mOsm (Wescor vapour-pressure osmometer). The modiolus was dissected out and the organ of Corti was removed to a 60 µl droplet containing 0.25 mg/ml trypsin (Sigma). After a 10 min incubation period the tissue was briefly tritiated and placed in a 200 µl bath on the stage of a Nikon TMD inverted microscope equipped for Nomarski DIC optics and real-time contrast enhanced (DSP-200, Dage-MTI, USA) video-microscopy. Images, obtained with either ×63 (1.3NA, Leitz, Germany) or ×100 (1.3NA, Nikon, Japan) oil immersion objectives, were stored on sVHS tape for off-line analysis using a 8 bit videoframe grabber (PC Vision Plus, Imaging Technologies, USA) controlled by image analysis software (Image-Pro Plus, Media Cybernetics, USA). Whole-cell current or voltage-recordings were made (after Hamill *et al.*, 1981) from the isolated OHCs using patch-pipettes filled with an internal solution of composition: KCl 144; $MgCl_2$ 2; NaH_2PO_4 1, Na_2HPO_4 8; D-glucose 3; EGTA 1; adjusted for pH 7.25 with 1 N KOH and with an osmolarity of approximately 308 mOsm. Voltage-clamp and current-clamp records were stored using a DAT recorder (TEAC RD-101T) and directly digitized via the patch–clamp interface (Tecmar TL-1, USA; Axon Instruments PClamp v5.5 software, USA). On achieving a stable whole-cell recording, capacitive charging transients were recorded in response to a 10 mV hyperpolarizing step (20 msec) from a holding potential of –60 mV, for subsequent determination of series resistance and membrane capacitance (after Housley and Ashmore, 1992). For experiments determining charge accumulation and volume responses to applied ATP, the charging transient was cancelled and 85–95% series resistance compensation achieved using the Axopatch 200. All data were corrected for junction potential and series resistance voltage errors. Drugs (adenosine 5'-triphosphate, sodium salt; Serva) or acetylcholine (Serva) were pressure-applied (via a microprocessor-controlled solenoid, Mac Valve Pacific, New Zealand) in external solution using a glass micropipette (typically 3–5 MΩ) positioned within 200 mm of the cell.

Results

OHCs were isolated from all turns of the cochlea and varied in length from 22 to 79 μm. Whole-cell recordings were obtained from cells after gigaohm seal formation. Cells were rejected if initial zero-current potentials (V_z) were less than –60 mV (mean = –68.4 ± 1.84 mV, SEM, n = 22). Membrane capacitance varied between 16.6 and 36.2 pF.

Electrophysiological localization of ATP receptors

In agreement with our previous localization studies (Housley *et al.*, 1992, 1993), application of ATP produced a rapid activation of an inward current, with minimum latency, when applied directly to the apical pole of the OHC. Additional support for the localization of the P_2-purinoceptors to this region was obtained using a double-barrel drug pipette. ACh (50 μM) applied to the base of an OHC, separately or together with ATP (10 μM), elicited a rapidly activating outward current, whereas the outward current was replaced by a similarly rapidly activating inward current (ATP response) when the ACh + ATP pipette was directed to the apical pole of the cell (Fig. 1).

ATP-induced electro-osmotic modulation of OHC volume

Application of ATP (10 μM; 0.2–15 sec) to voltage-clamped (V_h = –60 mV) OHCs resulted in a rapid and reversible activation of the OHC P_2-purinoceptor-mediated non-selective cation conductance (Naka-gawa *et al.*, 1990; Housley *et al.*, 1992). The ATP-gated inward current varied from <100 pA to >2.1 nA, with an increase in magnitude in the smaller OHCs (Fig. 2).

Associated with the recording or these ATP-activated inward currents was the observation that the OHCs, and in particular the shorter cells, showed a reversible swelling and shortening immediately after the application of ATP. The volume loading was also present under current-clamp (Fig. 3). In an earlier experiment, when an isolated OHC was deliberately bathed in an hyperosmotic solution (approximately

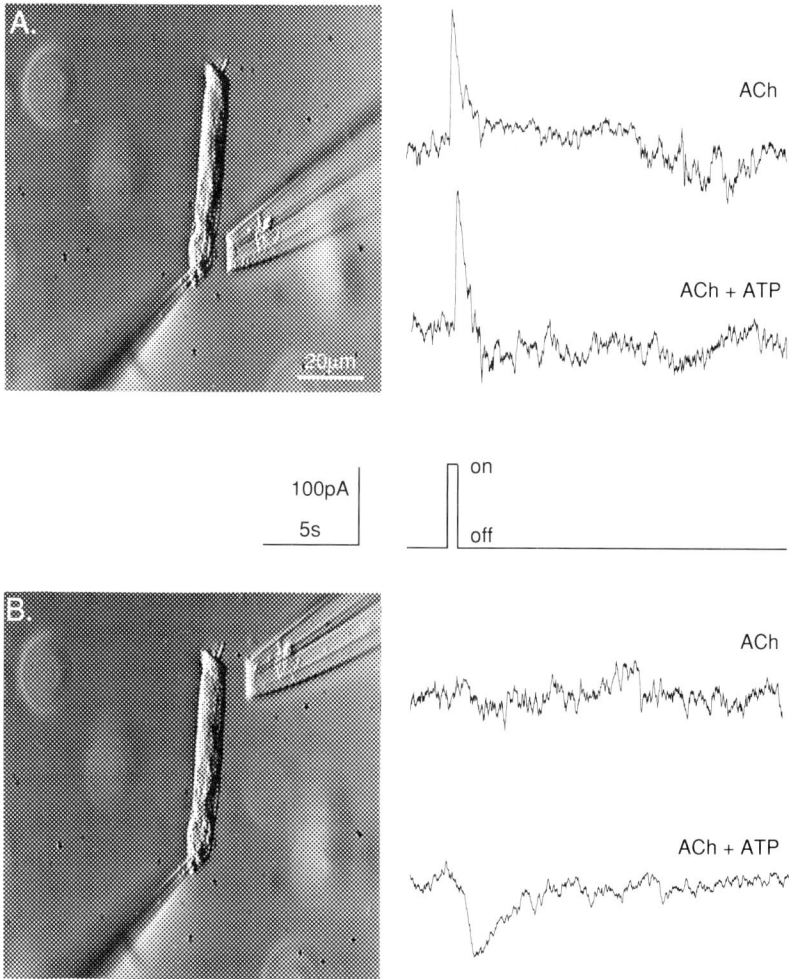

Fig. 1. Diametrically opposed fields for cholinergic and purinergic receptors are suggested by the activation of comparable outward currents (+115 pA and +109 pA) when ACh (50 μM) or ACh + ATP were applied (0.5 sec) to the base of an OHC, and activation of an inward current (–110pA) when ACh + ATP (10 μM) was applied to the apex of the cell. Holding potential = –60mV).

330 mOsm), application of ATP (100 μM, 3–5 sec) produced clearly discernible cell volume loading under both current- and voltage-clamp, although the effect was most pronounced under voltage-clamp. The latter observation was consistent with the removal of the self-limiting response to ATP associated with a depolarization towards the reversal

Active hearing

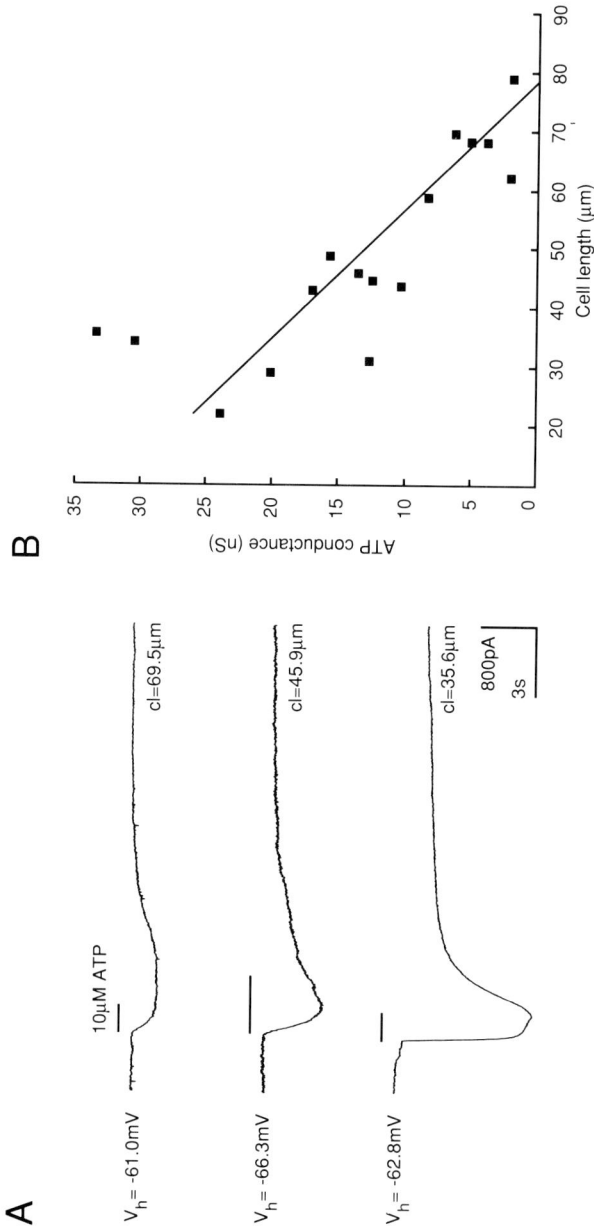

Fig. 2. Analysis of ATP-activated conductance in isolated OHC. (A) Examples of current responses of OHC to brief (1–2 sec) application of ATP (10 μM) under voltage-clamp (holding potential indicated) for small, intermediate and large OHC. (B) Net ATP conductances calculated from the ATP-activated currents assuming the reversal potential is 0 mV (after Nakagawa *et al.*, 1990). The progressive increase in conductance with decreasing cell length suggests that a tonotopic bias in the number of ATP-receptor gated ion channels exists towards the basal turn region of the cochlea. The regression line is fitted by the equation $y = -0.46x + 36.06$ ($r = 0.81$; P 0.5, one-way ANOVA).

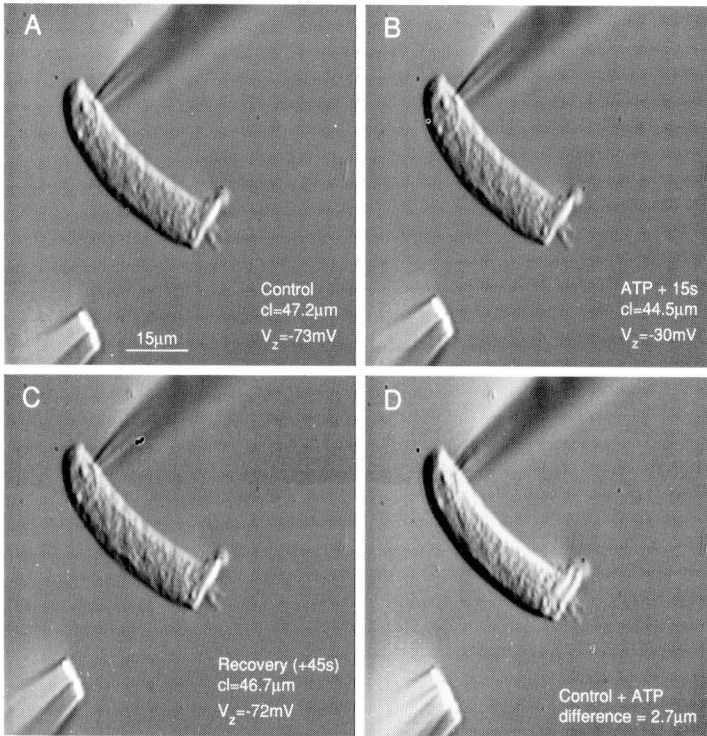

Fig. 3. Shortening of an OHC following application of ATP (10 μM, 1 sec) under current-clamp. (A) Control. (B) Fifteen seconds after the ATP the cell had depolarized by 43 mV and had shortened by approximately 6%. (C) Recovery of the membrane potential and cell length at $t = 45$ sec. (D). Digitial superimposition of control and t = 15 sec images (A, B) highlights the change in cell length. Associated with the shortening is an increase in cell width.

potential of the ATP-gated conductance under current-clamp, where Vz changed from –72 mV to –35 mV. Image analysis of the cell within seconds of the onset of the current response to ATP (under voltage-clamp) indicated that the initial increase in cell volume was localized to the infracuticular region of the cell (Fig. 4) — a result consistent with the localization of the P_2 purinoceptors to the apical surface of these cells.

Volume-dependent modulation of the gain of OHC electromotility

To determine the effect of the ATP-induced volume change on OHC electromotility, a series of experiments was performed which recorded

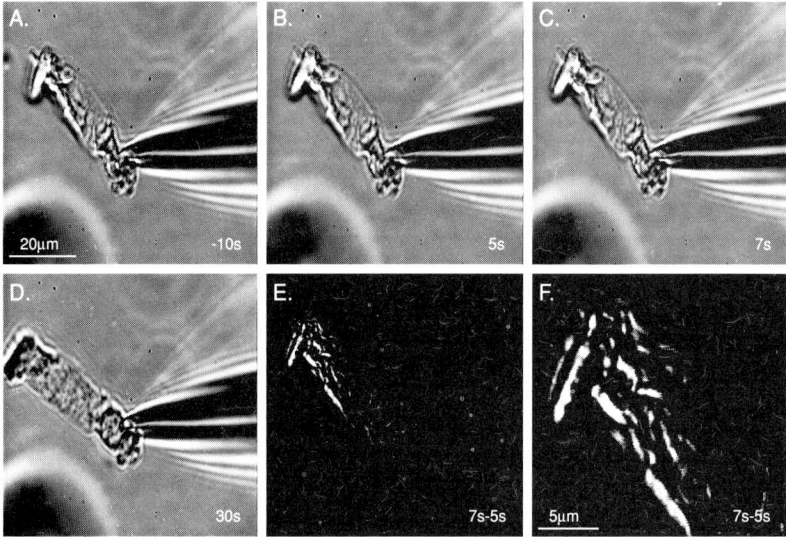

Fig. 4. Image analysis of an OHC showing the increase in volume associated with a transient exposure to ATP (100 μM, 3 sec) during voltage-clamp (V_h = –60 mV). The maximum current response (non shown) was –1.55 nA, returning to baseline within 2 min. (A) Control, 10 sec prior to application of ATP. (B)–(D) Images 5, 7 and 30 sec after ATP was applied to the cell. Note the substantial increase in cell volume by 30 sec. (E) Digital subtraction of images at 7 sec and 5 sec (B and C). Note the localization of the onset of the swelling in the cell membrane to the infracuticular region. (F). Magnified image of the infra cuticular region from (E).

membrane potential-driven length changes before and after exposure to brief (0.4–13 sec) applications of ATP (100 μM) under voltage-clamp (V_h=–60mV). To provide a measure of accumulating osmotic loading, the current records were integrated over 5 sec intervals for 50 sec, starting 10 sec prior to the ATP application. Cell volume was measured from images of the cell digitized over the same 5 sec intervals and determined from the equation volume = $\pi r^2 l$, where l = cell length and r = mean width/2 (mean width was determined from multiple measurements made from the infracuticular to supranuclear regions of the cell). Cell length change in response to changes in membrane voltage was determined from contrast-enhanced, digitally magnified images of the cuticular plate region, obtained by digitizing videoframes immediately before and during 600 msec voltage-steps (–120 mV to 0 mV, 3 sec delay between steps). The minimum resolution of the differential measurement was 58 nm. Cells typically exhibited a non-linear response, with rectification in length changes toward depolarizing potentials. Changes in membrane conductance and V_z showed no correlation with changes in the gain of the electromotile response, as determined from the slope of the voltage–length change relationship (Fig. 5), whereas a substantial increase in the slope of the length change–voltage relationship was associated with the volume loading accompanying the ATP response.

In six experiments, gain of the electromotile response increased after application of ATP (1–18 sec) by approximately 67% (control = 7.26 ± 1.88 nm/mV, mean ± SEM; after ATP = 12.11 ± 3.04 nm/mV; n = 6; P < 0.05; paired Student's t-test), measured 45 sec later. The range in gain prior to ATP was 3.1–14.1 nm/mV, increasing after exposure to an ATP to a range 4.5–22.1 nm/mV. This increase in efficiency in voltage-driven length change was associated with an increase in cell volume of approximately 32.1 ± 8.21% (P < 0.05, t-test) (control = 3.091 ± 0.452 pl; after ATP = 3.825 ± 0.518 pl; P < 0.05 paired t-test). There was a decrease in mean cell length from 50.59 μm (± 5.65) to 47.15 μm (7.01), corresponding to an average reduction of 8.9% (± 4.1%) but this was not significant because two out of six cells showed a marginal increase in cell lengths.

Analysis of cell volume with respect to time after application of ATP showed a consistent trend towards volume loading developing immediately after the onset of the current response to ATP, and increasing proportionally with the accumulating charge (Figs 6 and 7). In one of the experiments, successive applications of ATP (0.4–1 sec) produced initially reversible volume-loading of the cell, and then subsequently added to a constant inward flux of current which was associated with an almost doubling of cell volume from 3.12 to 5.74pl (Fig. 6). To analyse the relationship between the magnitude of the ATP-induced volume loading and gain of the electromotile response, a series of length change profiles to applied voltage-steps were obtained (as detailed in Fig. 5). In the case of the cell exposed to successive pulses of ATP, gain increased with initial cell volume increase, from 4.5 nm/mV to a maximum of 20.4 nm/mV after the fourth application of ATP (Fig. 6), subsequent applications of ATP were associated with continued volume loading and a drop in gain back towards the control levels (Figs 6 and 8), suggesting that electromotility is most efficient at an optimum volume/turgidity. Overall, cells which exhibited substantial ATP responses (500 pA) showed volume loading (Fig. 7) and an enhanced electromotility associated with the volume increase and cell shortening (Fig. 8). One cell (68.1 μm), which gave only a –400 pA response to 4 sec ATP (100 μM), had no detectable volume change and a minimal length change (less than +0.5 μm), accumulated charge over a 40 sec period after application of ATP was only –4.66 nC. This contrasts with cell 1 (Fig. 7) where a 13 sec application of ATP (10 μM) elicited a peak ATP-activated current of –1889 pA with a resulting –59.6 nC entry of charge in the 40 sec window; this was associated with an increase in cell volume from 2.72 to 4.09 pl.

Discussion

The ATP-mediated volume loading is consistent with a mechanism analogous to the previous demonstration of potassium depolarization-induced OHC shortening, which has been attributed to a current-mediated osmotic driving force, rather than the triggering of intracellular contractile processes (for a review see Brownell, 1990). Calculation of the net osmotic driving force produced by the influx of cations through the ATP-gated ion channels can be made from the data shown in Fig. 7. When the ATP-gated current (principally Na^+) entered the apex of the OHC under voltage-clamp, loading of osmotically active particles matched the integral of the measured current flux, although it arose in combination with K^+ efflux and Cl^- influx via the recording

Fig. 5 (opposite). An outline of the experimental protocol used to determine the influence of ATP-induced volume change on OHC electromotility. A. An example of the current responses to a voltage-clamp paradigm applied to an isolated OHC. The cell underwent four series (1–4) of voltage steps (–120 mV to 0 mV; 600 msec duration, 3 sec between steps), one preceding and three following a transient exposure to ATP (0.4 sec; 100 μM), which elicited an inward current response peaking at –2213 pA. The uppermost trace is the current record, the middle trace is the voltage-clamp record and the lower traces are higher-resolution records of the current responses to the voltage-steps shown in trace 1. The dashed line indicates zero current. (B) Current–voltage records obtained at 550 msec from the onset of the voltage-steps, from the records shown in (A). Note the shift in V_z from –67 mV to –73 mV subsequent to ATP, with slope conductance (–40 mV to –80 mV) remaining relatively unchanged (62 nS to 69 nS). (C) Length change–voltage relationships determined by comparing images (see D) of the cuticular plate region of the cell before and during the voltage-steps. Note the increase in gain (slope of the length change–voltage relationship measured between –80 mV and 0 mV), from 6.88 to 10.27nm/mV, immediately after the ATP induced volume-loading (see D). The gain subsequently dropped below the control level (3.41–2.49nm/ mV) as the cell volume returned to normal. (D) Video images of the OHC when the length change data was acquired (periods 1–4). Top panel shows the decrease in cell length and increase in cell volume subsquent to the application of ATP (cell length at 1= 30.2 μm, width = 7.99 μm, volume = 1.512 pl; cell length at 2 = 23.8 μm, width = 10.86 μm, volume = 2.203pl). Note the recovery of the cell shape over time (3–4). Bottom panel shows digitally subtracted images of the cuticular plate region of the cell obtained at –60 mV (holding potential) and 0 mV, from the voltage-step runs made at the time intervals indicated by the numbers in the top panel. The white region indicates change in cell position. To obtain the measurements shown in (C), the cuticular plate region of the cell was defined by a fixed pixel map for the resting position (at –60 mV) and then this was compared with the position of the cuticular plate in the image acquired after changing the membrane potential.

A

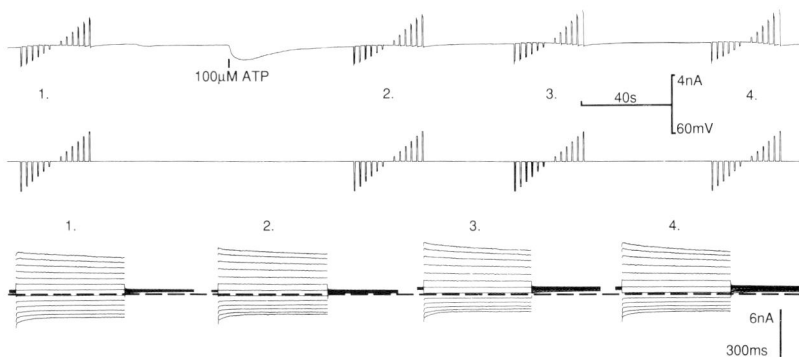

B

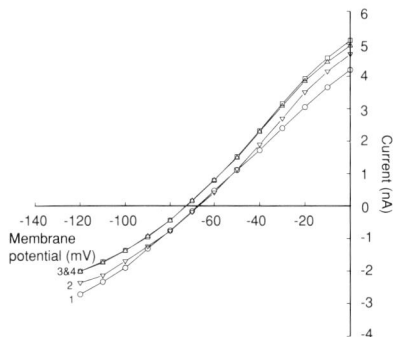

C

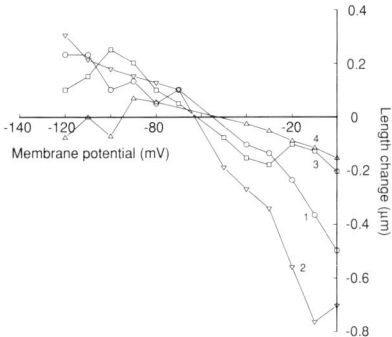

D

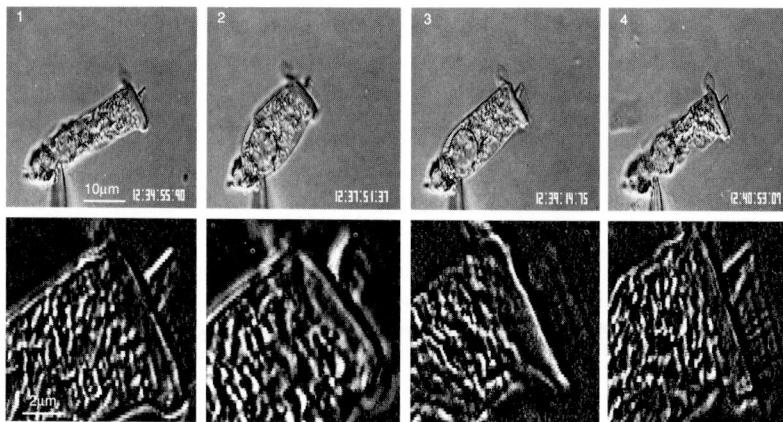

Fig. 5.

A

B

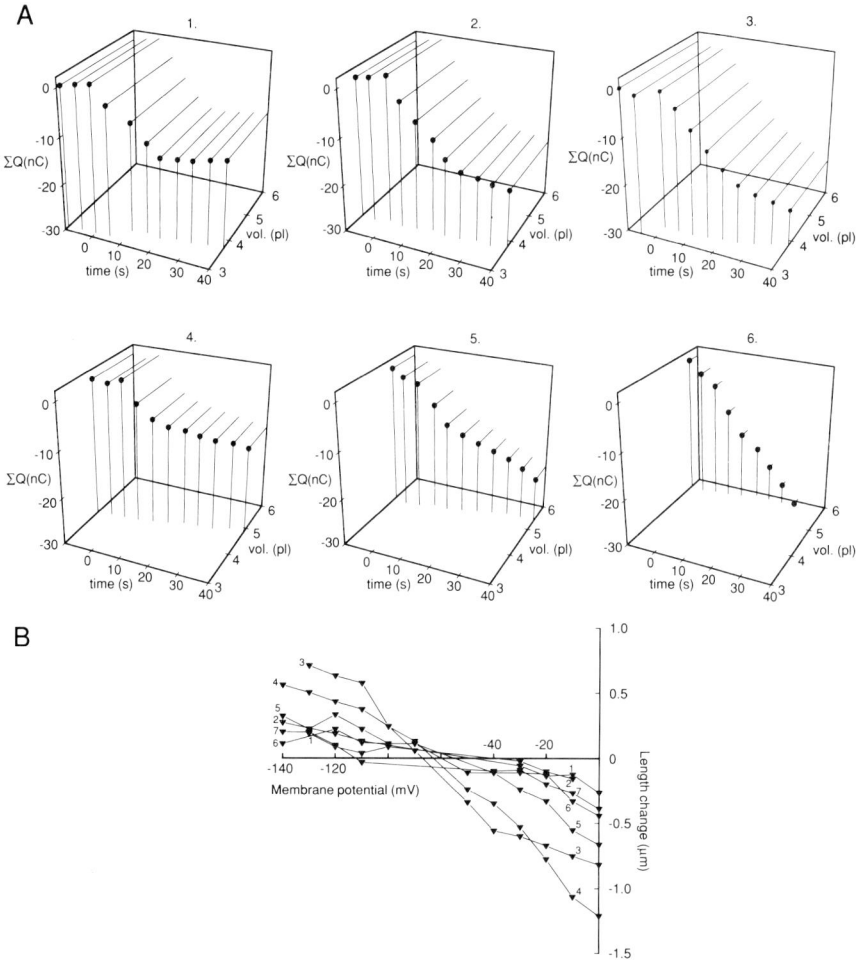

Fig. 6. The effect of the ATP-gated conductance on cell volume and hence efficiency of electromotility of an isolated OHC. Plots are with respect to time after successive applications of ATP (100 μM; 0.4 sec = 1 = peak current of −1058 pA; 1 sec = 2 = −746 pA; 1 sec = 3 = −891 pA; 0.4 sec = 4 = −780 pA; 0.4 sec = 5 = −724 pA; 0.4 sec = 6 = −691 pA), voltage-clamp (V_h) = −60 mV; V_z = −78 mV at 1. (A) Influx of charge was determined by integrating the current record over 5 sec intervals commencing 10 sec prior to application of ATP. Note the rapid inward flux after application of ATP at time 0. At this point cell volume increased in all cases, although towards the end of the experiment (runs 5 and 6) the ATP-induced inward current augmented a developing standing inward current. Note that volume changes were initially reversible but with time the OHC showed progressive volume loading, ultimately increasing from a baseline volume of 3.118 pl to 5.77 pl before lysis. (B) Determinations of electromotility obtained as outlined in the previous figure. Note that the gain increased from approximately 3 nm/mV (trace 1) to a maximum of 20.4 nm/mV (trace 4) after an ATP-mediated increase in volume to 3.9 pl (a 25% increase in volume from the start). Subsequent increase in volume (plots 4–6 in A) were associated with a reduction in electromotile gain back towards starting level (traces 5–7) (see also Figure 8, downward triangles). Trace 1 = control, traces 2–7 follow each determination of volume loading shown in (A).

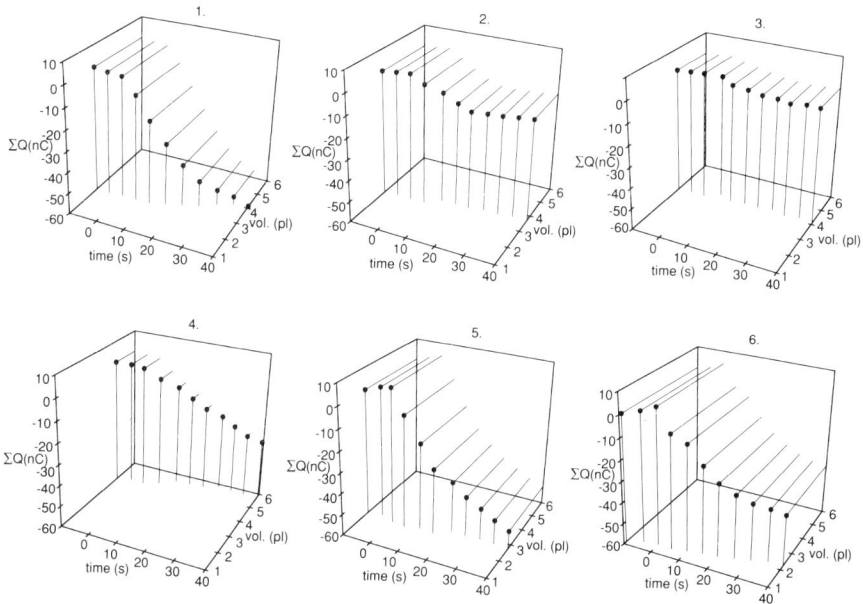

Fig. 7. The effect of ATP-gated influx of charge on cell volume change for six OHCs. Data were obtained as described for the previous figure. Note only cell 3 failed to exhibit a volume response associated with the influx of charge (principally Na$^+$), presumably because this was one of the larger cells (68.1 μm) and had the smallest response to ATP (–400 pA). Similarly all cells except cell 3 exhibited an increase in the slope of the length change–voltage relationship with the ATP-mediated increase in cell volume (see Fig. 8). Voltage-clamp ($V_h = -60$ mV) 1 = 13 sec, 10 μM ATP, peak ATP-activated current (I_p) = –1889 pA; 2 = 0.4 sec, 100 μM ATP, I_p= –1058 pA; 3 = 4 sec, 100 μM ATP, I_p= -400 pA; 4 = 1 sec, 100 μM ATP, I_p = –853 pA; 5 = 4 sec, 100 μM ATP, I_p = –2717 pA; 6 = 0.4 sec, 100 μM ATP, I_p = 2213 pA.

pipette which provided the voltage-clamp. Thus, because of the approximately equivalent mobilities of the K$^+$ and Cl$^-$ (Hille, 1992) in the internal solution, whole-cell recording was osmotically neutral, although re-equilibration of Cl$^-$ and Na$^+$ via the recording pipette, as well as potential membrane-bound ion exchange mechanisms, provide pathways for restoration of osmotic balance in addition to the influx of H$_2$O which was observed. A charge of 30–60 nC, driven in through the ATP-gated ion channels by the clamped driving force of –60 mV over 40 sec was sufficient to increase by 35% an average volume of 2.9 pl. Thus, based on elementary charge (1.60 × 10^{-19} C) and Avogadro's constant (6.0249 × 10^{23} mol^{-1}), an average 45 nC of charge would correspond to the influx of about 2.8 × 10^{11} osmotically active particles, or the equivalent of 160 mOsm (4.67 × 10^{-13} mol /2.9 × 10^{-12} l) of driving force,

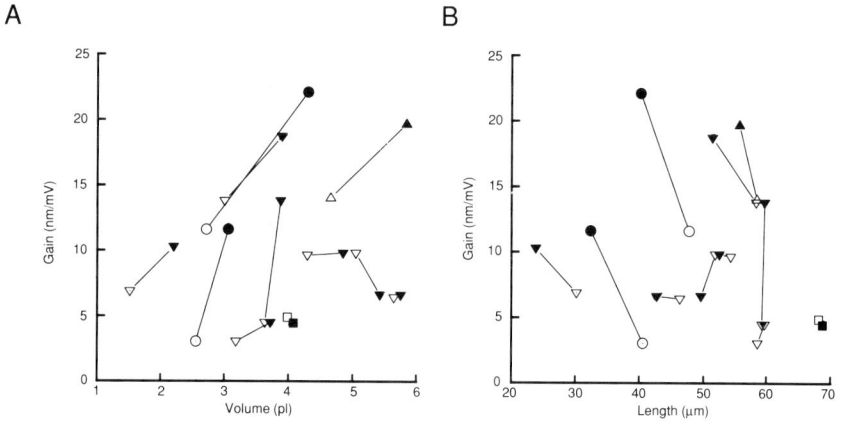

Fig. 8. Electromotile gain vs. cell volume and electromotile gain vs. cell length plots for the six OHCs described in the the previous figure before (open symbols) and after (closed symbols) exposure to ATP. Note that with the exception of a large cell (box), which had only a small response to ATP, failed to show volume or length changes, or a change in electromotile gain, the other OHC increase in gain after ATP. In the case of the OHC described in Fig. 6, which was given multiple exposures to ATP (shown here as inverted triangles), gain increased with increasing volume to an optimum and then subsequently decreased. A similar pattern is displayed with respect to cell length, with decreasing cell length being associated with the volume loading.

much of which must be negated by the rapid redistribuiton of ions by ion exchangers and by diffusion via the patch-pipette, as described above. Nevertheless, it is not surprising that the cell membrane can be seen to balloon out within seconds of exposure to ATP (Fig. 4). This compares with an earlier study which showed that an extracellular osmotic change from 300 to 280 mOsm produced a 5.8% length change, equivalent to a 14.8% volume change (Dulon et al., 1988). In one cell in the present study, a 19% volume change (3.12 to 3.71 pl) was produced by 10.23 nC of charge influx, equivalent to 34 mOsm of osmotic driving force. Therefore it does not seem unreasonable that the responses to ATP which these cells exhibit influence cell volume and hence cell turgidity, which has been linked to the efficient coupling of the electrokinetic proteins to the lateral wall membrane of the OHC (Brownell and Shehata, 1991; Santos-Sacchi, 1991). The gain of the electromotile response in the depolarizing direction was within the range reported previously (19.8 ± 8.3 nm/mV, Ashmore, 1987; 15nm/mV, Santos-Sacchi, 1989) with the non-linear attenuation at hyperpolarizing potentials being evident.

One outcome of the present study was the observation, both with current- and voltage-clamp, that sustained activation of the ATP-activated conductances in OHCs leads to rapid volume loading, potentially to destruction; this is an important finding, if under pathological conditions, such as cochlear ischaemia, significant concentrations of ATP are released into the cochlear fluids. Associated with this, determination of ATP receptor localization is very relevant to consideration of the physiological role of the P_2-purinoceptors in OHC function. The data shown in Figs 1 and 4 are consistent with the confinement of the ATP-gated ion channels to the endolymphatic surface of the OHCs. The consequence of this is that the K^+ would be the major cation entering through the ATP-gated channels, with a comparable driving force (150 mV) to that of mechanoelectrical transduction channels, but with clearly greater conductance. This has a fundamental implication with respect to clearance of an ATP-induced osmotic loading on the cell, as the K^+ ions would much more efficiently exit the cell via the basolateral membrane due to the activated I_{Kn} conductance present in these cells (Housley and Ashmore, 1992). In contrast to this, if the ATP receptors were on the basolateral surface, Na^+ loading of the cell could rapidly outpace the Na^+ exchange mechanisms, possibly even reversing the Na^+–Ca^{2+} exchanger (Ikeda *et al.*, 1992) and also producing an intracellular Na^+ block of the K^+ conductances. Ca^{2+} entry through ATP receptors located on the perilymphatic surface would also modulate the K^+-channels and non-selective cation channels present in these cells, whereas the more likely endolymphatic site for the receptors is subjected to a relatively low Ca^{2+} concentration (30 μM; Bosher and Warren, 1978) which may considerably reduce the impact of Ca^{2+} entry on the basolateral membrane conductances.

The degree to which ATP-mediated volume loading is likely to influence OHC electromotility *in vivo* and hence the active amplification of sound energy by the OHCs will largely depend upon the strength of the purinergic signal. Our evidence suggests that ATP may act as a local humoral factor within the endolymphatic compartment, a postulate supported by the detection of extracellular ATP in the endolymph at concentrations close to the physiological threshold for the ATP receptors (Thorne *et al.*, 1993) and more recently, the finding that subpicomole quantities of extracellular ATP injected into the endolymphatic compartment produce rapid and reversible attenuation in the endocochlear potential and cochlear microphonic (Thorne *et al.*, 1994).

Acknowledgements

The present study was supported by the Health Research Council of New Zealand, the Deafness Research Foundation of New Zealand, the

Wallath Trust and the New Zealand Lottery Grants Board. Thanks also to Denise Greenwood and Dr Bruce Mockett for expert technical assistance and to participants of the 'Active Hearing' symposium for useful discussions on the influence of the whole-cell patch-clamp on osmotic balance.

References

Ashmore, J. F. (1987). A fast motile response in guinea pig outer hair cells: the cellular basis for the cochlear amplifier. *J. Physiol.* **388:** 323–347.

Ashmore, J. F. and Ohmori, H. (1990). Control of intracellular calcium by ATP in isolated outer hair cells of guinea-pig cochlea. *J. Physiol.* **360:** 397–422.

Ashmore, J. F., Kolston, P. J. and Mammano, F. (1993). Dissecting the outer hair cell feedback loop. In *Biophysics of Hair Cell Sensory Systems* (H. Duifis, J. W. Horst, P. van Dijk and S. M. van Netten, eds), pp. 151–157. World Scientific, Singapore.

Barnard, E. A., Burnstock, G. and Webb, T. E. (1994). G protein-coupled receptors for ATP and other nucleotides: a new receptor family. *Trends Physiol. Sci.* **15:** 67–70.

Bosher, S. K. and Warren, R. L. (1978). Very low calcium content of cochlear endolymph, and extracellular fluid. *Nature* **273:** 377–378.

Brownell, W. E. (1990). Outer hair cell electromotility and otoacoustic emissions. *Ear Hearing* **11:** 82–91.

Brownell, W. E., Bader, C. R., Bertrand, D. and de Ribaupierre, Y. (1985). Evoked mechanical responses of isolated cochlear outer hair cells. *Science* **227:** 194–196.

Brownell, W. E. and Shehata, W. E. (1991). The effect of cytoplasmic turgor pressure on the static and dynamic mechanical properties of outer hair cells. In *Mechanics and Biophysics of Hearing* (P. Dallos, C. D. Geisler, J. W. Matthews, M. A. Ruggero and C. R. Steele, eds), pp. 52–60. Springer-Verlag, Berlin.

Dallos, P., Richard, H. and Evans, B. N. (1993). Theory of electrically driven shape changes of cochlear outer hair cells. *J. Neurophysiol.* **70:** 299–323.

Dubyak, G. R. (1991). Signal transduction by P_2-purinergic receptors for extracellular ATP. *Am. J. Respir. Cell Mol. Biol.* **4:** 295–300.

Dulon, D., Aran, J. M. and Schacht, J. (1988). Potassium-depolarization induces motility in isolated outer hair cells by an osmotic mechanism. *Hearing Res.* **32**, 123–130.

Dulon, D., Mollard, P. and Aran, J. M. (1991). Extracellular ATP elevates cystolic Ca^{2+} in cochlear inner hair cells. *NeuroReport* **2**: 69–72

Dulon D, Moataz, R., and Mollard, P. (1993). Characterisation of Ca^{2+} signals generated by extracellular nucleotides in supporting cells of the organ of Corti. *Cell Calcium* **14**, 245–254.

Eybalin, M. (1993). Neurotransmitters and neuromodulators of the mammalian cochlea. *Physiol. Rev.* **73**: 309–373.

Gordon, J. L. (1986) Extracellular ATP: effects, sources and fate. *Biochem. J.* **233**: 309–319.

Hamill, O. P., Marty, A., Neher, E., Sakmann, B. and Sigworth, F. J. (1981). Improved patch-clamp techniques for high resolution current recording from cells and cell-free membrane patches: *Pflüger's Arch. Eur. J. Physiol.* **391**: 85–100.

Housley, G. D. and Ashmore, J. F. (1992). Ionic currents of outer hair cells isolated from the guinea-pig cochlea. *J. Physiol.* **448**, 73–98.

Housley, G. D., Greenwood, D. and Ashmore, J. F. (1992). Localisation of cholinergic and purinergic receptors on outer hair cells isolated from the guinea-pig cochlea. *Proc. R. Soc. Lond.* B **249**, 265–273

Housley, G. D., Greenwood, D., Mockett, B. G., Muñoz, D. J. B. and Thorne, P. R. (1993). Differential actions of ATP-activated conductances in outer and inner hair cells isolated from the guinea-pig organ of Corti: a humoral purinergic influence on cochlea function. In *Biophysics of Hair Cell Sensory Systems* (H. Duifhuis, J. W. Horst, P. van Dijk and S. M. van Netten, eds), pp. 116–123. World Scientific, Singapore.

Hille, B. (1992). *Ionic Channels of Excitable Membranes.* Sinauer, Sunderland.

Ikeda, K., Saito, Y., Nishiyama, A. and Takasaka, T. (1992). Na^+–Ca^{2+} exchange in the isolated outer hair cells of the guinea-pig studied by fluorescence image microscopy. *Pflüger's Arch. Eur. J. Physiol.* **420**: 493–499.

Illes, P. and Nörenberg, W. (1993). Neuronal ATP receptors and their mechanism of action. *Trends Physiol. Sci.* **14**: 50–54.

Kakehata, S., Nakagawa, T., Takasaka, T. and Akaike, N. (1993). Cellular mechanism of acetlycholine-induced response in dissociated outer hair cells of guinea-pig cochlea. *J. Physiol.* **463:** 227–244.

Kolston, P. J. and Ashmore, J. F. (1992). Action of ATP on isolated Hensen's cells of guinea-pig cochlea investigated by calcium imaging and whole cell recording. *J. Physiol.* **446:** 389P.

Mockett, B. G., Housley, G. D. and Thorne, P. R. (1994). Fluorescence imaging of extracellular purinergic receptor sites and putative ecto-ATPase sites on isolated cochlear hair cells. *J. Neurosci.* **14:** 6992–7007.

Nakagawa, T., Akaike, N., Kimisuki, T., Komune, S. and Toshio, A. (1990). ATP-induced current in isolated outer hair cells of the guinea-pig cochlea. *J. Neurophysiol.* **63:** 1068–1074.

Nilles, R., Järlebark, L., Zenner, H. P. and Heilbronn, E. (1994). ATP-induced cytoplasmic [Ca^{2+}] increases in isolated cochlear outer hair cells involved receptor and channel mechanism. *Hearing Res.* **73:** 27–34.

Santos-Sacchi, J. (1989). Asymmetry in voltage-dependent movements of isolated outer hair cell motility and capacitance. *J. Neurosci.* **9:** 2954–2962

Santos-Sacchi, J. (1991). Reversible inhibition of voltage-dependent movements of isolated outer hair cells from the organ of Corti. *J. Neurosci.* **11:** 3096–3110.

Thorne, P. R., Muñoz, D. J. B., Housley, G. D. and Billett, T. E. (1993). Extracellular adenosine triphosphate (ATP) in the endolymph and perilymph of the guinea-pig cochlea. In *Biophysics of Sensory Hair Cell Systems* (H. Duifhuis, J. W. Horst, P. van Dijk, and S. M. van Netten, eds), p. 412. World Scientific, Singapore.

Thorne, P. R., Muñoz, D. J. B., Housley, G. D. and Battersby, J. (1994). Evidence for a humoral role for extracellular ATP in modulating cochlear function. *Drug Dev. Res.* **31** (4): 328.

Active Mechanisms in the Response of the Auditory System to Over- or Understimulation

RICHARD A. ALTSCHULER, YEHOASH RAPHAEL, HYUN HO LIM, JEROME DUPONT, KAZUO SATO AND JOSEF M. MILLER

Kresge Hearing Research Institute, University of Michigan, 1301 East Ann Street, Ann Arbor, MI 48109-0506, USA

Introduction

Active biomechanical and molecular mechanisms are important components of normal auditory processing. There may also be active mechanisms to enable the auditory system to react and adapt to changes in its input and the stresses they induce. In this chapter we will consider potential active processes involved in the responses to over- and understimulation of the auditory system. Some of these mechanisms may be specialized for reaction to stress, while others could utilize the same active processes used in normal hearing. Both overstimulation and understimulation are common stresses to the auditory system, the former occurring throughout life, the latter becoming greater with aging, although some hair cell loss is seen even in normal hearing young. It is therefore likely that active mechanisms have developed in the mammalian sensorineural epithelium and central auditory system (CAS) for reacting and adapting to over- and/or understimulation.

Overstimulation

High-intensity noise is a common stress to the auditory system, which can cause temporary (TTS) or permanent (PTS) threshold shifts/hearing loss. Active mechanisms appear to exist both to protect

239

the auditory system from noise-induced damage as well as to minimize the effects of damage when protective mechanisms fail. TTS, in fact, may in itself be a protective mechanism, decreasing the influence of continued noise. The mechanisms behind TTS and its recovery are not completely understood and there are likely to be several different processes involved. Mechanisms that have been proposed include: (i) effects on stereocilia rigidity (e.g. Canlon *et al.*, 1987; Saunders *et al.*, 1986; Tilney *et al.*, 1982); (ii) loss of receptor currents due to disruption of transduction and ion channels (e.g. Patuzzi, 1994); (iii) excitotoxicity related swelling and/or retraction of auditory nerve proximal processes and their recovery/regeneration (Eybalin, 1994; Robertson, 1983); (iv) disruption or uncoupling of the active biomechanical mechanism (e.g. Neely, 1993); (v) metabolic and homeostatic overload/fatigue (e.g. Hawkins, 1973; Lim and Melnick, 1971; Saunders *et al.*, 1986). Facets of these mechanisms could require active control(s). We will consider four mechanisms for active protection within the sensorineural epithelium: the heat shock proteins, the glutathione pathway, regulation of the excitatory synapses and actions of supporting cells. The potential role of the efferents will not be considered.

Heat Shock Proteins

Heat shock proteins (HSPs) are more appropriately termed 'stress shock proteins' since they are expressed in response to a wide variety of stresses in a large number of cell types. There are a number of families of HSPs, named by their denatured molecular weights – HSP 20s, HSP 60s, HSP 70s, HSP 90s, HSP 110s and the ubiquitins. HSPs are now known to be both constitutive (expressed without stress) with a role in normal cellular functions and perhaps protection, and induced in response to stress, with a protective function. Interference with expression of induced HSPs (e.g. by injection of antibodies into fibroblasts) decreases cell survival after stresses (Ciocca *et al.*, 1993; Riabowol *et al.*, 1988; Tytell *et al.*, 1993; Welch, 1992), while overexpression (often by priming with a pre-exposure) increases tolerance to stresses (Barbe *et al.*, 1988; Hutter *et al.*, 1994; Tytell *et al.*, 1993). Many mechanisms have been proposed to explain the protective function of HSPs, including folding, refolding and chaperoning of proteins, stabilization of receptors and proteins, and interaction with cytoskeleton (Ciocca *et al.*, 1993; Tytell *et al.*, 1993; Welch *et al.*, 1992). In the cochlea, the expression of HSP 72 has been shown with hyperthermia (Deschesne *et al.*, 1992; Myers *et al.*, 1992; Neely *et al.*,

1991) and hypoxia (Myers *et al.*, 1992). We examined the expression of HSP 27, 72 and 90 in the cochlea with and without noise.

Rats were exposed to 105 dB SPL broad-band noise for 90 min. This resulted in TTS averaging 50 dB for 3–5 hr. Four to six hours after the cessation of the exposure, expression of HSPs in the cochlea was assessed using antibodies to HSP 27 (Sigma), HSP 72 (Stressgen) and HSP 90 (Stressgen) with immunocytochemistry or Western blots. Other rats without noise exposure were similarly examined. A noise-induced expression of HSP 72 was seen in all three rows of outer hair cells (Fig. 1A) and in the stria vascularis (Lim *et al.*, 1993). This expression was greatest 5–6 hr after the noise exposure. HSP 27 also showed noise-induced expression in outer hair cells (Fig. 1B), with a similar temporal pattern (Ditto *et al.*, 1994). HSP 27 also had a constitutive expression, confined to the stereocilia of inner and outer hair cells (Fig. 1C). HSP 90 had a constitutive expression in both inner and outer hair cells (Fig. 1D), which was modulated by noise exposure (Lim *et al.*, 1994).

Each of the HSPs may play specific important roles in the protection of the cochlea. HSP 27 was originally described as an actin-binding protein (Miron *et al.*, 1991) and has been shown to influence actin polymerization (Lavoie *et al.*, 1993). Constitutive expression of HSP 27 in hair cell stereocilia and its up-regulation after noise exposure could play a role in regulation of stereocilia rigidity and therefore in TTS and its recovery. HSP 72 stabilizes spectrin in the cortical lattice of red blood cells (Gudi and Gupta, 1993) and could be playing a similar role in the outer hair cell cortical lattice. HSP 90 has also been shown to interact with the cytoskeleton. HSPs could also be playing a more general protective role in protein stabilization and refolding after noise induced denaturation. Constitutive levels could be involved in protection from the initial insult and its effects, while induced HSPs could help both with the longer-term effects and in recovery.

Expression of induced HSP 72 is also seen in auditory brain-stem nuclei after noise overstimulation (unpublished observations) and so HSPs may also be playing a protective function centrally.

Fig. 1 (following pages). Heat shock protein immunoreactive staining in surface preparations of the second turn of the rat cochlea. Noise-induced expression of HSP72 (A) and HSP 27 (B) is seen in outer hair cells. Constitutive expression of HSP 90 is seen in inner and outer hair cells (C), while constitutive expression of HSP 27 is seen in inner and outer hair cell stereocilia (D).

A

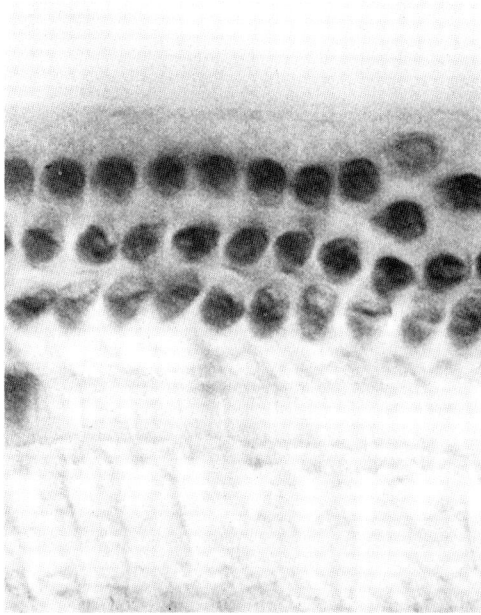

B

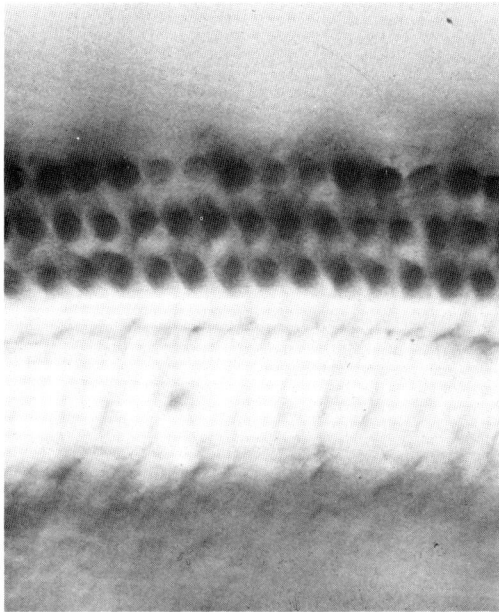

Fig.1.

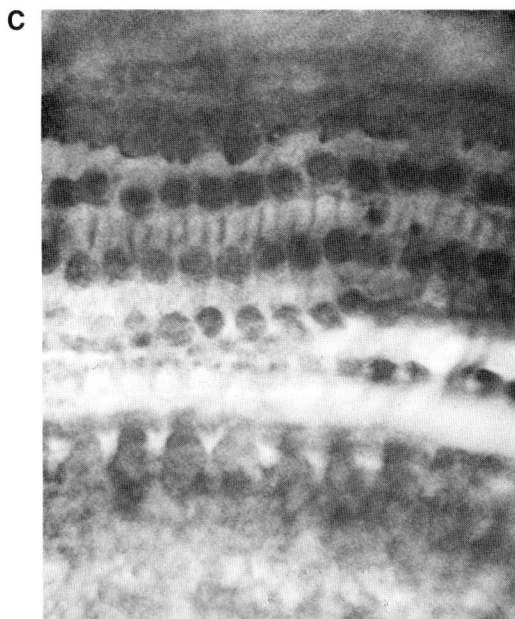

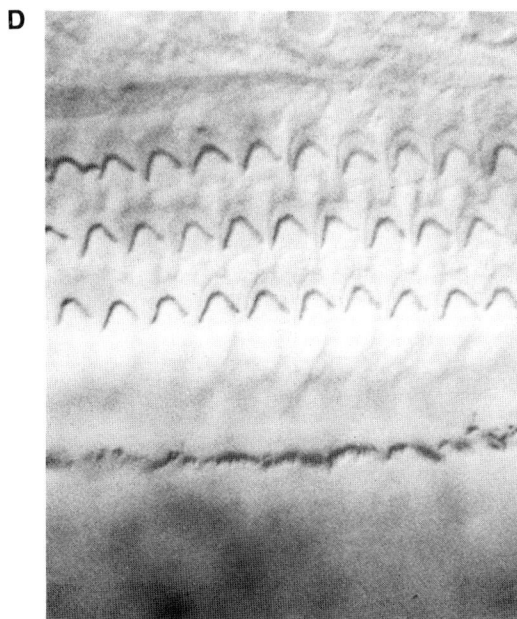

Fig. 1 (*continued*).

Glutathione and detoxification mechanisms

The glutathione pathway is another protective mechanism, used by a wide variety of cells that may also play a role in the cochlea. Recent immunocytochemical studies by El Barbary *et al.* (1993) showed the glutathione-*S*-transferase (GST) immunoreactive staining in hair cells, where this enzyme could play an important role in a general detoxification pathway. It is interesting that localization of GST in outer hair cells in an apical location by the centriole/basal body area coincides with where Aran *et al.* (1993) have shown the accumulation of aminoglycosides with long-term sublethal treatments, and where Kachar *et al.* (1993) have shown a vesicular pathway to endolymph. This detoxification pathway may play a role in preventing damage from many types of stresses involving free radical formation, including overstimulation and ototoxic insult (Hoffman *et al.*, 1988; Garetz *et al.*, 1994; Seidman *et al.*, 1993).

Regulation of the excitatory synapse

Another source of overstimulation-related damage in the cochlea is from excitotoxicity (Eybalin, 1994; Robertson, 1983). Protective mechanisms may exist to decrease its effects. One mechanism could be related to the swelling, retraction and regrowth of afferent proximal processes (Eybalin, 1994). Feedback mechanisms involving nitric oxide (NO) may also exist. Zdanski *et al.* (1994) have shown diaphorase activity in spiral ganglion cells, and Fessendon *et al.* (1994) have shown NO synthetase and diaphorase activity in lateral wall, supporting cells and nerve terminals. Humoral factors (such at ATP) may also regulate release (Housely, 1994). Another potential mechanism may involve changes in post-synaptic receptors, either in expression, placement or post-translational modifications to change either their general sensitivity or calcium permeability.

Supporting cells

Supporting cells may play a protective role in the cochlea. Unlike hair cells, supporting cells are in contact with the basement membrane, and are therefore able to communicate with the extracellular matrix. This has been shown to play important roles in maintenance of other epithelial tissues, and is most likely to be also true for the auditory epithelium. It has been shown that supporting cells in the organ of Corti

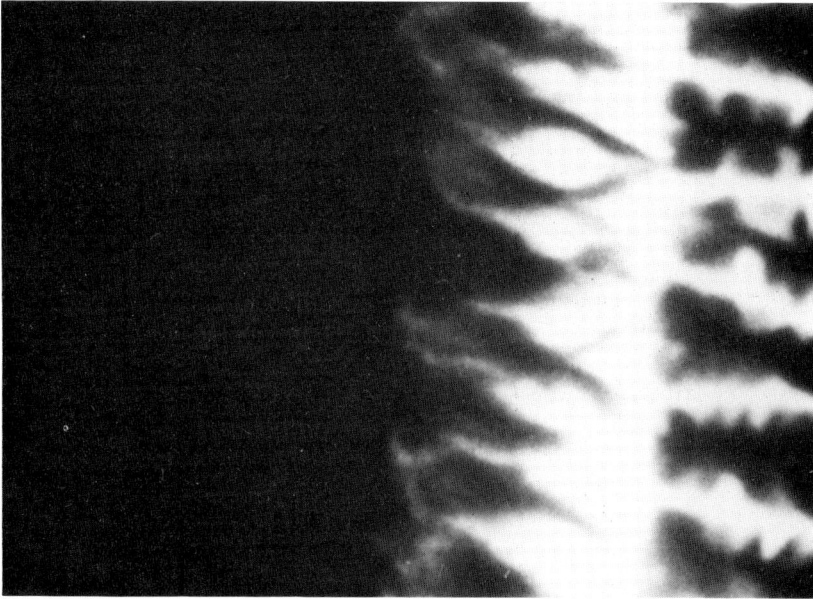

Fig. 2. Phalloidin labeling of f-actin in apical process(es) of pillar cell(s) reveals a spring-like structure that may be involved in medial stabilization of the outer hair cell motility.

transiently express fibronectin after acoustic trauma (Woolf, 1994), indicating that molecular changes occur in supporting cells in response to trauma and that matrix-related molecules are involved in this response.

Our recent observations show cytoskeletal structures that can potentially serve as active or resistive modulators of the forces generated by outer hair cell motility (Raphael *et al.*, 1994). These cytoskeletal elements were found in pillar cells and Deiters cells. Notably, a spring-like coil of actin was detected in the apical region of pillar cells in monkey and man, opposing medial motion of OHCs (Fig. 2). There are other potentially active processes, such as the tension fibroblasts in the spiral ligament, that could react and adapt to overstimulation of the basilar membrane (Henson and Henson, 1988).

There are therefore several mechanism in the cochlea that provide for protection and appear to represent active mechanisms for dealing with stresses such as overstimulation. Further studies will be needed to further characterize these and other mechanisms, how they operate, and how and why they fail.

Understimulation Deafness

Understimulation might seem to be less of a stress to the auditory system than overstimulation; however, it can have consequences that are as severe. The consequences are particularly striking if under-stimulation occurs during development, though our focus has been on the effects in the mature mammal. Very different effects are seen in the peripheral vs. the CAS. In the CAS a general 'down-regulation' associated with reduced activity occurs in neuronal elements with effects also seen in synapses and glia. Transient changes are also seen, as the neuronal elements react and adapt to changes in activity of their inputs. In the cochlea, however, there is degeneration rather than, or in addition to, down-regulation; a deafness-related loss of spiral ganglion cells. The latter change is irreversible, but may be preventable.

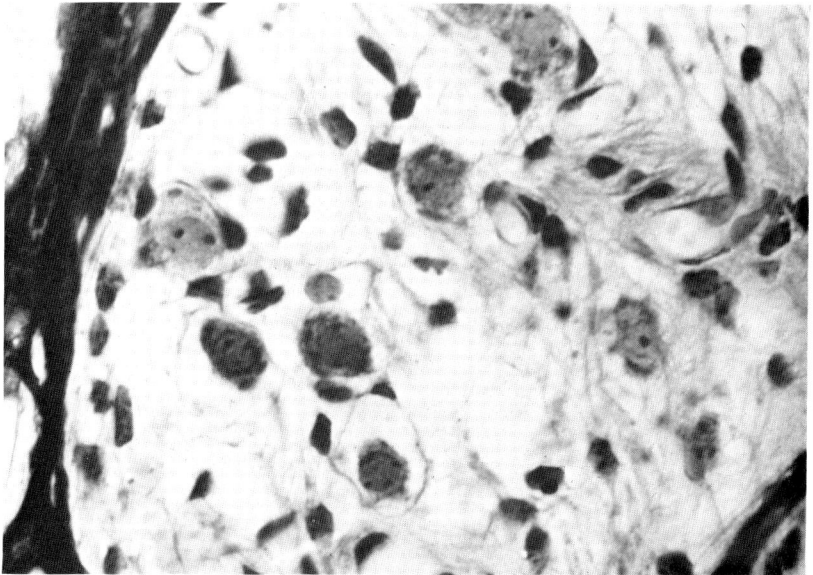

Fig. 3. Cross-section through Rosenthal's canal from the third turn of the cochlear spiral 50 days after deafening with kanamycin and ethacrynic acid. Two spiral ganglion cells with morphological features consistent with apoptosis are seen, while another spiral ganglion cell appears normal.

Deafness-related changes in peripheral auditory system

Spiral ganglion cell loss occurs under conditions where inner hair cells are lost, including noise, ototoxicity and genetic deafness. The time between inner hair cell loss and spiral ganglion cell loss varies between species, with a 50% loss taking about 9 weeks in guinea-pig (Jyung *et al.*, 1989; Webster and Webster, 1981; Zappia and Altschuler, 1989) and close to 2 years in dog (unpublished observations). Recent studies from our group and others show that spiral ganglion cell loss after inner hair cell loss can be reduced by electrically stimulating the remaining spiral ganglion cells with a cochlear implant (Hartshorn *et al.*, 1991; Leake *et al.*, 1992; Lousteau, 1983). This would suggest that the spiral ganglion cell death is related to loss of activity, rather than the initial stress or cause of inner hair cell loss. In preliminary studies, we examined the effects of chronic application of TTX (blocking AP in auditory nerve but not generator potentials in afferent dendrites) into perilymph. A month of application did not cause any significant spiral ganglion cell loss. It is also possible that factors other than activity or a combination of factors are necessary to induce cell death. This might also explain why some spiral ganglion cells are lost immediately and others are lost much more slowly, as well as variations between species. It is possible that apoptotic mechanisms are involved, initiated either by the lack of activity or by mixed signals to the spiral ganglion cell. We see many of the morphological signs of apoptosis, including condensation of nuclear material, accumulation of condensed chromatin along the margins of the nucleus and fragmentation of nuclear material in spiral ganglion cells after inner hair cell loss (Fig. 4); however, use of apoptosis-specific markers will be necessary to determine its role in deafness-related spiral ganglion cell loss.

Deafness-related central auditory system changes

In the mature mammal there is little cell loss in the CAS as a

Fig. 4 (following pages). Changes in glycine immunoreactivity and in glycine receptor expression in the lateral superior olive after deafness. (A) and (C) show side ipsilateral to lesion (where excitatory input from CN is lost and ipsilateral inhibitory input maintained). (B) and (C) are contralateral to lesion where excitatory input is maintained and inhibitory input lost. There is a decrease in glycine immunoreactive staining contralateral to the lesion (B) while expression of the alpha 1 subunit of the glycine receptor expression decreases ipsilaterally (C).

A

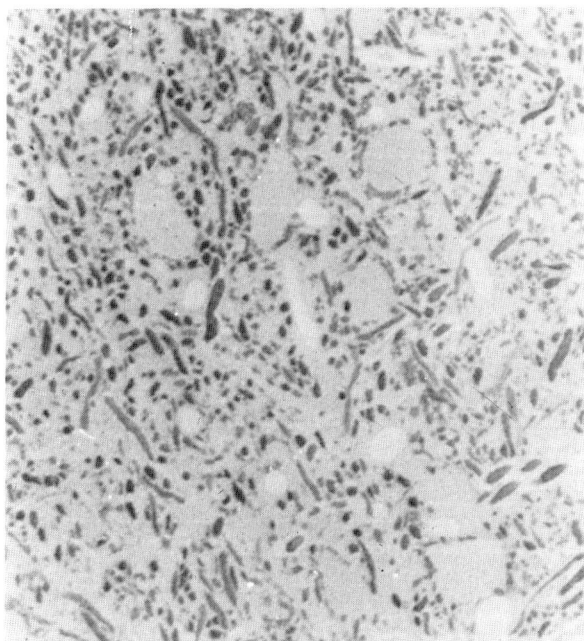

B

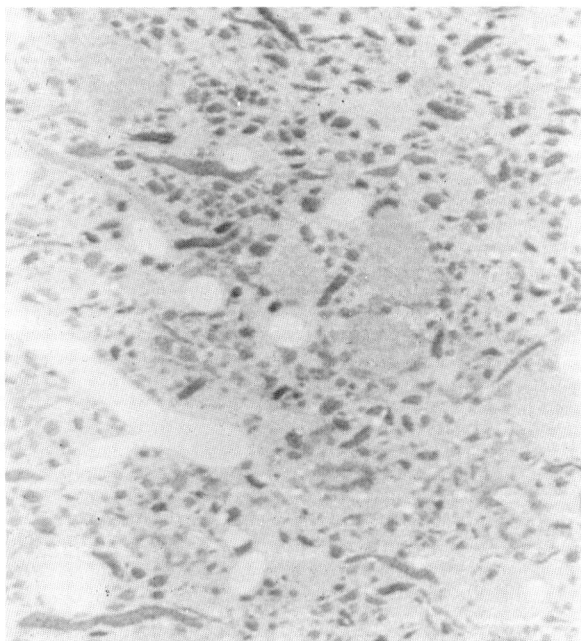

Fig. 4.

C

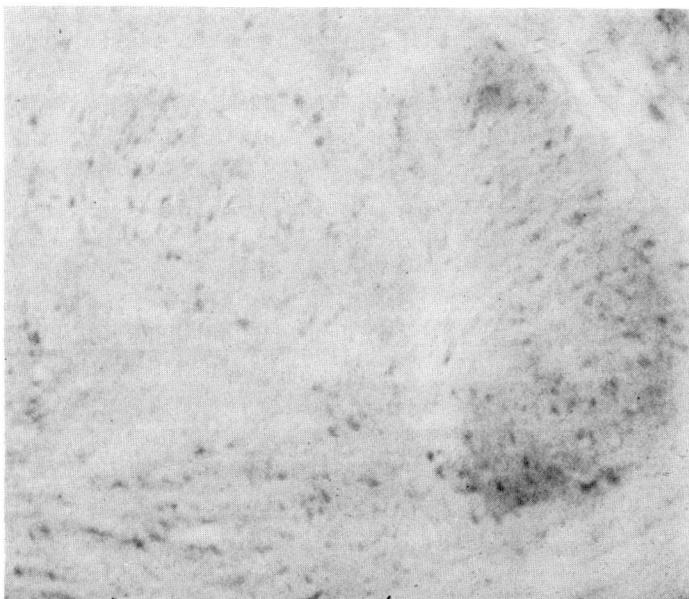

D

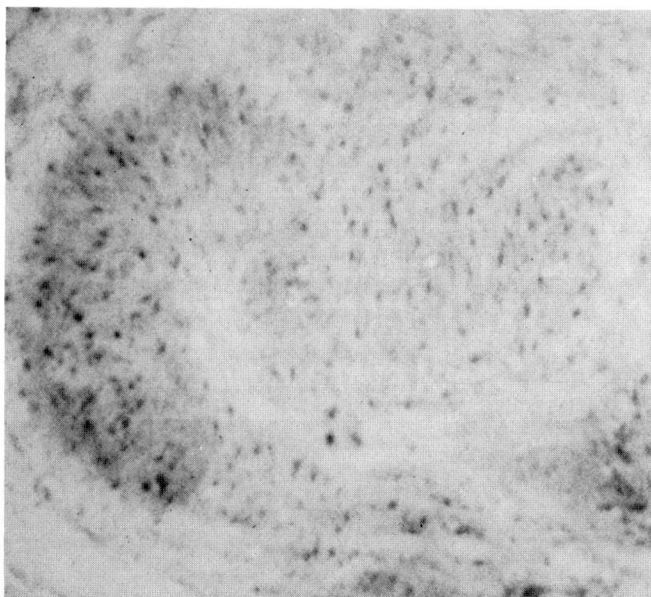

Fig. 4 (*continued*).

consequence of deafness, but there are are many deafness-related changes (Miller *et al.*, 1992; Rubel *et al.*, 1990) related to a decrease in activity and function. One of the most marked changes is in cell size. In the ventral cochlear nucleus we observe a deafness-related cell size decrease of about 20% (Lesperance *et al.*, 1994) along with changes in synaptic structure (Miller *et al.*, 1992). There is also a decrease in cell size in the lateral superior olive, though this change appears to be transient, with cell size returning to normal after several weeks. There are also changes in neurotransmitters and receptors, some of which may also be transient. Dupont *et al.* (1994) found deafness-related changes in GABA immunostaining of cell bodies in the ipsilateral LSO and in the contralateral central nucleus of the inferior colliculus 10 days after the unilateral deafening, with a progressive recovery towards normal by 25 days. In the ipsilateral inferior colliculus, GABA immunostaining showed a transient increase. We also have preliminary evidence of changes both in glycine immunoreactive terminals and in glycine receptor in the superior olivary complex following deafness. There is also a decrease of glycine positive terminals in the MNTB, DMPO and LSO 5–10 days after deafening, as well as decreases in intensity of staining (Fig. 4A,B). Our preliminary studies also show a decrease in expression of the alpha 1 subunit of the glycine receptor in the LSO 2–5 days following deafness (Fig. 4B,C) as well as up-regulation of the alpha 2 subunit which has its greatest expression during development (Dupont *et al.*, 1994).

There appear to be functional ramifications of these deafness-related changes. Four or nine weeks after deafness we found a marked decrease in electrical evoked excitability of central auditory nuclei, as reflected in 2-deoxyglucose (2DG) metabolic uptake (Schwartz *et al.*, 1993).

The activity dependence of cell size changes was demonstrated by Pasic and Rubel (1991) who showed a 20% decrease in cell size in the gerbil CN after blocking peripheral activity with TTX; this change was reversed after TTX administration ceased. We have also shown that many deafness related changes can be prevented or reversed by electrical stimulation with cochlear prostheses, although neither the prevention nor reversal is complete. Chronic electrical stimulation can largely prevent deafness-related flattening of auditory nerve synapses (Miller *et al.*, 1992), reverse deafness-related cell size changes in CN (Miller *et al.*, 1994) and reverse threshold changes for 2DG activity in IC (Schwartz *et al.*, 1993). We need to examine further the influence of the period of time of deafness and look at neurotransmitters and receptors.

It will be important to determine the reason for these deafness-related changes and why some are transient and others maintained. Transient changes in transmitters and receptors may be related to trying to enhance activation, and this might be most effective when lesion or loss of activity is only partial or 'patchy'. This is the case with age-related hearing loss and might explain why age-related changes in auditory brain-stem transmitters and receptors have been reported (Helfert *et al.*, 1994; Milbrandt *et al.*, 1994). It will also be important to understand why cell loss occurs in the periphery and only down-regulation centrally, and the mechanisms behind these changes.

Acknowledgements

This work was supported by NIDCD grants DC00383 and DC00274.

References

Aran, J. M., Dulon, D., Hiel, H., Erre, J. P. and Aurousseau, C. (1993). Ototoxicity of aminoglycosides: recent results on uptake and clearance or gentamycin. *Rev. Laryngol. Otol. Rhinol.* **114:** 125–128.

Barbe, M. F., Tytell, M., Gower, D. J. and Welsh, W. J. (1988). Hyperthermia protects against light damage in the rat retina. *Science* **241:** 1817–1820.

Canlon, B., Miller, J., Flock, A. and Borg, E. (1987). Pure tone overstimulation changes the micromechanical properties of the inner hair cell stereocilia. *Hearing Res.* **30:** 65–72.

Ciocca, D. R., Oesterreich, S., Chamness, G. C., McHuire, W. L. and Fuqua, S. A. W. (1993). Biological and clinical implications of heat shock protein 27000 (Hsp 27): a review. *J. Natl. Cancer Inst.* **85:** 1558–1570.

Dechesne, C. J., Kim, H. N., Nowak, T. S., Jr and Wenthold, R. J. (1992). Expression of heat shock protein, HSP 72, in guinea pig and rat cochlea after hyperthermia: immunochemical and *in situ* hybridization analysis. *Hearing Res.* **59:** 195–204.

Ditto, J. L., Raphael, Y., Lim, H. H., Wang, Y. and Altschuler, R. A. (1994). Stress shock response in the rat cochlea: noise induced expression of HSP 27. *Abstr. Assoc. Res. Otolaryngol.* **17:** 181.

Dupont, J., Bonneau, J. M. and Altschuler, R. A. (1994). GABA and glycine changes in the guinea pig auditory brainstem after total destruction of the inner ear. *Abstr. Assoc. Res. Otolaryngol.* **17:** 11.

El Barbary, A., Altschuler, R. A. and Schacht, J. (1993). Glutathione *S*-transferases in the organ of Corti of the rat: enzymatic activity, subunit composition and immunohistochemical localization. *Hearing Res.* **71**: 80–90.

Fessendon, J. D., Coling, D. E. and Schacht, J. (1994). Nitric oxide synthase in the cochlea. *Abstr. Assoc. Res. Otolaryngol.* **17**: 137.

Gudi, T. and Gupta, C. M. (1993). Hsp 70-like protein in erythrocyte cytosol and its interactions with membrane skeleton under heat and pathologic stress. *J. Biol. Chem.* **268**: 21344–21350.

Garetz, S., Altschuler, R. A. and Schacht J (1994). Attentuation of gentamycin ototoxicity by glutathione in the guinea pig *in vivo. Hearing Res.* **77**: 81–97.

Hartshorn, D. O., Miller, J. M. and Altschuler, R. A. (1991). Protective effect of electrical stimulation on the deafened guinea pig cochlea. *Otolaryngol. Head Neck Surg.* **104**: 311–319.

Hawkins, J. E., Jr (1973). Comparative otopathology: aging, noise and ototoxic drugs. *Adv. Otorhinolaryngol.* **20**: 125–41.

Helfert, R. H., Sommer, T. J., Hughes, L. F., Jeffery, C. M. and Caspary, D. M. (1994). Age related changes in the synaptic organization of the central nucleus of the inferior colliculus of the Fischer 344 rat. *Abstr. Assoc. Res. Otolaryngol.* **17**: 11.

Henson, M. M. and Henson, O. W. (1988). Tension fibroblasts and the connective tissue matrix of the spiral ligament. *Hearing Res.* **35**: 237–258.

Hoffman, D. W., Jones-King, K. L., Whitworth, C. A. and Rybak, L. P. (1988). Potentiation of ototoxicity by glutathione depletion. *Ann. Otol. Rhinol. Laryngol.* **97**: 36–41.

Hutter, M. M., Sievers,R. E., Barbosa, V. and Wolfe, C. L. (1994). Heat-shock protein induction in rat hearts. A direct correlation between the amount of heat shock protein induced and the degree of myocardial protection. *Circulation* **89**: 355–60.

Jyung, R. W., Miller, J. M. and Cannon, S. C. (1989). Evaluation of eighth nerve integrity using the electrically evoked middle latency response. *Am. Acad. Otolaryngol.* **101**: 670–682.

Kachar, B., Battaglia, A., Jaeger, R. and Fex, J. (1993). Vesicular traffic around the cuticular plate. *Abstr. Assoc. Res. Otolaryngol.* **16**: 455.

Lavoie, J. N., Hickey, E., Weber, L. A. and Landry, J. (1993). Modulation of actin microfilament dynamics and fluid phase pinocytosis by phosphorylation of heat shock protein 27. *J. Biol. Chem.* **268**: 24210–24214.

Lesperance, M. M., Helfert, R. H., Altschuler, R. A. (1994). Deafness-induced cell size changes in rostral AVCN of the guinea pig. *Hearing Res.*, in press..

Leake, P. A., Snyder, R. L., Hradek, G. T. and Rebscher, S. J. (1992). Chronic intracochlear electrical stimulation in neonatally deafened cats: effects of intensity and stimulating electrode location. *Hearing Res.* **64:** 99–117.

Lim, D. J. and Melnick, W. (1971). Acoustic damage of the cochlea. A scanning and transmission electron microscopic observation. *Arch. Otolaryngol.* **94:** 294–305.

Lim, H. H., Jenkins, O. H., Myers, M. W., Miller, J. M. and Altschuler, R. A. (1993). Detection of HSP 72 synthesis after acoustic overstimulation in rat cochlea. *Hearing Res.* **69:** 146–150.

Lim, H. H., Moon, A., Wang, Y., Moran, A. and Altschuler, R. A. (1994). Stress shock response in the rat cochlea: Noise induced expression of HSP 90. *Abstr. Assoc. Res. Otolaryngol.* **17:** 67.

Lousteau, R. J. (1983). Increased spiral ganglion cell survival in electrically stimulated, deafened guinea pig cochleae. *Laryngoscope* **97:** 837–842.

Milbrandt, J. C. and Caspary, D. M. (1994). Age-related decrease in GABAB receptor binding in the inferior collicukus of the Fischer 344 rat. *Abstr. Assoc. Res. Otolaryngol.* **17:** 11.

Miller, J. M., Altschuler, R. A., Niparko, J. K., Hartshorn, D. O., Helfert, R. H. and Moore, J. K. (1992). Deafness induced changes in the central nervous system and their reversibility and prevention. In *Noise Induced Hearing Loss* (D. Marshall, ed.), pp. 130–145. Mosby YearBook, St Louis, MO.

Miller, J. M., Finger, P. A., Prieskorn, D. M., Moran, A. and Altschuler, R. A. (1994). Reversal of deafness-related cochlear nucleus cell size changes with electrical stimulation of the guinea pig cochlea. *Abstr. Assoc. Res. Otolaryngol.* **17:** 144.

Miron, T., Vancompernolle, K., Vandekerckhove, J., Wilchek, M. and Geiger, B. (1991). A 25-kD inhibitor of actin polymerization is a low molecular mass heat shock protein. *J. Cell Biol.* **114:** 255–261.

Myers, M. W., Quirk, W. S., Rizk, S. S., Miller, J. M. and Altschuler, R. A. (1992). Expression of the major mammalian stress protein in the rat cochlea following transient ischemia. *Laryngoscope* **102:** 981–987.

Neely, S. T. (1993). A model of cochlear mechanics with outer hair cell motility. *J. Acoust. Soc. Am.* **94:** 137–46.

Neely, J. G., Thompson, A. M. and Gower, D. J. (1991). Detection and localization of heat shock protein in the normal guinea pig cochlea. *Hearing Res.* **52:** 403–406.

Pasic, T. R. and Rubel, E. W. (1991). Cochlear nucleus cell size is regulated by auditory nerve electrical activity. *Otolaryngol. Head Neck Surg.* **104:** 6–13.

Raphael, Y., Athey, B., Wang, Y., Lee, M. and Altschuler, R. A. (1994). F-Actin and spectrin in the organ of Corti: comparative distribution in different cell types and mammalian species. *Hearing Res.* **76:** 173–187.

Riabowol, K. T., Mizzen,L. A. and Welch, W. J. (1988). Heat shock is lethal to fibroblasts microinjected with antibodies against HSP 70. *Science* **242:** 433–436.

Robertson, D. (1983). Functional significance of dendritic swelling after loud sounds in the guinea pig cochlea. *Hearing Res.* **9:** 263–78.

Rubel, E. W., Hyson, R. L. and Durham, D. (1990). Afferent regulation of neurons in brain stem auditory system. *J. Neurobiol.* **21:** 169–196.

Saunders, J. C., Canlon, B. and Flock, A. (1986). Changes in stereocilia micromechanics following overstimulation in metabolically blocked hair cells. *Hearing Res.* **24:** 217–25.

Schwartz, D. R., Schacht, J., Miller, J. M., Frey, K. and Altschuler, R. A. (1993). Chronic electrical stimulation reverses deafness-related depression of electrically evoked deoxyglucose activity in the guinea pig inferior colliculus. *Hearing Res.* **70:** 243–249.

Seidman, M. D., Shivapuja, B. G. and Quirk, W. S. (1993). Protective effects of allupurinol and superoxide dismutase on noise induced cochlear damage. *Otolaryngol. Head Neck Surg.* **109:** 1052–1056.

Tilney, L. G., Saunders, J. C., Egelman, E. and DeRosier, D. J. (1982). Changes in the organization of actin filaments in the stereocilia of noise damaged lizard cochlea. *Hearing Res.* **7:** 181–97.

Tytell, M., Barbe, M. F. and Brown, I. R. (1993). Stress (heat shock) protein accumulation in the central nervous system. *Adv. Neurol.* **59:** 293–302.

Webster, M. and Webster, D. B. (1981). Spiral ganglion neuron loss following organ of Corti loss: a quantitative study. *Brain Res.* **212:** 17–30.

Welch (1992). Mammalian stress response: cell physiology, structure/function of stress proteins. and implications for medicine and disease. *Physiol. Rev.* **72**: 1063–1080.

Woolf, N. K. (1994). Inner ear fibronectin expression following viral infection and noise trauma. *Abstr. Assoc. Res. Otolaryngol.* **17**: 4.

Zappia, J. J. and Altschuler, R. A. (1989). Evaluation of the effect of ototopical neomycin on spiral ganglion cell density in the guinea pig. *Hearing Res.* **40**: 29–38.

Zdanski, C. J., Prazma, J., Petrusz, P., Grossman, G. and Raynor, H. C. (1994) Nitric oxide synthase is an active enzyme in the spiral ganglion cells of rats. *Abstr. Assoc. Res. Otolaryngol.* **17**: 137.

How are the Inner Hair Cells Stimulated Mechanically?

SHYAM M. KHANNA[1], MATS ULFENDAHL[3], ÅKE FLOCK[3] AND
CONOR HENEGHAN[2]

[1]*Department of Otolaryngology, College of Physicians and Surgeons;
and* [2]*Department of Electrical Engineering, Columbia University, New
York, NY 10032, USA*

[3]*Department of Physiology and Pharmacology, Division of
Physiology II, Karolinska Institutet, S-171 77 Stockholm, Sweden*

Introduction

The hearing organ has a complex cellular organization. The basic
pattern of organization is the same throughout the length of the organ
and is repeated in different mammalian ears, strongly suggesting that
this structural organization has important functional significance. In
order to understand the role of the structure in the function of the organ,
it is necessary to use non-invasive techniques, so that observations and
measurements can be carried out on different structural elements
without altering the function of the organ. The confocal slit microscope
(Koester *et al.*, 1994) and the heterodyne interferometer system
(Willemin *et al.*, 1988) were combined to visualize individual cells in the
intact organ of Corti and to measure their vibration in response to sound
applied to the ear (ITER, 1989). This system allows us to explore the
relationship between the vibrations of different cochlear structures and
to define their micromechanical roles.

Our initial studies of the cochlea were carried out in the apical turn of
the guinea-pig temporal bone immersed in tissue culture medium
(Ulfendahl *et al.*, 1989). This preparation has gone through a succession
of improvements during the last five years, including (i) oxygenation of
the immersion fluid, (ii) perfusion of oxygenated tissue culture medium

through scala tympani (Flock *et al.*, manuscript in preparation) and (iii) a non-immersed middle ear. These steps have resulted in more sharply tuned mechanical responses, higher characteristic frequencies for a given location and increased stability of the response.

An important issue is how well an isolated preparation represents the function of the hearing organ in the living animal. Several studies of the function of the sensory hair cells have been made using preparations ranging from isolated outer hair cells (OHCs) maintained in tissue culture medium (Brownell *et al.*, 1985; Canlon *et al.*, 1988; Hudspeth and Corey, 1977; Zenner, 1986) to the intact organ of a living animal, e.g. the fish lateral line (van Netten and Khanna, 1993) and the mammalian basilar membrane observed through the round window (Johnstone and Boyle, 1967; Khanna and Leonard, 1982; Rhode, 1970; Robles *et al.*, 1986; Ruggero *et al.*, 1991). Using the isolated preparations, important information about the sensory cells has been obtained, including electromotility (Ashmore, 1987; Brownell *et al.*, 1985; Dallos *et al.*, 1991), acoustically induced motile responses (Brundin *et al.*, 1991; Canlon *et al.*, 1988) and transduction mechanisms (Hudspeth, 1982).

The temporal bone preparation provides excellent optical access to the organ of Corti for studying cochlear function. The favorable optical conditions allow the visualization, positive identification and vibration measurement of many structures. Spontaneous vibrations of cells have been observed (Keilson *et al.*, 1993; Khanna *et al.*, 1993), and it has been shown that both summation potentials and cochlear microphonic responses are present in the isolated preparation (Brundin *et al.*, 1992). The vibratory and displacement responses have been extensively recorded (e.g. Brundin *et al.*, 1991; ITER, 1989; Khanna *et al.*, 1990; Ulfendahl *et al.*, 1993; Ulfendahl and Khanna, 1993). The displacement response and the receptor potentials can be reversibly altered by methylene blue administration (Brundin *et al.*, 1991). The vibratory response is altered by acoustic overstimulation in a way similar to that observed in intact ears (Ulfendahl *et al.*, 1993). These diverse observations suggest that many of the key responses observed by other investigators in the inner ears of living animals are present in the apical turns of the temporal bone preparation.

In the original temporal bone preparation (Ulfendahl *et al.*, 1989) the middle ear is fluid-filled. The immersion causes a sharp drop off in the middle ear response at frequencies above 1.5 kHz due to the fluid loading of the tympanic membrane. As a result the experiments were limited to the third and fourth turns where the frequencies of interest lie below 1.5 kHz and calibration is required to calculate what the response would be in an air-filled middle ear. However, when comparing measurements made in the same cochlea, immersion effects are common to all measurements and therefore cancel out.

To obtain the high sensitivity of hearing, the acoustic energy picked up by the external ear must be efficiently transferredthrough the middle ear and the inner ear fluids to the sensory cells in the organ of Corti. One consequence of this requirement is that tight coupling must exist all along the path through which the stimulus is transmitted. Thus, vibration induced at any point along this path will be observed both in a forward as well as in a backward direction towards the middle ear. This tight mechanical coupling between the adjacent elements makes it difficult to determine where a specific response originates in the organ.

It is well established that the main afferent pathway for the neural signals originates at the inner hair cells (IHCs) (Spoendlin, 1972). An important question therefore is: 'How are the inner hair cells stimulated?' The IHC is completely surrounded by supporting cells (inner border cells) except at the apex (Spoendlin, 1966). The apical end of the IHC is in close proximity to the inner pillar cell. The inner border cells rest on top of the outer edge of the osseous spiral lamina (OSL). The vibration amplitude of the OSL has been measured by us and found to be negligible (unpublished observations); the bases of the IHCs can thus be considered to be stationary.

It is generally accepted that the effective mechanical stimulus that activates the transduction channels of the hair cells is the bending of the stereocilia bundle (Flock, 1965; Hudspeth and Corey, 1977). The tall rows of the OHC stereocilia are embedded in the tectorial membrane and the shearing motion between the tectorial membrane and the cuticular plate bends the stereocilia.

The IHC stereocilia are freestanding or loosely coupled to the tectorial membrane (for review see Lim, 1986). If they are freestanding, their bending would be produced by the relative motion of the stereocilia and the subtectorial fluid. If they are loosely coupled to the tectorial membrane, the bending mechanism would be similar to that for the outer hair cells.

Methods

The method used in the preparation of the guinea-pig temporal bone has been described previously (Ulfendahl *et al.*, 1989). The preparation was immersed in oxygenated tissue culture medium and perfused through the scala tympani (Flock *et al.*, manuscript in preparation). The techniques used in the visualization of the cells, application of sound and measurement of the cellular vibration response have been described previously (ITER, 1989; Koester *et al.*, 1994; Ulfendahl *et al.*, 1989, 1991). Glass beads (diameter approx. 9 μm) were placed on the tectorial membrane after partly removing Reissner's membrane (Ulfendahl *et al.*, 1994). The effects of opening scala media on the tectorial membrane motion are not known and will be checked in future experiments. The velocity of the beads placed on the tectorial membrane and the velocity of the apical ends of hair cells were measured in response to sinusoidal signals applied to the ear at frequencies between 25 and 1500 Hz, at sound pressure levels between 94 and 104 dB. The signals were generated and the response was recorded using a 16 bit digital signal processing system (Ariel DSP16). The buffer size was 1024 bins and the sampling rate was 25 kilosamples/sec. The responses were averaged 20 times and the total signal duration was 0.82 sec for each frequency. Vibration measurements were carried out at a viewing angle from which the IHCs could be clearly seen (vertical goniometer angle setting 0 deg—all angles referred to in this paper are with respect to this zero) and after rotation of the vertical goniometer by –19.8 and +15 deg. The microscope is focused precisely at the center of rotation of the goniometers. Their rotation therefore changes the angle at which a selected cell is viewed and at which its vibration is measured. The radial axis of the cochlea through the points of measurement was in the vertical goniometer rotation plane. The angle of the plane of the reticular lamina with respect to the vertical goniometer zero was not measured, and therefore the direction of the measured motion cannot be related precisely to the cochlear anatomy.

   In some experiments beads were found on the basilar membrane. Velocity measurements were made at these beads, at the adjacent basilar membrane regions and at the Hensen's cells located along the same radial axis.

Observations

   Stability of response is important when measuring the differential vibration of structures with very similar amplitudes. Two tuning curves measured 86 min apart at the same Hensen's cell are shown in Fig. 1, illustrating the excellent repeatability of the response. The repeatability is generally even better than that shown in the figure, but repeated measurements at the same location are seldom made when the response is good. The high repeatability makes it possible to carry out some of the measurements described in this paper.

   The shape of the tuning curves observed from beads placed on the tectorial membrane and from the reticular lamina below the beads was

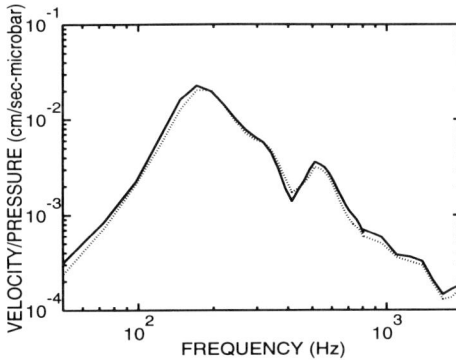

Fig. 1. Two tuning curves (velocity per unit sound pressure) recorded in the fourth turn from a Hensen's cell at the inner edge, in response to sinusoidal stimuli applied to the tympanic membrane. Curve measured near the beginning of the experiment (solid line) and 86 min later (dotted line). The response is highly stable.

similar. This was true for all radial locations of the beads. Tuning curves measured from a bead placed on the tectorial membrane (solid line) and at the reticular lamina, from a nearby second row OHC (dotted line) are shown in Fig. 2. The two curves are quite close and the shape of the tuning curves is quite smooth. These observations are typical of our experiments (over 100 cells) and lead to the conclusion that the placement of beads in this region of the cochlea does not alter the shape of the tuning curve.

When tuning curves are measured on the basilar membrane along a radial direction outwards from the Hensen's cell, the shape of the

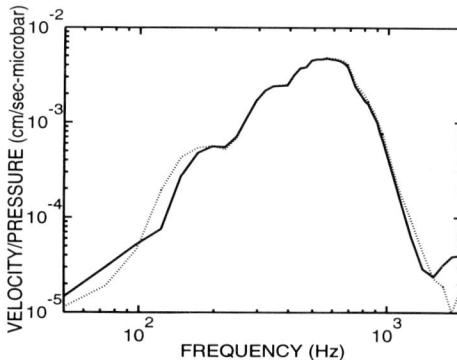

Fig. 2. Tuning curve measured in the third turn from a bead placed on the tectorial membrane (solid line) and from a second row OHC (dotted line) near the bead. The shapes and the magnitudes of the two curves are quite close.

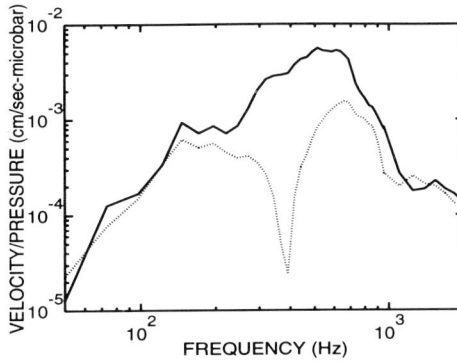

Fig. 3. Tuning curve measured in the third turn from a Hensen's cell (solid line) on the outer edge and from a bead (dotted line) placed on the basilar membrane just outside and below the edge of the Hensen's cells. The tuning curves are quite different in shape and magnitude and there is a sharp minimum at 390 Hz, in the response measured at the bead.

tuning curves observed is similar to that at the Hensen's cells, but their magnitude decreases with distance. However, when a bead is placed on the basilar membrane the shape of the tuning curve measured from the bead is altered. Figure 3 shows a tuning curve measured in the third turn from a Hensen's cell at the outer edge (solid line). A tuning curve recorded from a bead on the basilar membrane located about 20 μm from the edge of the Hensen's cells is also shown (dotted line). The curve from the bead shows a sharp notch centered at about 390 Hz. Similar notches were seen in other experiments; however, the frequency and the magnitude of the notch depend on the radial location of the bead and may also depend on the mass of the bead. The effect of bead placement on the basilar membrane was found to be local. Tuning curves measured directly on the basilar membrane adjacent to the location of the bead do not show a notch and the shape of the curves is similar to that measured at the Hensen's cell along the same radial axis.

The two-dimensional velocity at stimulus frequencies of 415, 586 and 927 Hz is shown for the OHC (top row), for the IHC (middle row) and for the bead (bottom row) in Fig. 4. The vertical axis chosen to illustrate the two-dimensional motion lies along the direction of maximum

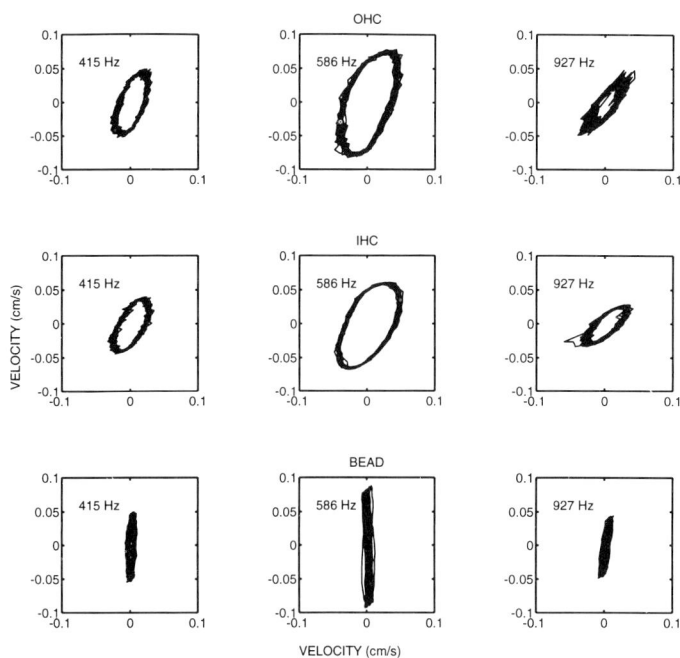

Fig. 4. Velocity plots in two dimensions: of a first row OHC (top), of an adjacent IHC (middle) and of a bead on the tectorial membrane just above the IHC (bottom). The vertical axis represents the direction of maximum response of the bead and the horizontal axis, a direction perpendicular to it. Velocity is shown at stimulus frequencies of 415, 586 and 927 Hz. The velocity plots are quite similar for the OHCs and IHCs, but different for the tectorial membrane where the vertical component is dominant.

vibration of the bead. At the characteristic frequency (586 Hz) the direction of maximum vibration was +31 deg for the bead, +57 deg for the OHC and +60 deg for the IHC (all with respect to 0 deg of the goniometer). The direction of maximum vibration is frequency dependent as seen in this figure.

The vibration of the bead is predominantly along the vertical axis, while the OHCs and IHCs have components in both horizontal as well as the vertical directions. The magnitude of vibration at the OHC is higher than that at the IHC as seen previously (ITER, 1989). The vertical components at the tectorial membrane and the OHC are quite similar, indicating that the tectorial membrane motion in the vertical direction simply follows the motion of the OHC beneath. The horizontal motion is, however, quite different at the two locations. The bead shows very little horizontal motion, while the magnitude of horizontal motion both at the IHCs and OHCs is comparable to its vertical motion. The relative

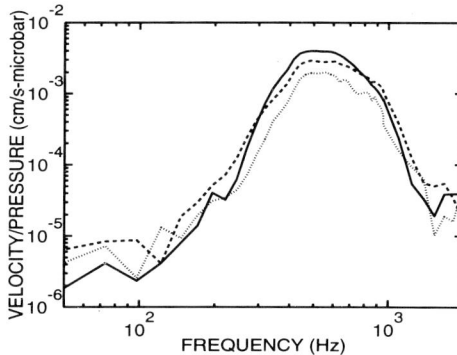

Fig. 5. The tuning curves for the difference velocity (per unit sound pressure) between a bead placed on the tectorial membrane in the third turn and the apical surface of a nearby first row OHC. The vertical component of the difference is shown as a dashed line and the horizontal component as a dotted line. Vertical and horizontal directions are defined as in Fig. 4. For comparison, the tuning curve for the vertical component of the absolute velocity (per unit sound pressure) of the OHC is shown as a solid line. Tuning curves for the difference in velocity have basically the same shape as that for the vertical vibration of the OHC.

horizontal motion between the tectorial membrane and the hair cell is thus mainly due to the horizontal motion at the apex of the hair cell.

Since the tips of the tallest stereocilia of the OHCs are embedded in

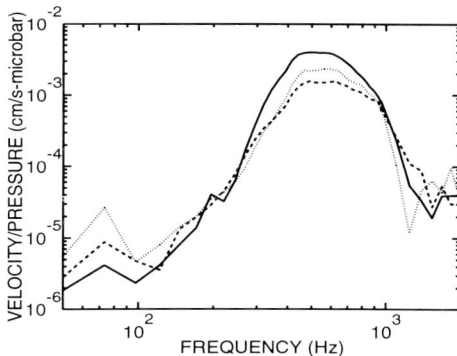

Fig. 6. The tuning curves for the difference velocity (per unit sound pressure) between a bead placed on the tectorial membrane in the third turn and the apical surface of a nearby first row IHC. The vertical component of the difference is shown as a dashed line and the horizontal component as a dotted line. Vertical and horizontal directions are defined as in Fig. 4. For comparison, the tuning curve for the vertical component of the absolute velocity (per unit sound pressure) of the adjacent first row OHC is shown as a solid line. Tuning curves for the difference in velocity have basically the same shape as that for the vertical vibration of the OHC.

the tectorial membrane, the stereocilia deflection is determined by the relative motion between the tectorial membrane and the OHCs. The tuning of this relative motion is plotted in Fig. 5. The differential vertical component (dashed line) is the difference between the vertical components of the OHC and the bead velocity that were shown in Fig. 4. Velocities are normalized for unit sound pressure. The horizontal component (dotted line) is similarly derived. For comparison the OHC (solid line) velocity tuning curve measured in the direction of maximal vibration of the bead is included. The difference components are smaller in magnitude but similar in shape to the solid curve. There appears to be no additional sharpening of tuning in the difference response.

A similar set of difference tuning curves derived from the two-dimensional motion of the bead and the IHC is shown in Fig. 6. The vertical component of relative motion is shown by the dashed lines and the horizontal component by the dotted line. For reference the solid line shows the velocity tuning curve of the OHC as in Fig. 5. The difference tuning curves for the IHCs and OHCs are very similar in shape.

Discussion

Bending of the stereocilia bundles is considered to be the fundamental mechanical stimulus for hair cell mechanosensory organs. The degree of bending of the stereocilia has not been measured in the intact organ of Corti and the determination of the stereocilia motion is therefore of fundamental importance for understanding auditory sensory transduction. The bundles are too small to be visualized directly with our present microscope. However, the bundle motion can be deduced from measurements of the motion of the tectorial membrane and of the apical ends of the OHCs below. Since the tectorial membrane is too transparent to be visualized directly with our system, it was necessary to attach microscopic glass beads to its surface in order to measure its vibration. Placing the beads on the tectorial membrane and Hensen's cell region did not appear to change the normal mode of vibration. This did not surprise us as, in our view, the OHCs amplify the basilar membrane motion and drive the reticular lamina (ITER, 1989). An incidental observation, however, was that the beads placed on the basilar membrane did alter its tuning. The beads load the basilar membrane and change the shape of the tuning curve locally. Therefore vibrations measured from beads placed on the basilar membrane may not reflect its true mechanical characteristics.

By viewing the two-dimensional motion of the tectorial membrane and OHCs, we can see that the vertical vibrationof the tectorial membrane follows the vertical motion of the OHCs below. However, the OHC vibration has both vertical and horizontal components. The horizontal component of motion between the tectorial membrane and the OHC is therefore produced mainly by the motion of the OHCs. The horizontal motion of the hair cells is not transmitted to the tectorial membrane due to the bending of the stereocilia.

The apical ends of the IHCs are driven by the OHCs via the mechanical coupling provided by the reticular lamina. The stimulation of the IHCs may take place either through bending of their stereocilia by shearing forces (if they are loosely coupled to the tectorial membrane), or by the movement of the stereocilia in the sub tectorial fluid (if they are free standing). Our present experiments do not allow us to distinguish between these two possibilities.

No additional sharpening is seen in the differential response as compared to the tuning curves measured at the apical end of the OHCs. This does not agree with some passive models of the cochlear mechanics (Allen, 1980; Zwislocki, 1979) in which tectorial membrane resonance is invoked to produce a sharpening of the mechanical response.

Acknowledgements

This research was supported by a program project grant from NIDCD, the Emil Capita Foundation, the Swedish Medical Research Council (0246, 9888) and the Karolinska Institute.

References

Allen, J. B. (1980). Cochlear micromechanics – a physical model of transduction. *J. Acoust. Soc. Am.* **68**: 1660–1670.

Ashmore, J. F. (1987). A fast motile response in guinea-pig outer hair cells: the cellular basis of the cochlear amplifier. *J. Physiol.* **388**: 323–347.

Brownell, W. E., Bader, C. R., Bertrand, D. and de Ribaupierre, Y. (1985). Evoked mechanical response of isolated outer hair cells. *Science* **227**: 194–197.

Brundin, L. and Russell, R. (1994). Tuned phasic and tonic motile responses of isolated outer hair cells to direct mechanical stimulation of the cell body. *Hearing Res.* **73**: 35–45.

Brundin, L., Flock, Å., Khanna, S. M. and Ulfendahl, M. (1991). Frequency specific position shift in the guinea pig organ of Corti. *Neurosci. Lett.* **128:** 77–80.

Brundin, L., Flock, B. and Flock, Å. (1992). Sound induced displacement of the guinea pig organ and its relation to cochlear potentials. *Hearing Res.* **58:** 175–184.

Canlon, B., Brundin, L. and Flock, Å. (1988). Acoustic stimulation causes tonotopic alterations in the length of outer hair cells from the guinea pig hearing organ. *Proc. Natl. Acad. Sci. USA* **85:** 7033–7035.

Dallos, P., Evans, B. N. and Hallworth, R. (1991). Nature of the motor element in electrokinetic shape changes of cochlear outer hair cells. *Nature* **350:** 155–157.

Flock, Å. (1965). Transducing mechanisms in the lateral line canal organ receptors. *Cold Spring Harbor Symp. Quant. Biol.* **XXX:** 133–145.

Hudspeth, A. J. (1982). Extracellular current flow and the site of transduction by vertebrate hair cells. *J. Neurosci.* **2:** 1–10.

Hudspeth, A. J. and Corey, D. P. (1977). Sensitivity, polarity and conductance change in the response of vertebrate hair cells to controlled mechanical stimuli. *Proc. Natl. Acad. Sci. USA* **74:** 2407–2411.

International Team for Ear Research (ITER). (1989). Cellular vibration and motility in the organ of Corti. *Acta Otolaryngol. Suppl.* **467:** 7–279.

Johnstone, B. M. and Boyle, A. J. F. (1967). Basilar membrane vibrations examined with the Mössbauer technique. *Science* **158:** 390–391.

Keilson, S. E., Khanna, S. M., Teich, M. C. and Ulfendahl, M. (1993). Spontaneous cellular vibrations in the guinea pig cochlea. *Acta Otolaryngol.* **113:** 591–597.

Khanna, S. M. and Leonard, D. G. B. (1982). Basilar membrane tuning in the cat cochlea. *Science* **215:** 305–306.

Khanna, S. M., Ulfendahl, M. and Flock, Å. (1990). Cellular mechanical responses in the cochlea. In *Cochlear Mechanisms and Otoacoustic Emissions* (F. Grandori, G. Cianfrone and D. T. Kemp, eds), *Adv. Audiol.* **7:** 13–26. Karger, Basel.

Khanna, S. M., Keilson, S. E., Ulfendahl, M. and Teich, M. C. (1993). Spontaneous cellular vibrations in the guinea pig temporal bone preparation. *Br. J. Audiol.* **27:** 79–83.

Koester, C. J., Khanna, S.M, Rosskothen, H. D., Tackaberry, R. B. and

Ulfendahl, M. (1994). Confocal slit, divided aperture microscope: applications in ear research. *Appl. Opt.* **33**: 702–708.

Lim, D. J. (1986). Functional structure of the organ of Corti: a review. *Hearing Res.* **22**: 117–146.

Rhode, W. S. (1970). Observations of the vibration of the basilar membrane in squirrel monkeys using Mössbauer technique. *J. Acoust. Soc. Am.* **49**: 1218–1231.

Robles, L., Ruggero, M. A. and Rich, N. C. (1986). Basilar membrane mechanics at the base of the chinchilla cochlea. I. Input output functions, tuning curves and response phases. *J. Acoust. Soc. Am.* **80**: 1364–1374.

Ruggero, M. A. and Rich, N. C. (1991). Application of a commercially-manufactured Doppler shift laser velocimeter to the measurement of basilar membrane vibration. *Hearing Res.* **51**: 215–230.

Spoendlin, H. (1966). The organization of the cochlear receptor. *Adv. Otorhinolaryngol.* **13**: 1–227.

Spoendlin, H. (1972). Innervation densities of the cochlea. *Acta Otolaryngol.* **73**: 235–248.

Ulfendahl, M. and Khanna, S. M. (1993). Tuning characteristics of the hearing organ in an *in vitro* preparation of the gerbil temporal bone. *Pflüger's Arch.* 424, 95–104.

Ulfendahl, M., Flock, Å. and Khanna, S. M. (1989). A temporal bone preparation for the study of cochlear micromechanics at the cellular level. *Hearing Res.* **40**: 55–64.

Ulfendahl, M., Khanna, S. M. and Flock, Å. (1991). Effects of opening and resealing the cochlea on the mechanical response in the isolated temporal bone preparation. *Hearing Res.* **57**: 31–37.

Ulfendahl, M., Khanna, S. M. and Löfstrand, P. (1993). Changes in the mechanical tuning characteristics of the hearing organ following acoustic overstimulation. *Eur. J. Neurosci.* **5**: 713–723.

Ulfendahl, M., Khanna, S. M. and Heneghan, C. (1994). Shearing motion between the auditory hair cells and the overlying tectorial membrane measured by laser heterodyne interferometry. Submitted.

van Netten, S. M. and, S. M. Khanna (1993). Mechanical demodulation of hydrodynamic stimuli performed by the lateral line organ. In *Progress in Brain Research* (J. H. J. Allum, ed.)., Vol.97, pp. 45–51. Elsevier, Amsterdam.

Willemin, J. F., Dändliker, R. and Khanna, S. M. (1988). Heterodyne interferometer for submicroscopic vibration measurements in the inner ear. *J. Acoust. Soc. Am.* **83**: 787–795.

Zenner, H. P. (1986). Motile responses in outer hair cells. *Hearing Res.* **22**: 83–90.

Zwislocki, J. J. and Kletsky, E. J. (1979). Tectorial membrane: a possible effect on frequency analysis in the cochlea. *Science* **204**: 639–641.

Micromechanics of Cellular Structures in the Mammalian Cochlea: Auditory and Electrical Stimulation

ANTHONY W. GUMMER, WERNER HEMMERT AND HANS-PETER ZENNER

Section of Physiological Acoustics and Communication, Department of Otolaryngology, University of Tübingen, Silcherstraße 5, 72076 Tübingen, Germany

Introduction

The mechanical impedances of the organ of Corti are unknown; of particular interest for cochlear tuning are those of the basilar membrane (BM), tectorial membrane (TM) and outer hair cells (OHCs). Theoretically, however, it should be possible to use the electromotile properties of the OHCs to determine the impedances. Thus, if sinusoidal current injected locally into the organ of Corti causes sinusoidal length changes of the OHCs and if forces were consequently exerted on the BM and TM, then under ideal conditions, measurement of the velocity of these structures would yield their mechanical impedance. Evidence in support of this approach is to be found in the most recent literature. First, in response to an externally applied electric field the evoked displacement of isolated OHCs in a microchamber is independent of stimulus frequency up to at least 24 kHz (Dallos, 1994). If the mechanical impedance of the OHC were stiffness controlled in the frequency range of interest, then the OHC would act as a constant force generator. Generally speaking, instead of the pressure generator – the loudspeaker – being located in the external ear, as it normally is for acoustic stimulation, it is located within the cochlear partition for electrical

271

stimulation – the OHCs are the electromechanical transducers. Second, the voltage-displacement sensitivity of OHC electromotility is maintained *in situ*, within the organ of Corti (Mammano and Ashmore, 1993). In a temporal bone preparation, Mammano and Ashmore (1993) showed that negative current injected into scala media causes OHC depolarization and contraction so that the BM is displaced towards scala media and the reticular lamina towards scala tympani. Third, the voltage-force sensitivity of OHC electromotility also appears to be maintained, as estimated from the *in vivo* BM response to current injection (Xue *et al.*, 1993). Fourth, electrical stimulation *in vivo* induces travelling wave motion (Nuttall and Dolan, 1993) and compound action potentials (Yates and Kirk, 1993) apical to the position of the stimulus electrode; the frequency selectivity is comparable to that of acoustic stimulation of healthy ears.

With the assumption that OHCs act as ideal electromechanical transducers in response to current injection, we estimated the frequency response of the BM, TM and Hensen's cells (HeC) by measuring their velocity in response to white-noise current. Vibration measurements were made with a laser Doppler velocimeter. The upper two turns of the guinea-pig cochlea were chosen for recordings because this region allows focusing of the laser beam on different structures without having to change the orientation of the beam. Recordings were made from an isolated temporal bone preparation with air-filled middle-ear cavities. By ensuring that the cochlea remained fluid filled and the middle ear moist in a water-saturated atmosphere, it has been shown (Gummer *et al.*, 1993) that the acoustically evoked vibration tuning curves have similar frequency selectivity to the threshold tuning curves of auditory nerve fibres and velocity tuning curves *in vivo* (Cooper and Rhode, 1993), both derived from this apical region of the cochlea. Nevertheless, for electrical stimulation the cochlea will be functioning in an open-loop condition, because under *in vitro* conditions active feedback is not intact in the absence of the *in vivo* values of the endocochlear potential.

Briefly summarized, we show that the acoustic anti-resonance at about 0.5 oct below acoustic resonance is due to mechanical resonance in the TM and propose that electromechanics of the TM may be an important feature of cochlear tuning.

Methods

Guinea-pigs (250–400 g) were decapitated after cervical dislocation, and the temporal bone was removed and cemented (Harvard) to a Delrin sound delivery cone. The temporal bone was removed ventrally to expose the cochlear apex. A sheet of parafilm was glued (Histoacryl) to the apical end of the cochlea to support mechanically a fluid droplet. A small hole was made in the cochlear wall under the droplet, leaving stria vascularis, Reissner's membrane and the helicotrema intact. The fluid droplet was Hanks' solution for those vibration measurements made before opening Reissner's membrane, and artificial endolymph for measurements made with an opened Reissner's membrane; the Hanks' solution was substituted for artificial endolymph before opening. The artificial endolymph was in mM: K 150, Na 1, Ca 2, Cl 130, HCO_3 25, with pH 7.4, 308 mOsm (Kronester-Frei, 1979).

The organ of Corti of the third and fourth turns were viewed with a microscope (Leitz Aristomet) through a Nikon ×20 water immersion objective (NA: 0.40; working distance: 3 mm; focal depth: 2 µm), positioned on the fluid droplet. Cochlear structures were discerned on a video display using a Hamamatsu (C3077) camera and controller (C2400).

Vibration was measured with a Polytec OFV-302 laser Doppler velocimeter coupled to the microscope. The diameter of the scattered laser beam (~5 µm) was less than that of an OHC (~7 µm). To make reliable measurements from the weakly reflecting surfaces of the BM and TM, reflective polystyrene microspheres (Polyscience, diameter 10 µm) were introduced onto the organ of Corti.

Sound was presented closed field from a Beyer DT48 loudspeaker and sound pressure was measured with a Brüel & Kjaer 1/4 in. condenser microphone inserted axially into the conical sound coupler. Sound pressure was flat (±5 dB) from 0.1 to 4 kHz, with measured second and third harmonics more than 40 dB and 55 dB, respectively, below the fundamental component at 110 dB SPL. For current injection, a silver wire (diameter 300 µm) was placed at the apical end of the cochlea in scala vestibuli. The ground electrode was placed in the fossa of the paraflocculus, which was filled with Hanks' solution. The amplitude spectrum of the injected current was flat (±1 dB) up to 2 kHz; harmonic distortion components were less than –60 dB for 5 µA. Acoustical or electrical stimuli consisted of band-limited white noise (0.05–2 kHz; 800 points) derived from the FFT Analyzer (AND AD-3525). The number of averages was 20 for the HeC and 100 for the TM and BM; the averaging time was about 1 sec per average.

Results

Thirty-two temporal bones were used; most data are from the fourth turn (n = 32); in three cases third-turn recordings were also made.

Acoustically and electrically evoked responses were measured from 0.1 to 2 kHz and were independent of Reissner's membrane being intact. Typically, 30 min elapsed between decapitation and the first vibration measurement. For the duration of the recording session (30–60 min), the middle-ear (acoustic) response was flat (±1 dB) up to 2 kHz and the acoustic responses of the organ of Corti remained unchanged (±2 dB) up to 2 kHz, so that acoustic responses have been referenced to constant sound pressure level near the tympanic membrane. No detectable difference was found between responses measured with or without polystyrene microspheres.

Electrically evoked resonance

Current injection evoked a resonance in the motion of the TM and HeC and an anti-resonance for the BM, all of which were vulnerable.

Figure 1 compares the acoustically and electrically evoked velocities of a HeC measured with Reissner's membrane intact. The maximum response for acoustic stimulation, called the acoustic CF, was 535 Hz. A 'sharp', acoustically evoked anti-resonance was found at 368 Hz, 0.54 oct below the CF. An anti-resonance is a common feature of vibration recordings from the fourth turn (Cooper and Rhode, 1993; Gummer *et al.*, 1993; Khanna, 1989). Centred on this anti-resonance is an electrically evoked resonance. The phase difference from the onset of the resonant region to the resonant frequency is 0.25 cycles, suggesting that the resonance is a simple mass-spring-damper. The Q_{3dB} was 2.1, implying that the resonance is underdamped with damping ratio of 0.43. Above the resonance, the velocity increased with frequency with a slope of 10 dB/oct. The resonance was vulnerable, vanishing irreversibly after 20 min (not illustrated); however, the amplitude of the HeC motion increased by at least an order of magnitude relative to the first recorded values, but the amplitude slope decreased to 4 dB/oct. The acoustic response was unchanged. In other words, the motion of the HeC became stiffness coupled. (The actual slope for a spring would be 6 dB/oct.) That the acoustic response remained robust should not be surprising because *in vitro* the organ of Corti is known to move as a single body for sound stimulation (Gummer *et al.*, 1993; Khanna, 1989) – the frequency responses of the HeCs, BM and hair cells have the same form, although their absolute magnitudes differ.

Electrically evoked responses of the TM and BM are compared in Fig. 2. The TM shows a resonance and the BM an anti-resonance, both located near the acoustic anti-resonance at 420 Hz. (The acoustic responses are not illustrated.) The fact that the TM and BM show

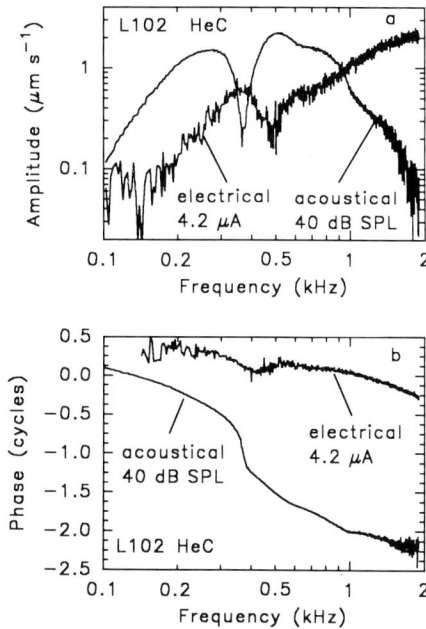

Fig. 1. Acoustically and electrically evoked motion in the organ of Corti of the isolated temporal bone preparation. (a) Amplitude, (b) phase of velocity of a Hensen's cell (HeC) in the fourth cochlear turn as a function of frequency. Reissner's membrane was intact. The stimulus spectrum was band-limited white noise with 60 dB SPL (per frequency) for sound and 4.2 µA (per frequency) for current stimulation (740 frequencies). For acoustic stimulation, velocity amplitudes have been linearly corrected to 40 dB SPL to allow easy comparison with the electrical responses. For example, at 930 Hz, a 4.2 µA current evoked an amplitude equivalent to that for sound at 40 dB SPL. Phase is negative for a lag. Velocity is positive towards scala media (towards the laser), pressure is positive for rarefraction and current is positive for injection into scala media.

opposing amplitude effects implies that the resonance is not due to a resonance in the (Thévenin equivalent) voltage source and impedance driving the OHCs, nor to a resonance in the force generator. Instead, it is more likely to be a mechanical resonance in the TM, the BM exhibiting anti-resonance because in this frequency range the TM shunts the greater part of the energy from the OHC.

For frequencies outside the resonant region, the TM and BM move in opposite directions, as evidenced by the 0.5 cycle differences in their phase responses (Fig. 2b). This is consistent with the hypothesis that electrical stimulation causes synchronous length changes of the OHCs within the organ of Corti and compares with the recent data of Mammano and Ashmore (1993). For positive current injection in scala

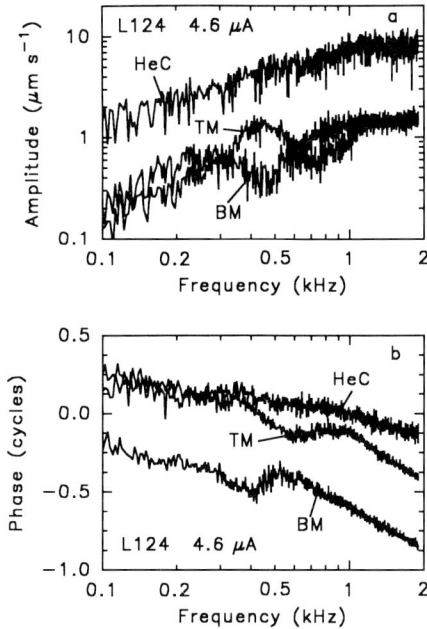

Fig. 2. Electrically evoked motion of the tectorial membrane (TM), basilar membrane (BM) and Hensen's cell (HeC) in the fourth cochlear turn. (a) Amplitude, (b) phase of velocity for 4.6 μA current injection into scala media. Reissner's membrane was opened. Recordings from the TM and BM are from polystyrene microspheres, made 10 and 11 min after the recording of the HeC. Acoustic CF: 640 Hz. Frequency of acoustic anti-resonance: 420 Hz.

media, the OHC hyperpolarizes and elongates, causing the BM to move towards scala tympani, consistent with the 0.25 cycle phase lag at low frequencies. The anti-phase motion of the TM and BM clearly excludes bulk fluid flow or electrophonics as the source of the electrically evoked resonance.

For low frequencies (<250 Hz) the slopes of the velocity amplitudes for both the TM and BM are 6 dB/oct, implying stiffness-controlled motion. The amplitude of TM velocity is 4.3 dB greater than that for the BM, implying that the BM is 1.6 times stiffer than the TM.

For frequencies below the resonant region, the HeC moves in phase with the TM, but does not exhibit resonance, presumably because the HeC motion has already become decoupled from the TM. Up to 1.2 kHz, the amplitude slope is 4 dB/oct, suggesting near stiffness-controlled motion of the HeC.

Non-resonant electrically evoked responses

Fifty-eight per cent of preparations did not show electrically evoked resonant phenomena; an example is illustrated in Fig. 3. Current

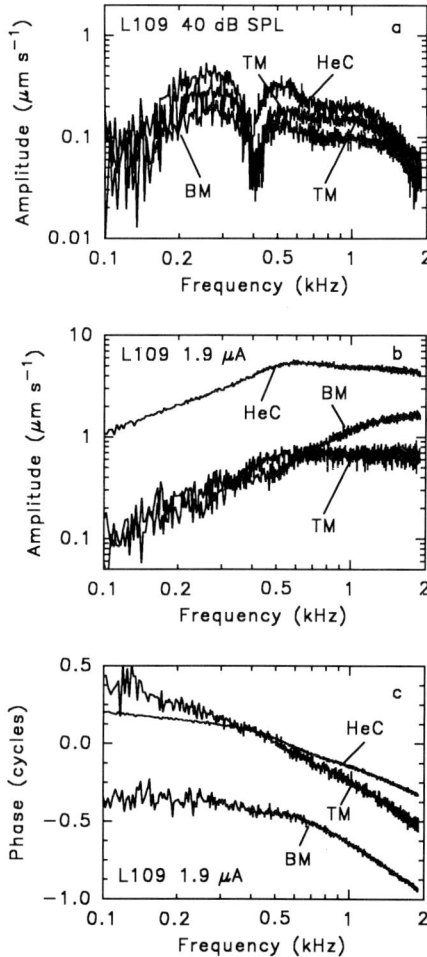

Fig. 3. Acoustically and electrically evoked motion of the tectorial membrane (TM), basilar membrane (BM) and Hensen's cell (HeC) in the fourth cochlear turn. (a) Velocity amplitude for acoustic stimulation referenced to 40 dB SPL; (b) amplitude and (c) phase of velocity for 1.9 μA current injection into scala media. Reissner's membrane was opened. Recordings from the TM and BM are from polystyrene microspheres. Acoustic CF, as determined from the HeC response, was 560 Hz and the acoustic anti-resonance was at 400 Hz.

injection caused the TM and BM to move in opposite phases (Fig. 3c), again implying synchronous length changes of the OHCs.

The electrically evoked responses of the HeC, TM and BM were 6 dB/oct below their respective 3 dB frequencies (Fig. 3b), implying stiffness-coupled motion. The 3 dB frequency served as an estimator of the resonant frequency; the estimates concurred with the frequency where the phase relative to the low-frequency asymptotic value was –0.25 cycle.

The $f_{3\,dB}$ for the TM was 385 Hz, which is approximately equal to the acoustic anti-resonance frequency (400 Hz), providing strong evidence that the TM is responsible for the acoustic anti-resonance. Above 385 Hz, the TM acts as a mass, i.e. in the acoustic CF region the OHCs deliver kinetic energy to the TM. For the BM, however, $f_{3\,dB}$ is 1033 Hz, which is well above the acoustic CF (560 Hz), i.e. near the acoustic CF, the OHCs deliver potential energy to the BM.

HeC motion is strongly coupled to the TM motion with similar $f_{3\,dB}$ (408 Hz); at this frequency they move in phase.

At low frequencies (<300 Hz), the electrically (and acoustically) evoked motion of the TM is 3 dB greater than that of the BM, meaning that the BM is 1.4 times stiffer than the TM. This, together with the ratio of their cut-off frequencies (2.68), means that the TM is five times heavier than the BM. This ratio should be treated with caution, however, because it is dependent on the measurement positions on the TM and BM. The TM location was over the inner hair cells and the BM location was just radial from the projection of the HeCs onto the Claudius cells in the pectinate zone. In order to gain insight into the possible size of the effect, one may use the electrically evoked HeC response as a measure of the upper limit of the TM response. The TM response was always less than the HeC response and the two responses exhibited similar 3 dB frequencies. The ratio of the low-frequency HeC and BM responses is 10 and the ratio of the 3 dB frequencies for the BM and the HeC is 2.53. The BM is thus estimated to be 10 times stiffer than the TM (instead of 1.4) and the mass ratio of TM to BM becomes 0.8 (instead of 5).

Electrically evoked otoacoustic emissions

Current injection evoked acoustic emissions; an example is illustrated in Fig. 4. The spectrum contains two distinct resonances, one at 610 Hz and the other an octave higher. For 4.2 µA, the resonant amplitudes were 7.5 and 5 dB SPL. The emission amplitude was linearly dependent on current amplitude up to 300 µA, implying that OHC

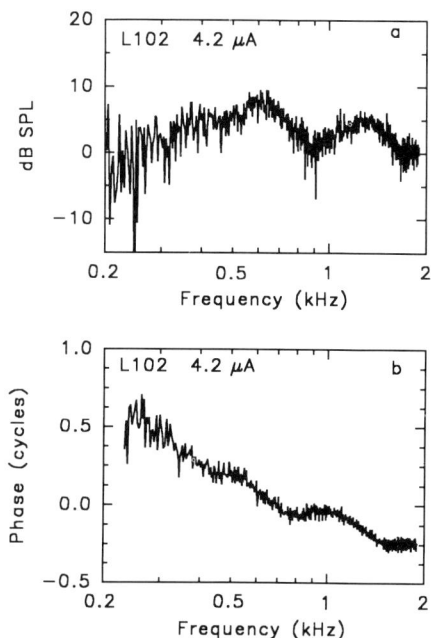

Fig. 4. Electrically evoked otoacoustic emissions measured near the tympanic membrane for 4.2 μA current injection into scala media. (a) Amplitude, (b) phase of sound pressure.

electromotility was linear for the currents used in determining the frequency responses. Consistent with the postulated role of OHCs in generating emissions, a resonant frequency of 600 Hz suggests that this spectral region was derived from vibrations initiated by OHCs in the fourth turn. Tonotopy suggests that the second peak may have been generated by OHCs of the third turn.

Discussion

Advantage has been taken of the electromotile properties of OHCs to elucidate micromechanics of the organ of Corti. In this analysis we have assumed that the OHC electromotility and its driving source voltage and impedance are flat within the frequency range of interest. Two principle findings were described. First, the acoustic anti-resonance, always found at about 0.5 oct below the acoustic CF, is almost certainly due to a radial resonance in the TM, which shunts energy away from the transverse motion of the cochlear partition. From experiments in which electrically evoked resonance was not detected, current injection showed that the mechanical resonant frequency of the TM was located at the acoustic anti-resonance. Moreover, the mechanical resonant frequency of the BM was about one octave above the acoustic CF, meaning that near acoustic resonance, TM motion is mass coupled and BM motion is stiffness coupled. The same conclusion for the TM was reached by Zwislocki and Cefaratti (1989), based on their *in vivo* stiffness measurements of the TM. It also supports Allen's resonant TM model of enhanced cochlear tuning, in which the TM is tuned to about 0.5 oct below the CF, where it introduces an anti-resonance (Allen, 1980; Allen and Fahey, 1993).

The second principle finding was an electrically evoked resonance in the TM response located at the acoustic anti-resonance. Since HeC motion is coupled to that of the TM, HeCs also exhibited the resonance phenomena. The BM, however, exhibited an electrically evoked anti-resonance. The electrically evoked resonant phenomena seemed to be a feature of 'good' preparations; it was vulnerable, vanishing about half an hour after beginning the electrical stimulation experiments. It cannot be decided from these experiments whether the effect is an epiphenomenom related to the experimental configuration, or whether it is a manifestation of hitherto unexplored electromechanical processes in the organ of Corti.

It is proposed that the electrically evoked resonance is due to electromechanical processes in the TM. The TM was resonant but the BM anti-resonant, suggesting that the TM shunted energy away from the BM. Moreover, the electrically evoked resonance and anti-resonance were both located at the acoustic anti-resonance, which in turn was probably due to a mechanical TM resonance. There is evidence for the presence of negative fixed charges on the fibrils of the TM – *in vitro* experiments have shown that the TM supports a Donnan equilibrium, with the TM negative to extracelluar fluid (Steel, 1983). The osmotic pressure in the TM would then be sensitive to changes of ionic concentrations. Consequently, current injection may well have caused fluctuations in the TM osmotic pressure, which may have significantly affected both its mass and elasticity. Given the extreme sensitivity of the TM to changes of ionic concentrations (Kronester-Frei, 1979), it is hardly surprising that the electrically evoked resonance was vulnerable, vanishing irreversibly. On the other hand, whether osmotic pressure changes are sufficiently fast remains to be determined.

These experiments suggest that electromechanics of the TM might be an important feature of cochlear tuning.

Acknowledgements

This work was supported by the Deutsche Forschungsgemeinschaft, SFB 307, Teilprojekt C10.

References

Allen, J. B. (1980). Cochlear micromechanics – a physical model of transduction. *J. Acoust. Soc. Am.* **68:** 1660–1670.

Allen, J. B. and Fahey, P. F. (1993). A second cochlear-frequency map that correlates distortion product and neural tuning measurements. *J. Acoust. Soc. Am.* **94:** 809–816.

Cooper, N. P. and Rhode, W. S. (1993). Nonlinear mechanics at the base and apex of the mammalian cochlea: *in vivo* observations using a displacement-sensitive laser interferometer. In *Biophysics of Hair Cell Sensory Systems* (H. Duifhuis, J. W. Horst, P. van Dijk and S. M. van Netten, eds), pp. 249–257. World Scientific, Singapore.

Dallos, P. (1994). Biophysics of outer hair cell motility. In *Association for Research in Otolaryngology*, February 6–10, p. 166.

Gummer, A. W., Hemmert, W., Morioka, I., Reis, P., Reuter, G. and Zenner, H.-P. (1993). Cellular motility in the guinea-pig cochlea. In *Biophysics of Hair Cell Sensory Systems* (H. Duifhuis, J. W. Horst, P. van Dijk and S. M. van Netten, eds), pp. 229–236. World Scientific, Singapore.

Khanna, S. M. (1989). Cellular vibration and motility in the organ of Corti. *Acta Otolaryngol. Suppl.* 467.

Kronester-Frei, A. (1979). The effect of changes in endolymphatic ion concentrations on the tectorial membrane. *Hearing Res.* **1:** 81–94.

Mammano, F. and Ashmore, J. F. (1993). Reverse transduction measured in the isolated cochlea by laser Michelson interferometry. *Nature* 365, 838–841.

Nuttall, A. F. and Dolan, D. F. (1993). Basilar membrane velocity responses to acoustic and intracochlear electric stimuli. In *Biophysics of Hair Cell Sensory Systems* (H. Duifhuis, J. W. Horst, P. van Dijk and S. M. van Netten, eds), pp. 288–294. World Scientific, Singapore.

Steel, K. P. (1983). Donnan equilibrium in the tectorial membrane. *Hearing Res.* **12:** 265–272.

Xue, S., Mountain, D. C. and Hubbard, A. E. (1993). Direct measurement of electrically-evoked basilar membrane motion. In *Biophysics of Hair Cell Sensory Systems* (H. Duifhuis, J. W. Horst, P. van Dijk and S. M. van Netten, eds), pp. 361–369. World Scientific, Singapore.

Yates, G. K. and Kirk, D. L. (1993). Electrically evoked travelling waves in the guinea pig cochlea. In *Biophysics of Hair Cell Sensory Systems* (H. Duifhuis, J. W. Horst, P. van Dijk and S. M. van Netten, eds), pp. 352–359. World Scientific, Singapore.

Zwislocki, J. J. and Cefaratti, L. K. (1989). Tectorial membrane II: Stiffness measurements *in vivo*. *Hearing Res.* **42:** 211–228.

Basilar Membrane Motion and Position Changes Induced by Direct Current Stimulation

A. L. NUTTALL, W. J. KONG, T. Y. REN AND D. F. DOLAN

Kresge Hearing Research Institute, University of Michigan, 1301 East Ann Street, Ann Arbor, MI 48109-0506, USA

Introduction

The study of *in vivo* electromotility has been facilitated by the experimental procedure of electric current stimulation of the cochlea. Hubbard and Mountain (1983) showed that alternating current applied to the cochlea produces otoacoustic emissions and, when mixed with an acoustic stimulus, produces otoacoustic distortion products. Nuttall (1985) has shown that direct current when applied across the cochlea duct altered the frequency tuning curve of inner hair cells (IHCs), increasing and decreasing their characteristic frequency and sensitivity. The mechanism for these changes was hypothesized to be outer hair cell (OHC) dependent.

We have reported recently (Nuttall *et al.*, 1994) that direct current applied to the cochlea enhances or decreases the velocity responses of the basilar membrane (BM) to acoustic pure tones. The changes in the mechanical response of the BM are consistent with and account for the earlier findings of Nuttall (1985). In general, studies that show current-induced sounds of cochlear origin or BM vibration support the hypothesis of a functional electromotility in the organ of Corti. Cellular forces induced by electric current stimulation must be sufficient to move organ of Corti structures, resulting in otoacoustic emissions or tonal-like traveling wave responses on the BM. Current stimulation also modifies the responsiveness of the inner ear by electrically influencing the electromechanical 'gain' of the organ of Corti. The current work investigates the dynamic response of organ of Corti gain changes. An important goal of this study is to provide direct evidence of electromotility *in vivo* and to begin to characterize the micromechanical motion of cellular structures produced by this organ of Corti electromotility.

Methods

Young pigmented guinea-pigs with body weights between 300 and 500 g were used in this study. They were anesthetized with pentobarbital (15 mg/kg i.p.)/ Innovar-vet (0.4 ml/kg i.m.) and prepared surgically to measure BM velocity responses. Surgery was performed to create an opening of the ventral lateral surface of the auditory bulla to expose the cochlea with preservation of the ossicles and the tympanic membrane. The tensor tympani and the stapedius muscles were sectioned. The cochlea was opened in the scala tympani area of the first turn to allow visualization of the BM. Gold-coated glass microbeads of approximately 20 μm in diameter were dropped onto the scala tympani surface of the BM to serve as reflectors for the laser beam of the laser Doppler velocimeter (LDV) (Polytec OFV 1102 Vibrometer). The velocimeter was coupled to a compound microscope and focused onto the microbeads (the surgical details and velocity method are described more fully in Nuttall *et al.*, 1991).

A wire electrode placed on the round window (RW) and a ground electrode on the neck were used to monitor cochlear sensitivity. The compound action potential of the cochlea was measured for tone bursts between 2 and 40 kHz. Platinum–iridium wire electrodes were placed on the RW and into a hole drilled in the scala vestibuli of the first turn. Electric current pulses were passed from a BAK (BSI-1) opto-isolated stimulator under computer control. Current pulses ranged to 600 μA generated as rectangular pulses of 1.5 msec duration with a repetition rate between 10 and 40 per second. A microcomputer was used, with custom software, to average the velocity responses of the BM (256–2048 sweeps) sampled at 8 μsec for 512 points. The LDV was used to derive both velocity and displacement responses of the BM. In this instrument, displacement responses are obtained from a standard Mach–Zehnder interferometer with a sensitivity of 1/80 of a wavelength. Signal averaging increases the usable displacement range by at least a factor of 10. The effective displacement noise floor is less than 0.5 nm.

The animals were cannulated and maintained under anesthesia without artificial respiration. The EKG was monitored. Body temperature was controlled with a servo-regulated heated pad, and head and cochlea temperature was controlled independently by a head-holder

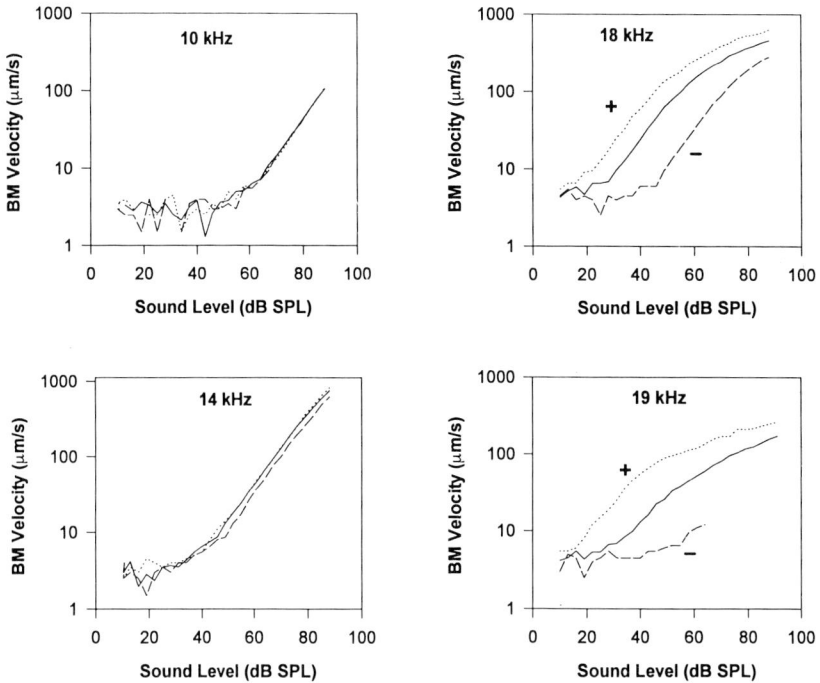

Fig. 1. The four panels of this figure show BM velocity in responses to continuous tones of increasing levels at four different frequencies. The solid curve in each panel is the control I/O function taken in the absence of current stimulation. Direct current application (±200 μA) shifts the I/O functions for frequencies near the characteristic frequency of 18 kHz.

heater and heat lamp. The animal experimental procedures were reviewed and approved by the University of Michigan Committee for the Use and Care of Laboratory Animals.

Results

Figure 1 shows the influence of direct current applied across the cochlea duct on the velocity responses of the BM to pure tone (sound). Each panel of this figure shows input/output (I/O) functions of BM velocity for varying sound pressure levels. The current has very little influence on acoustic frequencies well below the characteristic frequency of approximately 18 kHz (Note: 10 and 14 kHz). At frequencies above the characteristic frequency the influence of the current is relatively greater (compare 18 to 19 kHz). Positive current enhances the

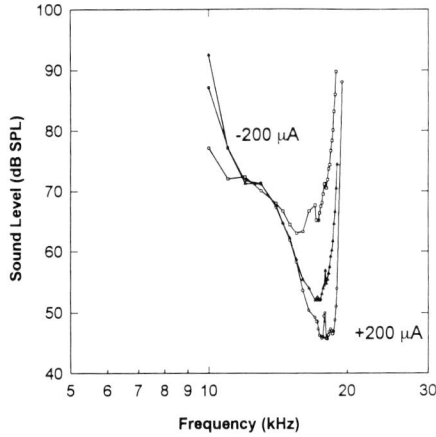

Fig. 2. Frequency tuning curves derived from I/O functions shifted by ±200 μA current (as shown in Fig. 1). A 50 μm/sec criterion velocity was used.

velocity response; it is less effective than is negative current in decreasing the response of the BM. The changes in the I/O functions result in dramatic shifts (in this case 50 μsec) of the isovelocity frequency tuning curves derived from the I/O function data. Figure 2 shows that a positive current of 200 μA both increases the sensitivity and raises the characteristic frequency of the isovelocity tuning curve. In contrast, negative current shifts the characteristic frequency lower, reduces the sensitivity and broadens the tuning curve. The changes induced by negative current are similar to that produced by hypoxia or other variables that reduce cochlear sensitivity. For example, hypoxia reduces IHC tuning (e.g. Brownell, 1983) and furosemide reduces BM tuning (Ruggero and Rich, 1991).

We have shown (Nuttall and Dolan, 1993a,b) that in the absence of acoustic stimulation the electric current itself produces a ringing response of the BM. The inset figure in Fig. 3 shows a transient ringing response to a current pulse of approximately 1.4 msec duration at 300 μA. The ringing response *during* the time of the positive current injection is enhanced (i.e. made larger) by the current. The later (in time) ringing response follows the offset of current. This ringing response is smaller in magnitude. A time–frequency distribution (the exponential distribution of Williams and Jeong, 1992) of the transient signal is shown in the main portion of Fig. 3. The frequency of the peak energy during positive current application is 19 kHz. The later-in-time energy peak is smaller and has a frequency of 18.1 kHz. This frequency (18.1 kHz) is the natural frequency for this tonotopic location on the BM

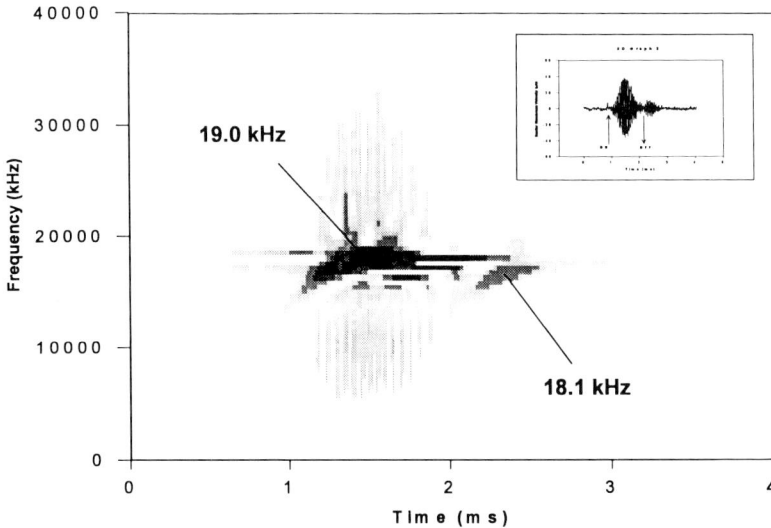

Fig. 3. A time–frequency distribution of the energy in the transient velocity response of the BM produced by a +300 μA rectangular pulse of current. The energy peak during positive current application is higher in level and in frequency than for the offset response following current application.

when measured with steady-state signals. Figure 4 shows a similar response for negative current application. During the time of negative current, the ringing responses of the BM are reduced as can be seen in the inset of Fig. 4. The time–frequency distribution shown in Figure 4 indicates that the frequency of the energy peak during negative current application is approximately 16 kHz, while the offset response energy peak was again 18.1 kHz. These changes in the resonant frequency of the organ of Corti are consistent with the shifts in the frequency tuning curve obtained with steady-state tones.

It is instructive to examine the dynamic change of an acoustic response during the application of electric current to the cochlea. Figure 5 shows that the increase in BM velocity which occurs during a positive current application is a complex function in time. There is an onset overshoot which is larger in size and earlier in time than can be accounted for by the BM velocity response due to the current application alone. For comparison, the velocity responses of the BM due to the current alone are shown as a solid line curve within the acoustic toneburst waveform.

Figure 5 also shows that positive current application results in a shift

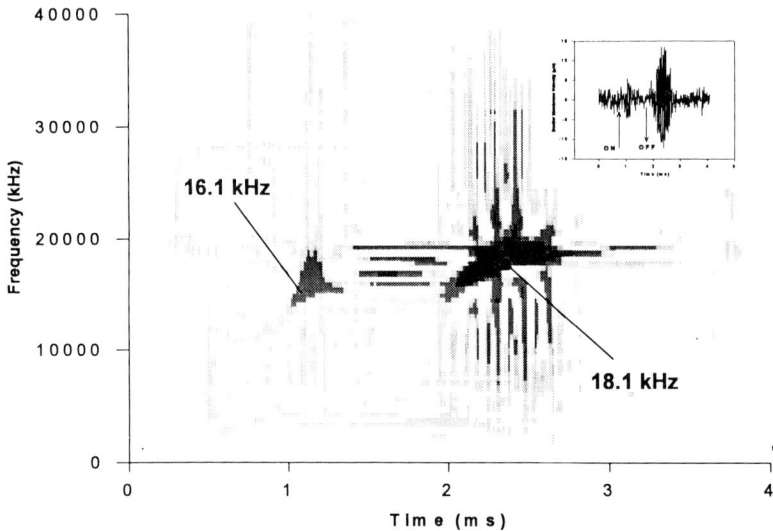

Fig. 4. A time–frequency distribution of the energy and the transient velocity response of the BM caused by a –300 µA current pulse. The energy of the transient response during negative current application is lower in level and in frequency than the offset response.

in position of the BM. This displacement was approximately 1 nm (in this case, with a +300 µA current) and in the scala tympani direction.

Other irregularities in the waveform are due to the large amount of 'baseline shift' due to animal motion and uncontrolled vibration of the LDV/ microscope. Figure 6 illustrates the influence of negative current injection on the acoustic velocity response. Negative current caused profound reduction in BM velocity, which was also accompanied by a shift in BM position of approximately 1 nm in the scala tympani direction. This direction of motion (i.e. scala tympani) is the expected direction for the hypothesized OHC elongation with hair cell hyper-polarization from negative current. The onset and offset time constants were not analyzed in this study.

Discussion

Alternating and direct current applied into the cochlea cause ringing responses of the BM (Nuttall and Dolan, 1993a,b). We have also shown that direct current alters the sensitivity of the organ of Corti in a frequency specific way (Nuttall *et al.*, 1994). Hubbard and Mountain (1983) have determined that alternating current applied across the organ of Corti produces otoacoustic emissions at the frequency of the alternating current stimulus as well as distortion product otoacoustic emissions, if the electrical stimulus is mixed with an acoustic pure tone (Hubbard and Mountain, 1983). Xue *et al.* (1993a,b) has also directly observed BM velocity responses that are altered by current stimulation.

The influence of direct current passed across the cochlea duct is clearly quite complex. In this study, current levels to approximately 600 μA were applied between the scala vestibuli and the scala tympani in the RW area. According to the electroanatomical models of the cochlea (Honrubia *et al.*, 1976), approximately 10% of the current passed from these electrodes will follow a path through the cochlea partition. A positive voltage applied with respect to the scala tympani should lead to increased current flowing through the hair cells of the organ of Corti.

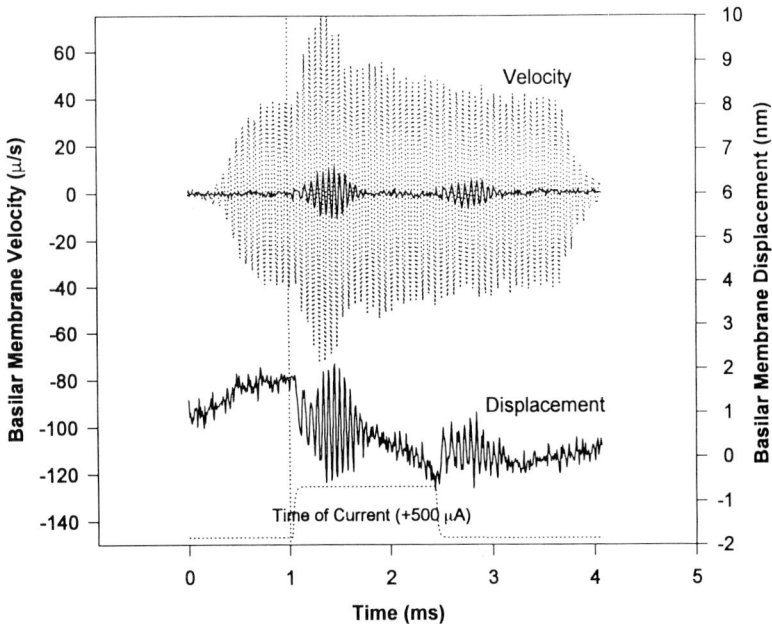

Fig. 5. The upper dotted curve shows the velocity response of the BM to a tone burst of 16.5 kHz at approximately 50 dB SPL. One millisecond following onset of the tone burst, a +500 μA direct current is passed across the cochlear duct. The upper solid curve is the transient response of the BM produced by the current alone. The time of current injection is given by the dotted rectangular pulse at the bottom of the graph. The displacement response to the current alone shows that there is a shift in position of the BM toward the scala tympani during current injection.

The distribution of potential across the network of resistors modeling the hair cells (and the shunt resistance paths of the scala media) predicts an increase in the endocochlear potential and a relative depolarization of hair cells. Using reasoning set forth by Roddy *et al.* (1994), a current of 10% of 600 µA applied across the cochlear partition would result in many tens of millivolts of depolarization of each OHC along one to several millimeters of BM in the basal turn area.

The influence of the direct current on the steady-state velocity responses of the BM to acoustic tone is clearly frequency and level dependent for the acoustic tone. It is also non-linear and asymmetrical with regard to current level and polarity. Figure 1 shows the relatively stronger effect of current on the I/O function at 19 kHz, a frequency above the characteristic frequency. The negative current can effectively abolish velocity responses for low sound pressure levels, whereas positive current increased the BM response to a tone. Positive current effects are more saturating, an effect not shown in the figures. Direct current applied across the cochlea duct therefore manipulates the effective gain of the organ of Corti. The data indicate that gain can be

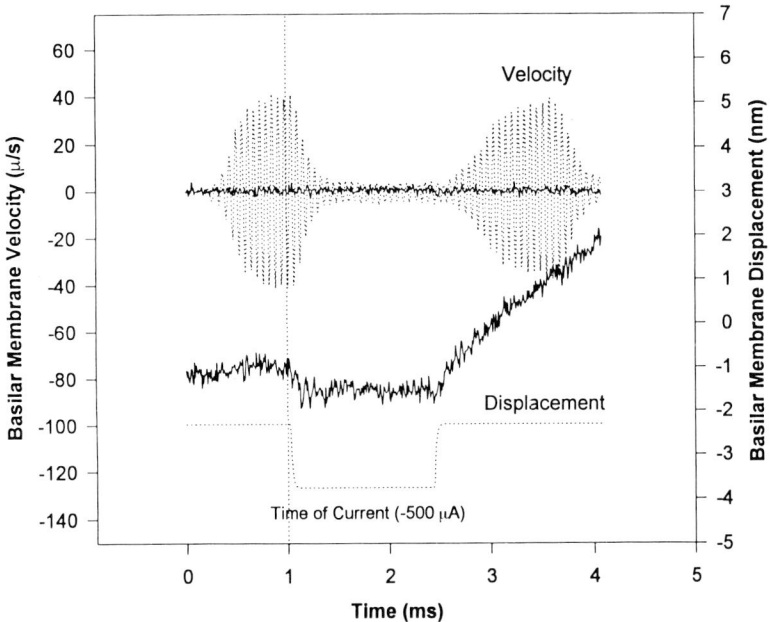

Fig. 6. An acoustic tone burst of 16.5 kHz combined with direct current application of the BM as described in Figure 5. In this example, the current is –500 µA. The lower portion of the lower dotted trace shows the time of the current injection. The displacement response shows that there is a shift in the position of the BM toward the scala tympani during passing of negative current.

both increased and decreased from its 'set point'. (For guinea-pig studies so far, a caveat must be made. Auditory sensitivity is usually somewhat compromised by the surgical procedure of the experiment, i.e. frequency tuning and sensitivity in animals of this report are not as sensitive as normal unoperated animals. It is not known if the gain increase can occur in completely normal animals.)

The change of the organ of Corti gain is a complex function in time, particularly for positive current application where the velocity responses contain a strong onset overshoot. The nature of this overshoot, as well as the onset and offset time constant for negative and positive current injections, should provide interesting new data about the active mechanical process in the organ of Corti in future analysis. Moreover, current application in the absence of tonal acoustic stimulation gives rise to transient responses. These transients are not of sufficient size to account for the offset overshoot phenomenon seen for positive current injection. In fact, it is likely that the acoustic response is quite dominant, and very little, if any, of the onset response is attributable to the current transient itself. We have shown that current pulses always generate these strong ringing responses in sensitive cochleas. The group delay of steady-state, current-induced velocity responses produced by alternating current are very similar to that of the natural acoustic responses.

This implies that current produces an electromotile response that then becomes acoustic-like as energy impinging on cochlea fluids is coupled back into a traveling wave. The transient responses represent the natural frequency of this location on the BM as it is stimulated by an energy step. Figures 3 and 4 show that the gain increase produced by positive current application is accompanied by a slight increase in the resonant frequency of this tonotopic location. Negative current application decreases the resonant frequency toward what might be called the passive natural frequency of this location along the BM.

Depolarization of OHCs has been shown to produce an electromotile contraction response *in vitro* (Brownell, 1983). At the expected membrane potential of OHCs (–70 mV), OHCs are not at an optimal point on their transfer function relating membrane voltage change to a length change (e.g. Evans *et al.*, 1989, 1991; Santos-Sacchi, 1989). Depolarization of hair cells is, therefore, associated with an increase in sensitivity of length change to membrane potential change. Further-more, the sensitivity of the length change of isolated OHCs is several nanometers per millivolt (Evans *et al.*, 1991). The data of Figs 5 and 6, therefore, present a paradox. Both negative and positive current resulted in displacement of the BM toward the scala tympani, and that displacement reached a level of only approximately 2 nm, although one may estimate OHC potential change to be tens of millivolts, as above. The displacement is constant during the duration of the current injection. These displacements may suggest the elongation of some organ of Corti element(s) during both polarities of current application.

The size of the displacements are consistent with those reported by Xue *et al.* (1993b). The motion is much smaller than found *in vitro* but may be consistent with the inherent stiffness of the BM/organ of Corti system (Olson and Mountain, 1994) and the estimated force generated by OHCs (Mountain and Hubbard, 1989; Xue *et al.*, 1993b). The polarity of the motion appears to require a more complex model of cellular activity in the organ of Corti than elongation/contraction of OHCs. A multiple component system, possibly combined with asymmetric or rectifying responses of length changes in cells of the organ of Corti, may be indicated.

Acknowledgements

This work was supported by NIH grants RO1 DC00141 and PO1 DC00078. The authors thank Maroula Bratakos for help with data analysis.

References

Brownell, W. E. (1983). Observations on a motile response in isolated hair cells. In *Mechanisms of Hearing* (W. R. Webster and L. M.Aitken, eds), pp. 5–10. Monash Press, Australia.

Evans, B. N., Dallos, P. and Hallworth, R. (1989). Asymmetries in motile responses of outer hair cells in simulated *in vivo* conditions. In *Cochlear Mechanisms* (J. P. Wilson and D. T. Kemp, eds), pp. 205–206. Plenum, New York.

Evans, B. N., Hallworth, R. and Dallos, P. (1991). Outer hair cell electromotility: the sensitivity and vulnerability of the DC component. *Hearing Res.* **52:** 288–304.

Honrubia, V., Strelioff, D. and Sitko, S. (1976). Electroanatomy of the cochlea: its role in cochlear potential measurements. In *Electrocochleography* (R. J. Ruben, C. Elberling and G. Salomn, eds), pp. 23–39. University Park, Maryland.

Hubbard, A. E. and Mountain, D. C. (1983). Alternating current delivered into the scala media alters sound pressure at the eardrum. *Science* **222:** 510–512.

Mountain, D. C. and Hubbard, A. E. (1989). Rapid force production in the cochlea. *Hearing Res.* **42:** 195–202.

Nuttall, A. L. (1985). Influence of direct current on dc receptor potentials from cochlear inner hair cells in the guinea pig. *J. Acoust. Soc. Am.,* **77:** 165–175.

Nuttall, A. L. and Dolan, D. F. (1993a). Basilar membrane velocity responses to acoustic and intracochlear electric stimuli. In *Biophysics of Hair Cell Sensory Systems* (H. Duifhuis, J. W. Horst, P. van Dijk and S. M. van Netten, eds), pp. 288–293. World Scientific, Singapore.

Nuttall, A. L. and Dolan, D. F. (1993b). Electrically induced traveling waves in the cochlea. *30th Inner Ear Biology Workshop,* Budapest, Hungary, September 1993.

Nuttall, A. L., Dolan, D. F. and Avinash, G. (1991). Laser Doppler velocimetry of basilar membrane vibration. *Hearing Res.* **51:** 203–213.

Nuttall, A. L., Zhang, M. and Dolan, D. F. (1994). Basilar membrane frequency tuning and sensitivity altered by direct current applied to the cochlea. *Abstr. 17th Midwinter Meeting of the Association for Research in Otolaryngology,* p. 355.

Olson, E. S. and Mountain, D. S. (1994). Mapping the cochlear partition's stiffness to its cellular architecture. *J. Acoust. Soc. Am.* **95:** 395–400.

Roddy, J., Hubbard, A. E., Mountain, D. C. and Xue, S. (1994). Effects of electrical biasing on electrically-evoked otoacoustic emissions. *Hearing Res.* **73:** 148–154.

Ruggero, M. A. and Rich, N. C. (1991). Furosemide alters organ of Corti mechanics: evidence for feedback of outer hair cells upon the basilar membrane. *J. Neurosci.* **11:** 1057–1067.

Santos-Sacchi, J. (1989). Asymmetry in voltage-dependent movements of isolated outer hair cells from the organ of Corti. *J. Neurosci.* **9:** 2954–2962.

Williams, W. and Jeong, J. (1992). Reduced interference time-frequency distributions. In *Time–Frequency Signal Analysis* (B. Boashash, ed.). Longman Cheshire, Halsted.

Xue, S., Mountain, D. C. and Hubbard, A. E. (1993a). Direct measurement of electrically-evoked basilar membrane motion. *Abstr. 16th Midwinter A.R.Otolaryngol.* 122.

Xue, S., Mountain, D. C. and Hubbard, A. E. (1993b). Acoustic enhancement of electrically-evoked otoacoustic emission on basilar membrane tuning: experiment results. *Hearing Res.* **70**: 121–126.

Measurements of the Basilar Membrane Resonance in the Cochlea of the Mustached Bat

I. J. RUSSELL[1] AND M. KÖSSL[2]

[1]School of Biology, University of Sussex, Falmer, Brighton BN1 9QG, UK

[2]Zoologisches Institut, Universität München, 80333 München, Germany

Introduction

Mustached bats, *Pteronotus parnellii*, use echolocation to hunt for insects in the forest canopy. They can detect, identify and locate their prey against acoustic reflections from the foliage on the basis of slight modulations in the frequency of the return echoes to the constant frequency (CF) component of their echolocating call caused by the insects' wing beats. This task is achieved largely through a cochlear frequency analyser of remarkable resolution. The 60 kHz region of the mustached bat cochlea, which responds to the dominant component of the CF call frequency, is greatly expanded to occupy about half a turn of the cochlea (Kössl and Vater, 1985a) and the bat modulates the frequency of its call so that the echo falls within the so-called acoustic fovea (Neuweiler, 1990). Evidence of the remarkable frequency resolution of the acoustic fovea can be seen in the sharp threshold minima and resonance of the cochlear microphonic potentials (CM) (Henson *et al.*, 1990; Suga and Jen, 1977) and strong spontaneous and evoked otoacoustic emissions (OAEs) at about 61 kHz (Kössl, 1994). In the present study OAEs were measured in the auditory meatus and a laser interferometer was used to measure displacements of the basal half-turn of the basilar membrane (BM) where frequencies above the range of the acoustic fovea are represented. From these measurements we have shown that the BM is indeed the carrier of the resonant oscillations which may reverberate between the 61 kHz foveal region and the basal end of the cochlea, and that the source of these reverberations can be modelled by a labile but simple resonator. We conclude that the BM resonance probably shapes the amazing frequency tuning of the cochlear microphonics and nerve fibers and leads to the emission of the OAEs.

Materials and Methods

Preparation

Mustached bats, *Pteronotus parnellii*, were anaesthetized with an initial dose of 2 mg/100 g Pentabarbital and either 1.2 mg/100 g Ketavet or 0.13 mg/100 g Hypnorm (Fentanyl derivative). The bats were maintained at 37 deg C, the level of anesthesia was continuously monitored and every 40 min the initial Ketavet or Hypnorm dose was administered. Tone stimuli were delivered through a calibrated, closed sound system under computer control. The sound system incorporated a measuring microphone which was used to measure OAEs (for detailed methods see Kössl, 1994).

The middle ear was opened to expose the round window of the cochlea. The round window was left intact and the beam from the laser diode interferometer was focused through the transparent round window membrane to form a 10 μm diameter spot in the middle of the basal half-turn of the BM. In these experiments it was relatively easy to direct the beam to any position on the BM which was visible through the round window and to compare vibrations of the BM with those of a fixed structure such as the spiral lamina which did not produce tuned displacements to tones presented at the tympanic membrane.

Interferometry

Displacements of the BM were measured by the self-mixing effect of a diode laser. The technique is similar to that for gas lasers (O'Neill and Beardon, 1993) but the interferometer is more compact and easier to implement for *in vivo* measurements. The self-mixing in a laser diode depends on reflecting back a small proportion of the light emitted by the laser into the laser cavity (Koelink *et al.*, 1992). The mixing of the back-reflected light with the light actually produced in the cavity has a cosine dependence of the laser intensity with the target position. As the target is displaced along the beam axis, the intensity of the diode laser varies sinusoidally with a period of $\lambda/2$. For the case of a target oscillating with an amplitude ρ about a fixed position P_0, the displacement $P(t)$ is given by:

$$P(t) = P_0 + \rho \sin(\omega t). \tag{1}$$

The self-mixing effect is essentially linear for changes in ρ of less than about 30 nm. The displacement-dependent effect was measured in the signal of the photodiode which is located behind the laser crystal in the photodiode laser housing. The bandwidth of the displacement-dependent signal was 200 kHz. Calibration of the signal was achieved by displacing the interferometer a known amount by a piezo electric driver which moved 12.5 nm/V.

The output of the interferometer was fed into a spectrum analyzer and a pair of lock-in amplifiers set in quadrature, and the phase and magnitude of the signal was calculated on-line by the computer.

Results

Evoked otoacoustic emissions (EOAEs), which are an indirect indicator of cochlear resonance in *Pteronotus* (Kössl, 1994) were measured at the beginning of each experiment to determine the cochlear resonance frequency. A stimulus-frequency EOAE was consistently emitted from each ear and its frequency was a few hundred hertz above the dominant, constant-frequency component of the echolocation call in the range of 60–63 kHz. Figure 1(A) shows an example of an EOAE which appears as a sharp maximum/minimum pattern in the frequency response measured acoustically at the tympanum. After opening up the middle ear cavity, the EOAEs varied in amplitude and often completely vanished although the cochlea remained untouched.

To monitor the sensitivity of the cochlear mechanics we applied two pure tone stimuli of different frequency (f_1, f_2) and measured the acoustic distortion product $2f_1 - f_2$ at the tympanum. Acoustic distortion products are produced by non-linear mechanical processing in the cochlea and correlate with hearing sensitivity (Gaskill and Brown, 1990; Kössl, 1994). Figure 1(B) shows the course of a typical experiment. Maximum distortion level was recorded at the OAE frequency. After anaesthetizing the bat, the level of distortion decreased by about 10–15

A

B

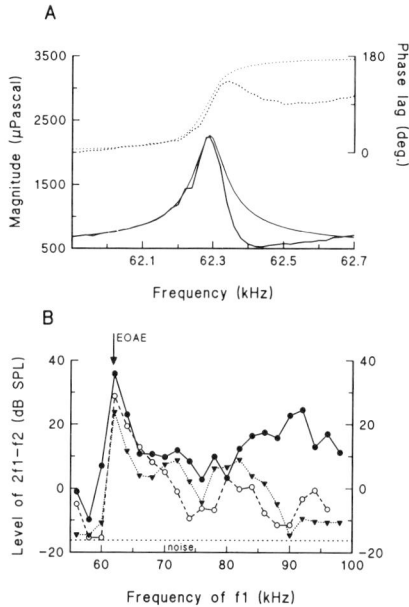

Fig. 1. (A) The magnitude of an evoked otoacoustic emission (EOAE) measured as a function of frequency in the outer ear canal during stimulation with a continuous pure tone of 32 dB SPL that was swept upwards in frequency. Interference between the EOAE and the incoming stimulus causes a sharp maximum/minimum pattern which is evident in the measured frequency response at about 62.3 kHz (lower trace) and is associated with a steep phase lag (upper trace). For comparison, the frequency dependency of the magnitude and phase of a resonator (equations 1 and 2) with a damping coefficient (k) of 0.00065 and a Q_{3dB} value of 719 is plotted (thin lines). (B) Acoustic distortion products measure in the ear canal. Two pure tone stimuli of different frequency (f_1, f_2) with a ratio f_2/f_1 of 1.02 were delivered at 70/60 dB SPL. The level of the recorded $2f_1 - f_2$ distortion product was used to monitor the mechanical sensitivity of the cochlea at all stages of the experiment: awake (closed circles, anaesthetized (open circles); middle ear open (triangles). (A) from Kössl and Russell (1995).

dB for stimulus frequencies up to 85 kHz. For higher frequencies the level of distortion decreased by up to 30 dB. When the middle ear cavity was opened, the distortion level changed further and frequency-dependent increases and decreases were observed. The data was taken to indicate that the mechanical processing of ultra high frequencies is extremely vulnerable to changes in the physiological state of the animal.

The laser diode interferometer was used to measure BM displacements through the round window membrane. The region of the BM through the round window membrane that was accessible to the beam of the laser diode was that of the first half turn of the cochlea. For sound pressure levels between 50 and 105 dB SPL, this region of the BM

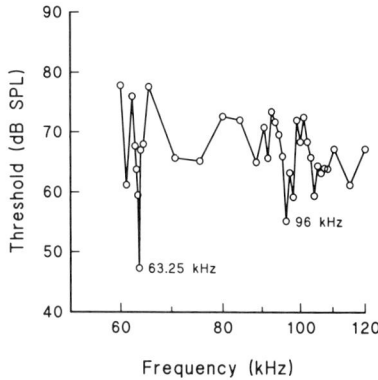

Fig. 2. Isoresponse tuning curve (0.1 nm criterion) measured at the 96 kHz region of the basilar membrane. Note minimum thresholds at 63.25 kHz (Q_{10dB} = 536) and at 96 kHz (Q_{10dB} = 30). (From Kössl and Russell, 1995.)

responded to frequencies of 88–98 kHz in agreement with the cochlear frequency map of *Pteronotus* (Kössl and Vater, 1985a). Recordings of BM responses to swept tones were made at constant stimulus levels. Recordings at frequencies in the 90 kHz range were made in only two preparations because the high-frequency BM responses were labile and deteriorated during the course of an experiment; an example of one of these recordings is shown in Fig. 3. Threshold minima occur at the frequency of the place of the recording (96 kHz) and at the frequency close to the dominant 63 kHz call frequency. Figure 4 shows data from a very sensitive individual with an initial EOAE frequency of 62 kHz. Sharply tuned maxima of BM displacement at frequencies between 61.3 and 62 kHz were measured from the first half turn of the BM over a period of more than 3 hr. The frequency of maximum BM displacement was associated with a phase lag of about 90 deg which is typical of a simple resonator. An attempt was made to fit the data according to the characteristics of a damped resonator with the displacement (D) and phase (P) given by:

$$D = D_0 \, / \, \{[1 - (f \, / \, f_0)^2]^2 + (2kf \, / \, f_0)^2\}^{0.5} \tag{2}$$

$$P = P_0 \, \arctan^{-1} \, [(2kf \, / \, f_0) \, / \, (1 - (f \, / \, f_0)^2] \tag{3}$$

where f is the driving frequency, f_0 is the resonance frequency of the undamped system, k is the damping coefficient and D_0 and P_0 are the displacement and phase values when f is infinitesimal (Bendat and Piersol, 1971).

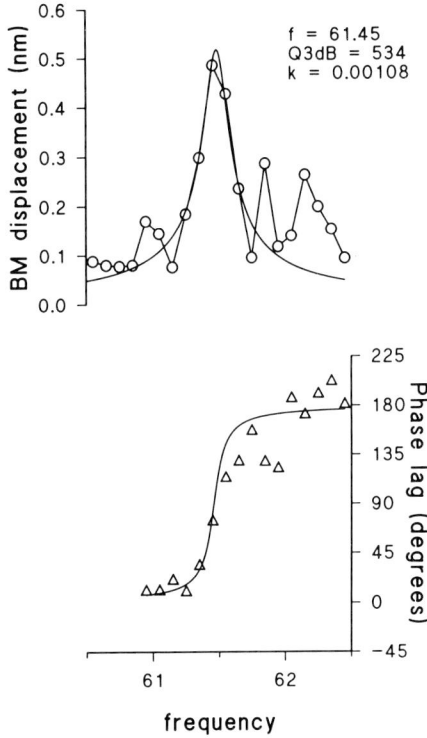

Fig. 3. BM displacement maximum and associated phase change measured in the first cochlear half-turn of a single individual to tones incremented by 100 Hz steps at frequencies around 61.4–61.5 kHz for constant sound pressure levels of 70 dB SPL. The signals were too small to obtain accurate phase measurements for frequencies below 60.95 kHz. The data were fitted, as a first approximation, according to the characteristics of a simple resonator (Eqns 1 and 2). The fitted displacement maxima had a Q_{3dB} values of 534, damping coefficient $k = 0.0011$.

From the fitted displacement data, Q_{3dB} (resonant frequency/ bandwidth at 3 dB of the peak magnitude) and resonant frequency values were derived for frequencies between 61.3 and 62 kHz. When the resonant frequency was between 61.4 and 61.5 kHz, the derived damping coefficients were extremely small and varied between 0.0015 and 0.00048 and the Q_{3dB} could exceed 900. Similar values were found for the damping coefficients of the resonant filter functions which were fitted to the EOAEs (see Fig. 1). The frequency and Q_{3dB} of the resonance varied slightly and continuously between measurements. In fact, following repetitive or intense tones around the resonance, both the Q_{3dB} and the magnitude of the BM response decreased, the phase

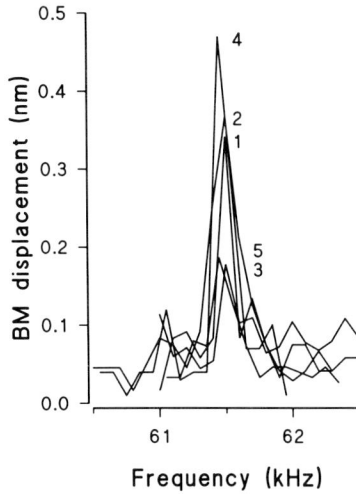

Fig. 4. The variability of BM displacement maxima for five successive measurements (labelled 1–5) over a 10 min period from the same point in the first cochlear half-turn of a single individual to tones incremented by 100 Hz steps at frequencies around 61.4–61.5 kHz for constant sound pressure levels of 70 dB SPL.

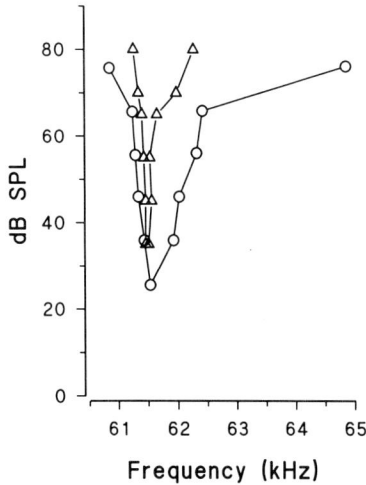

Fig. 5. Iso-displacement tuning curve (0.1 nm criterion) of the BM response (circles) and threshold neural tuning curve (triangles) to frequencies in the 60 kHz range. Q_{10dB} of BM tuning curve: 610; Q_{10dB} of neural tuning curve: 123.

change became more gradual, and in extreme cases the resonance even disappeared altogether for some minutes. This variability in the response of the BM to tones for frequencies between 60 and 63 kHz is shown in Fig. 4 for successive measurements taken over a 10 min period from a single location in the basal half turn of the cochlea.

Sharp resonant displacement maxima were observed for stimulus levels as low as 35 dB SPL. An isoresponse tuning curve (shown in Fig. 5) was derived from the resonance maxima at different sound pressure levels and is characterized by very steep slopes with a Q_{10dB} (center frequency/bandwidth 10 dB from tip) value around 610, which is sharper than that obtained for nerve fibers which innervate the same region of the basilar membrane. Although the BM response to constant stimulus levels can be modelled as a linear resonance, the peak displacements of the resonance change non-linearly with stimulus level (Kössl and Russell, in preparation). The large variations which were observed in the peak displacement for different measurements at the same SPL may indicate that the resonator is probably close to instability and, hence, liable to generate spontaneous oscillations.

Discussion

Enhanced tuning in the 60 kHz range can be explained by a simple mechanical resonance that is apparent in the displacement of the BM at constant stimulus level. Resonant oscillations at about 61 kHz are not restricted to the cochlear representation of this frequency and, in fact, all were recorded in more basally located regions of the cochlea. This finding provides direct evidence for the proposal that EOAEs arise as a consequence of propagated mechanical signals in the reverse direction down the cochlear partition (Kemp, 1981; Kössl and Vater, 1985b; Kössl and Russell, 1995). The Q_{3dB} values, the damping coefficients and the frequency of the BM resonance are similar to that of evoked OAEs and, accordingly, the BM resonance can account for the generation of otoacoustic emissions in *Pteronotus*. The very high frequency separation which is achieved at the level of cochlear mechanics is also indicated by measurement of acoustic two-tone distortions where the optimum frequency separation between the stimuli reaches minimum values of only about 30 Hz at the frequency of the resonance (Kössl, 1994).

At frequencies close to the resonance frequency and hence the OAE frequency, the responses of auditory nerve fibers and brainstem neurons are very sharply tuned with Q_{10dB} values up to 400 (Kössl and Vater, 1990; Suga *et al.*, 1975). Although this tuning is remarkable when compared with typical values of between 1 and 10 for the Q_{10dB} of auditory nerve fibers in the cochleas of non-echolocating mammals (Evans, 1972), the tuning of the afferent fibers of the acoustic fovea is not as sharp as the resonance we have measured in BM displacements (Q_{10dB} of 610, see Fig. 5). This may be because the neurons see the resonance after it has been filtered by the local mechanics of the cochlear partition and by the inner hair cells.

The 61 kHz BM resonance in the cochlea of *Pteronotus* is thought to emanate from discontinuities in the structure of the cochlea in the region of the acoustic fovea. For example, the thickness of the BM changes abruptly in the 60 kHz region (Kössl and Vater, 1985a). It has been proposed that this discontinuity could lead to reflection of the travelling wave and to reverberant oscillations along the BM between the 60 kHz place of the acoustic fovea and the basal end of the BM. Thus, standing-wave like phenomena are generated which lead to resonant 61 kHz oscillations along the whole extent of the BM in the basal region of the cochlea (Kemp, 1981; Kössl and Vater, 1985b). At present, it is not possible to distinguish between this hypothesis and another where the passive resonance has been attributed to different radial oscillation modes of the BM (Kolston *et al.*, 1989; Wilson and Bruns, 1983). However, the variability of both the 61 kHz BM resonance and the OAEs, their level dependency and, indeed, their momentary complete disappearance indicate that the mechanism of the 61 kHz resonance depends not only on the macromechanical properties of the BM but also on active and labile structural elements in the cochlear partition. In this respect, the motile outer hair cells (OHCs) (Brownell *et al.*, 1985) have been attributed as the sources of active, mechanical feedback to the cochlear partition in many models of frequency tuning in the peripheral auditory system. If the remarkable tuning of the BM in the 61 kHz region is determined through interaction between the passive resonant properties of the cochlea and the gain and time constant of mechanical feedback from the OHC into the cochlear partition, then maximum Q_{3dB} might be expected when the parameters of the active and passive systems are matched. A mismatch in the parameters of either one of the components might be expected to give rise to a broadening of the tuning and a slight shift in the best frequency of the resonance (see Fig. 3). Strong spontaneous OAEs may arise when the macromechanical resonator is driven by the force generated by the OHC feedback system.

The 61 kHz resonator in the cochlea of the mustached bat appears to have properties that depend critically on the mechanical state of a labile element in the organ of Corti, presumably the OHCs. Through the implementation of a labile, actively controlled macromechanical resonator in the cochlea, *Pteronotus* achieves a cochlear frequency resolution over a limited frequency range that is far beyond the scope of normal mammals and which may represent the quintessence of sharpness.

Acknowledgements

We thank J. Hartley for the design and construction of electronic equipment and T. S. Collett, C. J. Kros, G. Neuweiler, G. P. Richardson and M. Vater for helpful comments on various early drafts of the manuscript. The work was supported by grants from the DFG and MRC.

References

Bendat, J. S. and Piersol, A. G. (1971). *Random Data: Analysis and Measurement Procedures*. Wiley-Interscience, New York.

Brownell, W. E., Bader, C. R., Bertrand, D. and Ribaupierre, Y. (1985). Evoked mechanical responses of isolated hair cells. *Science* **227:** 194–196.

Davis, H. (1983). An active process in cochlear mechanics. *Hearing Res.* **9:** 79–90.

Evans, E. F. (1972). The frequency response and other properties of single fibers in the guinea pig cochlear nerve. *J. Physiol.* **226:** 263–287.

Henson, O. W., Jr, Koplas, P. A., Keating, A. W., Huffmann, R. F. and Henson, M. M. (1990). Cochlear resonance in the mustached bat: behavioural adaptations. *Hearing Res.* **50:** 259–274.

Kemp, D. T. (1981). Physiologically active micromechanics: one source of tinnitus. In *Tinnitus: Ciba Foundation Symposium 85* (D. Everet and G. Lawrenson, eds), pp. 54–81. Pitman, London.

Koelink, M. H., Slot, M., de Mul, F. F., Greve, J., Graaff, R., Dassel, A. C. M. and Adrnoudse, J. G. (1992). Laser Doppler velocimeter based on the self-mixing effect in a fiber-coupled semiconductor laser: theory. *Appl. Opt.* **18:** 3401–3408.

Kolston, P. J., Viergever, M. A., deBoer, E. and Diependaal, R. J. (1989). Realistic mechanical tuning in a micromechanical cochlear model. *J. Acoust. Soc. Am.* **86:** 136–140.

Kössl, M. (1994). Otoacoustic emissions from the cochlea of the 'constant frequency' bats *Pteronotus parnellii* and *Rhinolophus rouxi*. *Hearing Res.* **72:** 59–72.

Kössl, M. and Russell, I. J. (1995) Basilar membrane resonance in the cochlea of the mustached bat. *Proc. Natl. Acad. Sci. USA* **92:** 276–280.

Kössl, M. and Vater, M. (1985a). The cochlear frequency map of the moustache bat, *Pteronotus parnellii. Hearing Res.* **19:** 157–170.

Kössl, M. and Vater, M. (1985b). Evoked acoustic emissions and cochlear microphonics in the moustache bat, *Pteronotus parnellii. J. Comp. Physiol.* **A157:** 687–697.

Kössl, M. and Vater, M. (1990). Resonance phenomena in the cochlea of the mustached bat and their contribution to neuronal response characteristics in the cochlea nucleus. *J. Comp. Physiol.* **A166:** 711–720.

Neuweiler, G. (1990). Auditory adaptations for prey capture in echolocating bats. *Physiol. Rev.* **70:** 615–641.

O'Neill, M. P. and Beardon, A. (1993). The amplitude and phase of basilar membrane motion in the turtle measured with laser-feedback interferometry. In *Biophysics of Hair cell Sensory Systems* (H. Duifhius, J. E. Horst, P. van Dijk and S. M. van Netten, eds), pp. 398–405. World Scientific, Singapore.

Suga, N. and Jen, P. H.-S. (1977). Further studies on the peripheral auditory system of CF-FM bats specialized for fine frequency analysis of Doppler shifted echoes. *J. Exp. Biol.* **69:** 207–232.

Suga, N., Simmons, J. A. and Jen, P. H.-S. (1975). Peripheral specializations for fine frequency analysis of Doppler-shifted echoes in the auditory system of the CF-FM bat, *Pteronotus parnellii. J. Exp. Biol.* **63:** 161–192.

Wilson, J. P. and Bruns, V. (1983). Basilar membrane tuning properties in the specialized cochlea of the CF-FM bat, *Rhinolophus ferrumequinum. Hearing Res.* **10:** 15–35.

Mechanical Activity in the Apex of the Cochlea

W. S. RHODE AND N. P. COOPER

Department of Neurophysiology, University of Wisconsin, Madison, WI 53706, USA

Introduction

Mechanical measurements in the apical turn of the cochlea are crucial to our understanding of the auditory process for several reasons. Much of human hearing experience occurs for sounds with frequencies less than 5000 Hz. Glottal pulse periods and the first formant frequencies of speech are all less than 1000 Hz in frequency. The cochlear region most sensitive to these frequencies has been little studied because of its relative inaccessibility and one of the few studies that has been conducted, that of the International Team for Ear Research (ITER, 1989), has resulted in a set of controversial observations. Using a laser heterodyne interferometer that records from a small volume of space, they found that the amplitudes of movement of outer hair cells (OHCs) and Hensen cells were much larger than those of the basilar membrane. This is contrary to expectations based on numerous observations of cochlear mechanics in the mid- and basal- region of the cochlea (Rhode, 1971, 1973, 1978; Sellick *et al.*, 1982; Robles *et al.*, 1986; Cooper and Rhode, 1992). These latter observations have led to the conclusion that basilar membrane mechanics adequately account for auditory-nerve frequency threshold curves (FTCs).

Békésy's (1960) observations of cochlear mechanics suggested that the entire cochlear partition (Reissner's membrane to basilar membrane and everything in between) vibrates as a single unit with the motion principally determined by the properties of the basilar membrane. In support of this latter hypothesis, Rhode (1978) measured Reissner's membrane (RM) motion in the apical turn of the squirrel monkey's cochlea and found that the mechanical tuning properties were nearly the same as neural tuning in auditory nerve fibers with the same characteristic frequencies (=300–600 Hz).

The characterization of the vibration of individual cells by the ITER group has the potential to alter our concept of cochlear mechanics. In contrast to the ITER results, Gummer *et al.* (1993) reported that the basilar membrane vibrated with the same amplitudes as the cells of Hensen. Thus there appears to be a dispute over the behavior of the organ of Corti in the apical region which might also have implications for studies in the basal region of the cochlea.

We reported on some preliminary studies in the apex that have a number of features in common with the ITER results (Brundin *et al.*, 1991): large DC displacements (10 μm) with the AC response superimposed (Cooper and Rhode, 1993). These displacements were seen only at very high intensities (110 dB SPL). However, further studies suggest that they are the result of artefacts in our phasemeter (Cooper and Rhode, 1995).

We also observed non-linear phenomena in a notch region of the RM frequency transfer function including two-tone suppression. The notch appeared to be fairly robust with some of the features lasting for hours after the death of the animal. This latter result is very different from that found in the basal region were it is very difficult to maintain non-linear behavior regardless of how careful one tries to be with the preparation.

Additional experiments in the apex of the cochlea now confirm these earlier results. The notch appears to be a repeatable feature of the apical RM vibration. In a new observation, the basilar membrane lateral to the Hensen cells has been shown to vibrate with the same amplitude as RM and Hensen cells. It appears that a substantial portion of apical mechanics are linear and independent of an active process.

Methods

Substantial data were collected from 37 pigmented guinea-pigs. The animals were anesthetized and ventilated to maintain expired CO_2 levels of 4–5%. Body temperatures were maintained at 38.5 deg C. CAP and/or CM recordings were monitored to determine the viability of the cochleae.

A displacement-sensitive laser interferometer (Cooper and Rhode, 1992) was used to observe the sound-evoked patterns of various structures in the apical turn of the cochlea. The displacements, which include both AC and DC information, were sampled at 8 μsec intervals (=125 kHz) with a resolution of 0.47 nm. Highly reflective microbeads (diameters ~20–30 μm) were used to reflect the laser light back from the

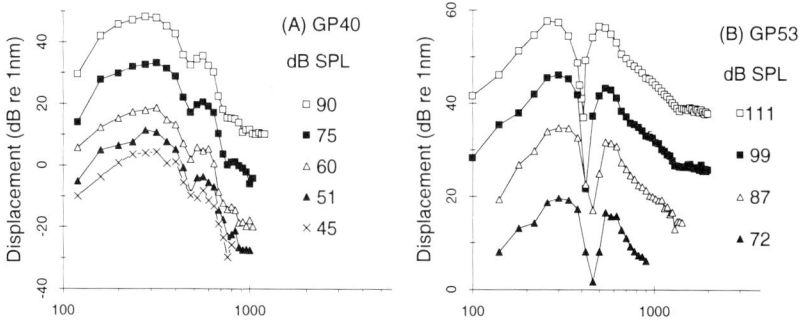

Fig. 1. Reissner's membrane transfer functions for two GPs. (A) Iso-intensity tuning curves for GP40 at the indicated intensities. (B) Iso-intensity curves for GP53. Adapted from Cooper and Rhode (1995).

normally transparent structures of the cochlear partition. The beads were introduced through a small hole (~200 × 500 μm) on the scala vestibuli side of the cochlear partition. In most of our experiments, observations were made at the level of RM. However, a number of recordings were also made from the underlying organ of Corti and basilar membrane — in these instances the reflective microbeads were pushed through a small hole in Reissner's membrane. A small glass-topped chamber was sealed over the apex of the cochlea to maintain near normal hydromechanics in all experiments. A complete description of the methods is given elsewhere (Cooper and Rhode, 1995).

Results

Reissner's membrane displacement

The access to the apical cochlea is through the scala vestibuli which results in exposing RM, a transparent, two-cell-thick membrane which forms one boundary of the scala media. The other boundary is formed by the basilar membrane upon which the organ of Corti rests. Because opening the scala media results in the loss of the endocochlear potential and the subsequent death of the organ of Corti, we chose to perform measurements on RM before attempting more difficult observations of the structures of the organ of Corti.

Figure 1 shows a series of RM displacement transfer functions at several intensities in two guinea-pigs. They illustrate the range of transfer function shapes that we observed in this series of experiments. The curves are characterized by a large, broad peak near 300 Hz, a notch of varying depth near 450 Hz, and a secondary peak in the 500–600 Hz region. The transfer curves in Fig. 1(b) exhibited the largest notch found in any animal. The notch frequency decreased with increasing intensity. The notch was a robust feature that persisted even after the animal was dead.

The vibration of RM is linear over nearly the entire set of stimulus frequencies to which it responds. However, in the region of the notch a number of non-linear phenomena have been observed. These include non-linear input/output curves, two-tone suppression and DC shifts. The set of I/O curves shown in Fig. 2 illustrate the general features of RM response: linear I/O curves for most of the frequencies (e.g. 200, 300 and 500 Hz), compressive non-linearities at 360 and 380 Hz, expansive non-linearity at 420 Hz, and a complex I/O curve at 400 Hz that has

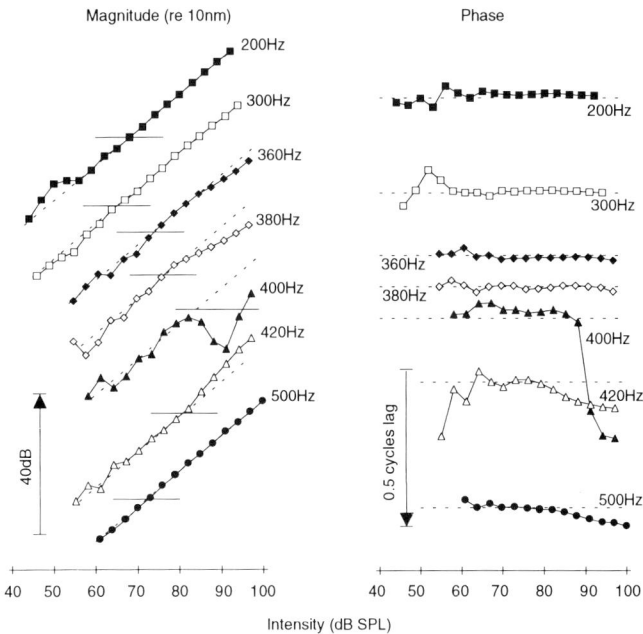

Fig. 2. Input/output functions for RM vibration in GP118. The individual curves are displaced vertically for clarity. The horizontal lines through each curve correspond to a 10 nm displacement in the left column. The corresponding phase relations are shown in the right-hand column.

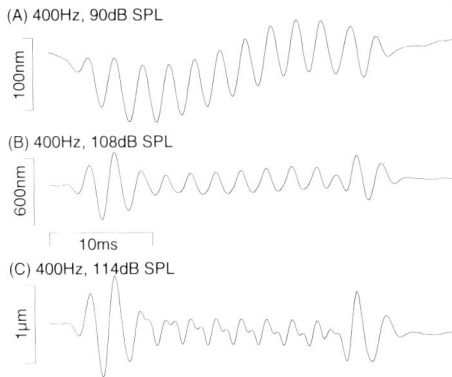

(A) 400Hz, 90dB SPL

100nm

(B) 400Hz, 108dB SPL

600nm

10ms

(C) 400Hz, 114dB SPL

1µm

Fig. 3. RM response waveforms at 400 Hz in GP53. The stimulus level is indicated in each panel.

compressive and expansive properties. The right-hand column illustrates the phase properties of RM, which are well behaved for most frequencies but show a rapid 180 deg phase lag at 400 Hz, the notch frequency. The other phase shifts were limited to frequencies between the peak frequency and the plateau region (defined by an unchanging phase characteristic in the RM/incus transfer function). In general, RM phase increasingly lagged that of the incus as intensity was increased.

Two modes of vibration

In the frequency region of the notch, there was often a short tone response pattern that suggests the presence of two vibration modes. Figure 3 shows the waveform at three intensities that cover a typical linear response (Fig. 3A), to one where the center is compressed (Fig. 3b) while the onset/offset portion is linear, to one at 114 dB SPL (Fig. 3c) where the compressed region shows a great deal of distortion. The distortion in the latter response could result in the peak-splitting phenomena that has been seen in auditory nerve responses at high levels (Kiang *et al.*, 1986; Ruggero and Rich, 1989).

When the non-monotonic I/O curve at 400 Hz is considered along with the rapid 180 deg phase change (Fig. 2b), the data suggest the possibility of two vibration modes which interact to give a minimum when they partially cancel each other. This is similar to the two-mode explanation offered for Kiang's notch.

High-level intensity dependence

Unlike the non-linearities that have been seen in the basal region of the cochlea, all the non-linear effects observed at the level of RM were at high stimulus levels (>80 dB SPL). Some of these non-linear phenomena were noted in Figs 2 and 3. At very high levels (e.g. 110 dB SPL) even more dramatic non-linearities are observed. In Fig. 4(a), a time-dependent adaptation is seen as a decrease in amplitude over the duration of the 30 msec stimulus along with peak splitting. DC or base-line shifts are observed (Fig. 4b,c). However, after due consideration, it is our interpretation that many of these high-level DC responses are due to the behavior of our phasemeter in response to rapidly changing phase changes. When we compensated for a presumed 0.5 cycle phase error at appropriate places in the responses, the DC shift disappeared as shown in Fig. 4(d).

While small DC shifts occur for stimulus levels 100 dB SPL, they are usually confined to values less than 100 nm. We have never recorded DC displacements as large as those required for some hypothesized automatic gain control function in the cochlea (LePage, 1987).

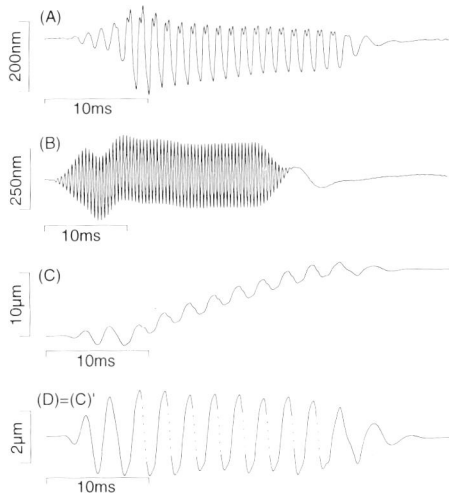

Fig. 4. Distorted RM responses evoked by intense levels of stimulation. (A) Peak-splitting and adaptation to a 110 dB SPL tone at 800 Hz. (B) Baseline shift to a 120 dB SPL tone at 2400 Hz. (C) A rapidly accumulating baseline shift at 120 dB SPL to a 400 Hz tone. (D) The waveform in (C) after it has been corrected for a presumed 0.5 cycle phase shift due to a phasemeter response error.

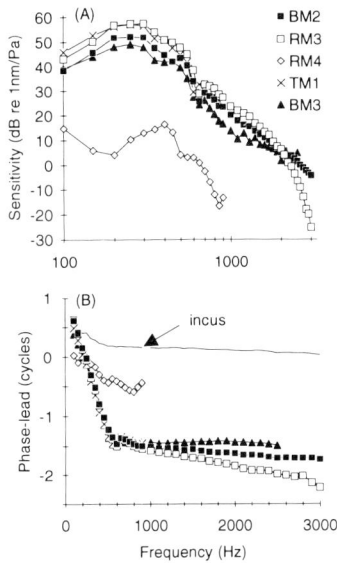

Fig. 5. Frequency response functions for various structures and locations in GP119. (A) Acoustico-mechanical sensitivities. (B) Phase transfer functions.

Basilar membrane vs. RM transfer functions

A key question about cochlear function is what is driving the inner hair cells (IHCs). As noted earlier there is some controversy as to whether the reticular lamina is largely driven by the motile OHCs or whether the basilar membrane bears the same relation to the organ of Corti as it does in the base of the cochlea. Several attempts were made to measure the vibration of structures after RM transfer function had been determined.

An opening was made in RM to allow the deposit of a reflective sphere/microbead on structures of the organ of Corti. This was a hit-or-miss situation since the final resting place of any sphere was not well controlled. The transfer functions for RM, the tectorial membrane and the basilar membrane are shown in Fig. 5. The sensitivity and the shape of the transfer functions are remarkably similar irrespective of the position of the measurement. The exception is curve RM4 which corresponded to a location near the inner edge of the spiral limbus. The interesting aspect of these measurements is the small difference between them. This could easily be explained by the difference in the radial position at which the measurements were made.

The phase characteristics near the center of the partition all show the same general features. A rapid decrease to about 1.5 cycles of phase-lag followed by a nearly constant value over a considerable range of frequencies. This is essentially the same phase behavior that has been found in the basal region of the cochlea and in the earlier studies of RM motion.

Discussion

Transfer functions in the apex of the cochlea – similarities

To date, the pronounced non-linearities at low intensities observed in the basal region of the cochlea (Rhode, 1971; Sellick *et al.*, 1982; Robles *et al.*, 1986; Cooper and Rhode, 1992) have not been duplicated in the apical region of the cochlea. In contrast, the agreement between various investigators in their measured RM transfer functions is rather remarkable in spite of differences of time, species and the preparation. We compare Békésy's RM tranfer function results obtained in guinea-pig with those of ITER (1989) obtained in a guinea-pig temporal bone preparation and Rhode's obtained in live squirrel monkey (1978) in Fig. 6. For stimulus frequencies below the peak the slopes are

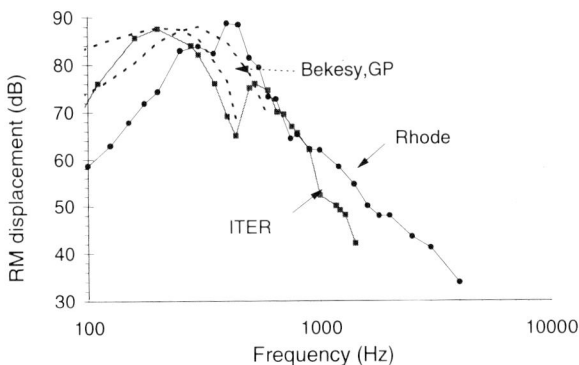

Fig. 6. A comparison of the displacement transfer functions in three preparations using three different measurement techniques for RM vibration. The stylized curve is adapted from Békésy (1960). The ITER (1989, p. 154) data was obtained in their guinea-pig temporal bone preparation in the fourth turn of the cochlea. The squirrel monkey (SM) data was obtained *in vivo* with a loose plastic cover over the opening into the apical turn (Rhode, 1978).

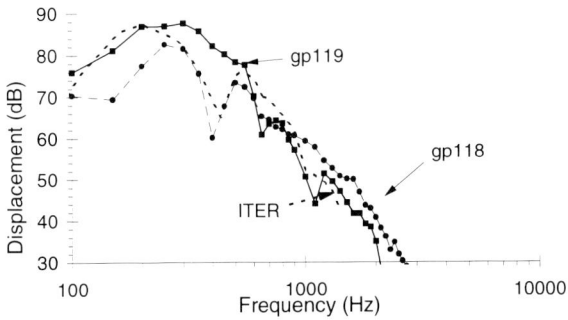

Fig. 7. Two further examples of RM displacement in guinea-pig apical turn (Cooper and Rhode, 1995) are compared with the ITER data of Fig. 6.

somewhat steeper than Békésy's (whose data were obtained from guinea-pigs that were in a questionable condition). Above the peak frequency, the slopes are all relatively close in magnitude. The principal difference from Békésy's transfer functions is the presence of the notch at 400 Hz in the ITER curve and the dip (notch?) at 300 Hz in the SM curve.

Two representative RM transfer functions from the present studies are compared with the ITER results in Fig. 7. The gp118 curve is remarkably similar to the ITER curve in both shape and slopes. The gp119 curve illustrates the range of variation in the notch region, i.e. the notch is absent. Nevertheless, the shapes and slopes are comparable.

A key question is whether the mechanical transfer ratios can explain

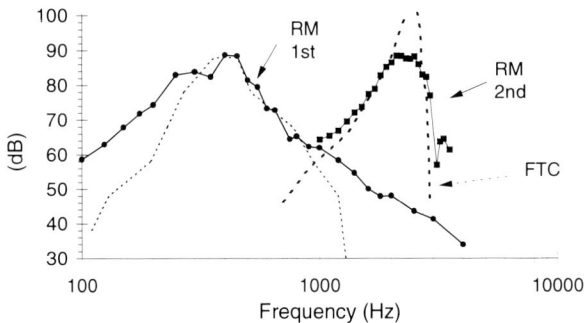

Fig. 8. A comparison of Reissner's membrane transfer curves (solid lines) to inverted and vertically shifted frequency threshold curves (FTCs). The CFs are near 400 Hz and 1500 Hz respectively and were collected in the apical turns of the squirrel monkey cochlea. Adapted from Rhode (1978). The vertical scale is arbitrary.

auditory nerve FTCs. This certainly is the case in the basal region of the cochlea. Figure 8 illustrates a comparison between RM transfer functions and FTCs in the squirrel monkey in the two apical turns of its cochlea. The slopes and shapes are in reasonable agreement, suggesting that there is a relation between the two. There are of course differences: first, the tip region is missing in the second turn RM transfer function. The FTC is a threshold curve (data at tip corresponds to 10 dB SPL) and would be considerably broader if an isointensity curve at 80 dB SPL were used as a comparison. Second, there is a plateau in the mechanical curves that has not been seen in the neural FTCs. Third, the low-frequency limb deviates somewhat from the FTC. However, the two curves could be brought into better agreement if velocity is taken to be the effective stimulus (at low frequencies) to the IHCs, thereby increasing the slope 6 dB/oct (Dallos, 1986). However, Ulfendahl *et al.* (1991) demonstrated that the effect of opening the cochlea on the transfer curves in the low frequency region is to reduce the amplitude for stimuli below a few hundred hertz. As the squirrel monkey RM data were obtained from unsealed cochlea, these two effects could tend to cancel each other. The differential stimulus to the IHCs could be substantially different than RM or basilar membrane transfer functions would indicate. For example, the movement of the limbus could be substantial (possibly equal to that of the basilar membrane) for frequencies in the plateau region and thereby reduce or eliminate the movement of the IHC cilia that results from the shear motion between the tectorial membrane and the reticular lamina. Certainly there have been arguments advanced for multiple modes of vibration in the cochlea that could explain this result (e.g. Békésy, 1960; Allen, 1980; Zwislocki, 1980).

Transfer functions in the apex of the cochlea — differences

The principal difference in transfer functions obtained by various groups to date is whether there is substantially more movement in the reticular lamina–OHC complex than in the basilar membrane. The ITER results indicate that the OHCs vibrate (with amplitudes) as much as 1000 times more than the basilar membrane. Confirmation fo this result would substantially alter our view of cochlear mechanics. Contrasted with this view of apical cochlear mechanics are the results of Békésy (1960), Gummer *et al.* (1993) and our data. These data suggest that the motion of individual structures is nearly the same whether one is looking at RM, the basilar membrane or the tectorial membrane A similar conclusion was reached by Khanna *et al.* (1992; but see commentary on Gummer *et al.*, 1993).

One possible explanation for the ITER result is that the indicated position of the BM measurement was near the lateral wall of the cochlea. It is clearly to be expected that more central regions would vibrate much more than lateral positions. The basilar membrane transfer function certainly bears a strong resemblance to the vibration of the outer bony shell of the cochlea (Fig. 3, p153, ITER).

Non-linear phenomena in the apex of the cochlea

While Rhode (1971, 1978) had demonstrated that there was a pronounced compressive non-linearity in the midfrequency region (6–7 kHz) of the squirrel monkey cochlea, he found RM in the apical cochlear turn to vibrate linearly. The marked difference between the apex and base results suggests that there is possibly something different about the apical cochlear region. This difference is also clear in intracellular recordings in hair cells (Cheatham and Dallos, 1993). The present results indicate that even the apical motion is more complicated than a linear system would allow.

The non-linearities occur over a narrow frequency range in the region of the transfer function notch for which compressive and expansive non-linearities were observed. There was also two-tone suppression, peak-splitting and DC shifts. Each of these were seen in the vibration of RM where one presumes that its motion merely reflects the motion of the organ of Corti. In contrast to the basal region where the transfer functions are extremely vulnerable, apical region mechanics seem to be relatively more robust. The non-linearities we observed were only for intensities >80–90 dB SPL and were only observed in live preparations in good condition.

Observations by Brundin *et al.* (1991) indicated that there is a DC response that is better tuned than the AC response. With a 25 Hz shift in the stimulus frequency, the DC response that could be as large as 20 μm, disappeared. Q^{10} for the DC response at 350 Hz is 7. It is argued that the position shift must be produced by an active non-linear amplification, which could be due to tuned length changes that have been seen in isolated OHCs (Brundin and Russell, 1993; Canlon *et al.*, 1988).

Our results are contaminated by the behavior of our phasemeter which makes errors in the face of rapid phase changes that result in waveforms that look very similar to the ITER results for DC shifts. We feel that at most there are small DC shifts (tens of nm) that we are presently confident of and only at high intensities. We do not believe that large DC shifts exist at normal physiological levels.

Understanding the details of how the neural FTCs are generated by the IHCs will require the effort of further measurements and cochlear models.

Acknowledgements

This work was funded by NIH grant DC01910.

References

Allen, J. B. (1980). Cochlear mechanics – a physical model for transduction. *J. Acoust. Soc. Am.* **68**: 1660–1670.

Békésy, G. von. (1960). *Experiments in Hearing*. McGraw-Hill, New York.

Brundin, L. and Russell, I. (1993). Sound-induced movements and frequency tuning in outer hair cells isolated from the guinea pig cochlea. In *Biophysics of Hair Cell Sensory Systems* (H. Duifhuis, J. W. Horst, P. van Dijk and S. M. van Netten, eds), pp. 182–191. World Scientific, New Jersey.

Brundin, L., Flock, Å., Khanna, S. M. and Ulfendahl, M. (1991). Frequency-specific position shift in the guinea pig organ of Corti. *Neurosci. Lett.* **128**: 77–80.

Canlon, B., Brundin, L. and Flock, Å. (1988). Acoustic stimulation causes tonotopic alterations in the length of isolated outer hair cells from the guinea pig hearing organ. *Proc. Natl. Acad. Sci. USA* **85**: 7033–7035.

Cheatham, M. A. and Dallos, P. (1993). Longitudinal comparisons of IHC ac and dc receptor potentials recorded form guinea pig cochlea. *Hearing Res.* **68**: 107–114.

Cooper, N. P. and Rhode, W. S. (1992). Basilar membrane mechanics in the hook region of cat and guinea-pig cochlea: Sharp tuning and non-linearity in the absence of baseline position shifts. *Hearing Res.* **63**: 163–190.

Cooper, N. P. and Rhode, W. S. (1993). Nonlinear mechanics at the base and the apex of the mammalian cochlea: *In-vivo* observations using a displacement sensitive laser interferometer. In *Biophysics of Hair Cell Sensory Systems* (H. Duifhuis, J. W. Horst, P. van Dijk and S. M. van Netten, eds), pp. 249–256. World Scientific, New Jersey.

Cooper, N. P. and Rhode, W. S. (1995). Nonlinear mechanics at the apex of the guinea-pig cochlea. *Hearing Res.* Submitted.

Dallos, P. (1986). Neurobiology of cochlear inner and outer hair cells: intracellular recordings. *Hearing Res.* **22:** 185–198.

Gummer, A. W., Hemmert, W., Morioka, I., Reis, P. Reuter, G. and Zenner, H.-P. (1993). Cellular motility measured in the guinea-pig cochlea. In *Biophysics of Hair Cell Sensory Systems* (H. Duifhuis, J. W. Horst, P. van Dijk and S. M. van Netten, eds), pp. 229–239. World Scientific, New Jersey.

International Team for Ear Research (ITER) (1989). Cellular vibration and motility in the organ of Corti. *Acta Otolaryngol.*, Suppl. 467.

Khanna, S. M., Flock, Å. and Ulfendahl, M. (1989). Comparison of the tuning of outer hair cells and the basilar membrane in the isolated cochlea. *Acta Otolaryngol.* Suppl. 467, 151–156.

Khanna, S. M., Flock, Å. and Ulfendahl, M. (1992). Mechanical analysis in the cochlea at the cellular level. In *Noise Induced Hearing Loss* (A. L. Dancer, D. Henderson, R. J. Salvi and Hammernick, eds), pp. 3–10. B. C. Decker, Philadelphia.

Kiang, N. Y. S., Liberman, M. C., Sewell, W. F. and Guinan, J. J. (1986). Single unit clues to cochlear mechanisms. *Hearing Res.* **22:** 171–182.

LePage, E. L. (1987). Frequency-dependent self-induced bias of the basilar membrane and its potential for controlling sensitivity and tuning in the mammalian cochlea. *J. Acoust. Soc. Am.* **82:** 139–154.

Rhode, W. S. (1971). Observations of the vibration of the basilar membrane in squirrel monkeys using the Mössbauer technique. *J. Acoust. Soc. Am.* **49:** 1218–1231.

Rhode, W. S. (1973). An investigation of post-mortem cochlear mechanics using the Mössbauer effect. In *Basic Mechanisms of Hearing* (A. R. Møller, ed.), Academic Press, New York.

Rhode, W. S. (1978). Some observations on cochlear mechanics. *J. Acoust. Soc. Am.* **64:** 158–176.

Robles, L., Ruggero, M. A. and Rich, N. C. (1986). Basilar membrane mechanics at the base of the chinchilla cochlea I. Input–output functions, tuning curves and response phases. *J. Acoust. Soc. Am.* **80:** 1364–1374.

Ruggero, M. A. and Rich, N. C. (1989). 'Peak-splitting': intensity effects in cochlear afferent responses to low frequency tones. In *Cochlear Mechanisms* (J. P. Wilson and D. T. Kemp, eds), pp. 259–267. Plenum, New York.

Sellick, P. M., Patuzzi, R. and Johnstone, B. M. (1982). Measurements of basilar membrane motion in the guinea-pig using the Mössbauer technique. *J. Acoust. Soc. Am.* **72:** 131–141.

Ulfendahl, M. Khanna, S. M. and Flock, Å. (1991). Effects of opening and resealing the cochlea on the mechanical response in the isolated temporal bone preparation. *Hearing Res.* **57:** 31–37.

Zwislocki, J. J. (1980). Two possible mechanisms of the second cochlear filter. In *Psychophysical, Physiological and Behavioral Studies in Hearing* (G. van Den Brink and F. A. Bilsen, eds). Delft University Press, Delft, The Netherlands.

What is the Mechanical Stimulus for the Inner Hair Cell? Clues from Auditory-Nerve and Basilar-Membrane Responses to Low-Frequency Sounds

MARIO A. RUGGERO

Department of Communication Sciences and Disorders, Northwestern University, 2299 North Campus Drive, Evanston, IL 60208-3550, USA

Introduction

In the study of sensory transduction the logical starting point should be to define the adequate stimulus to the sensory cell. It is remarkable that the stimulus for the inner hair cell (IHC) of the mammalian cochlea has not yet been fully specified in terms of the vibration of the cochlear partition. This is not a trivial gap in knowledge, since IHCs provide almost the entire afferent input to the auditory nerve. There is little doubt that mechanoreceptive hair cells, in general, are specialized to detect the displacement of their stereocilia. In the case of both IHCs and outer hair cells (OHCs) in *in vitro* preparations of the mammalian organ of Corti, displacement of the stereocilia toward the longer stereocilia causes depolarization (Russell *et al.*, 1986), in agreement with the response polarity of hair cells in other sensory organs. What is not known is the manner in which displacements of IHC stereocilia arise from the interactions of the basilar membrane (BM), the tectorial membrane the organ of Corti and the surrounding fluid spaces. This paper reviews evidence, based on the responses of hair cells, auditory-nerve fibers and the BM to low-frequency sound, bearing on the mechanisms that encode motion of the BM into the receptor potentials of IHCs.

Historical Background

Almost from the moment that the pioneers of cochlear electrophysiology recognized the separate identities of compound action potentials and cochlear microphonics, it was thought that, at least for low-frequency stimuli, there must be a unique relationship between vibration of the BM (and its electrical correlate, cochlear microphonics) and initiation of spikes in auditory-nerve fibers, regardless of their site of innervation. Derbyshire and Davis (1935) proposed that action potentials are generated at a time corresponding to the negative-to-positive transition of the cochlear microphonics recorded at the round window. Once it was shown that cochlear microphonics reflect low-frequency or static displacements of the BM (Békésy, 1951), it became generally accepted that auditory-nerve 'excitation occurs when the cochlear partition is deflected toward scala vestibuli' (Goldstein, 1968). More precisely, excitation synchronous with the negative-to-positive transition of round-window microphonics implied that the auditory-nerve fibers are excited when BM velocity, rather than displacement, is maximal. There was much to recommend this view. Firstly, almost self-evidently, the stereocilia of OHCs, which are firmly embedded in the tectorial membrane, must be tilted toward the basal body upon BM displacement toward scala vestibuli (Davis, 1958). Such a tilt was then suspected (and now known) to cause depolarization in the hair cells of the lateral line, the frog sacculus and the mammalian cochlea (Flock, 1971; Hudspeth and Corey, 1977; Russell *et al.*, 1986). Since cochlear microphonics are dominated by the responses of OHCs (Dallos and Cheatham, 1976), a maximally positive potential at the round window or in the basal region of scala tympani (indicating maximal depolarization of OHCs) occurs when the BM is maximally displaced toward scala vestibuli (e.g. Ruggero *et al.*, 1986). On the other hand, since the attachment of IHC stereocilia to the tectorial membrane appears to be flimsy or non-existent (see reviews by Lim, 1980, 1986), it was thought reasonable that their deflection should be velocity-coupled to BM displacement via frictional forces arising from the motion of fluids in the subtectorial space (Dallos *et al.*, 1972; Billone and Raynor, 1973).

Hair-Cell Receptor Potentials: Responses to Low-Frequency Tones

Intracellular recordings from both IHCs and OHCs support the notion that, at least for low-frequency stimulation at low and moderate intensities, OHCs act as displacement detectors, while IHCs act as

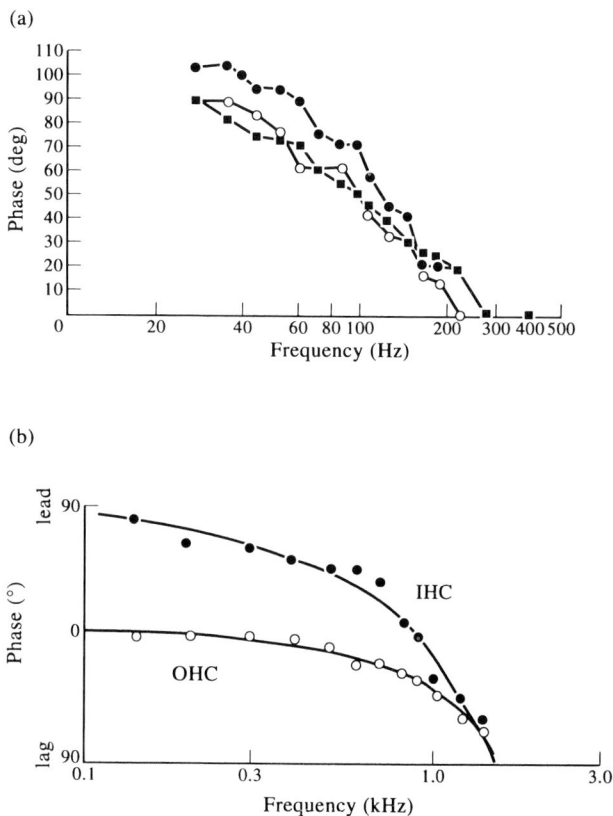

Fig. 1. The phases of IHC depolarization, relative to cochlear microphonics, in scala tympani. (A) Recordings from an IHC at the base of the guinea-pig cochlea. The symbols – open circles, filled circles and squares – indicate responses to tones presented at 80, 90 and 100 dB SPL, respectively. (B) Recordings from an IHC and an adjacent outer hair cell (OHC) at the third turn of the guinea-pig cochlea. [Reproduced, with permission, from (A) Fig. 6(A) of Russell and Sellick (1983) and (B) Fig. 2 of Dallos (1984).]

velocity detectors. At the base of the guinea-pig cochlea, where BM response phases are known most precisely (e.g. Wilson and Johnstone, 1975), OHCs are maximally depolarized when the BM is maximally displaced toward scala vestibuli or, equivalently, when the cochlear microphonic in scala tympani is maximally positive (Russell and Sellick, 1983). In contrast, IHCs innervating the same basal cochlear regions respond to very-low-frequency stimuli (below 50 Hz or so) in phase with maximal BM velocity (Fig. 1A; Nuttall *et al.*, 1981; Russell and Sellick, 1983). At frequencies above 40 Hz, responses progressively phase lag, so that at 200–300 Hz, the IHC appears to be responding to maximal BM displacement toward scala vestibuli. At apical cochlear regions, where *in vivo* recordings of BM vibrations are still unavailable, comparison of intracellular receptor potentials with extracellular micro-phonics suggest that response phases relative to BM motion are much the same as at the cochlear base: in response to low-frequency tones presented at low or moderate intensities, OHCs and IHCs signal maximum BM displacement or velocity, respectively, toward scala vestibuli (Dallos and Santos-Sacchi, 1983; Fig. 1B).

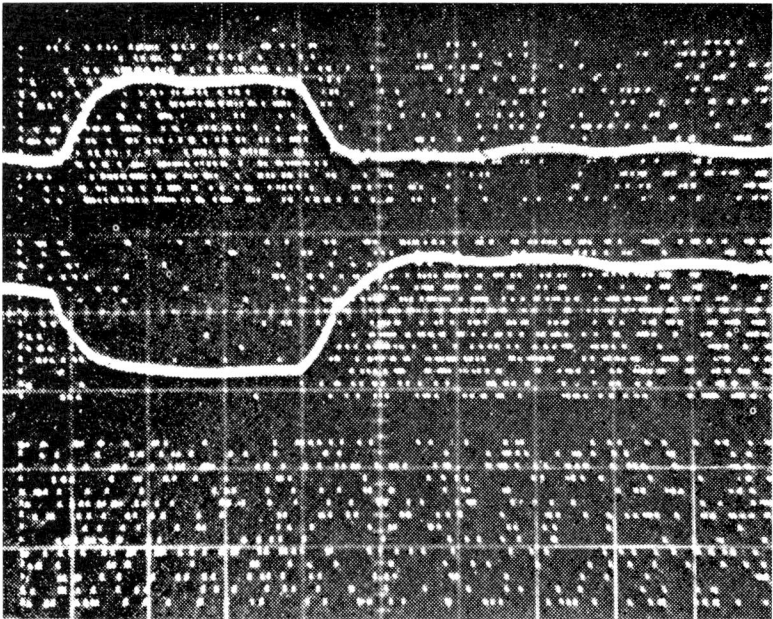

Fig. 2. Response of an auditory-nerve fiber (CF: 17.7 kHz) to trapezoidal displacements of the BM. Each dot represents one action potential. The continuous traces indicate the endocochlear potential. Bottom section: spontaneous activity; middle section: displacement toward scala vestibuli; top section: displacement toward scala tympani. [Reproduced, with permission, from Fig. 5(A) of Konishi and Nielsen (1978).]

Response Phases of Auditory-Nerve Fibers: Relation to Characteristic Frequency

In 1973, two reports (Konishi and Nielsen, 1973; Zwislocki and Sokolich, 1973) challenged current notions of auditory-nerve excitation

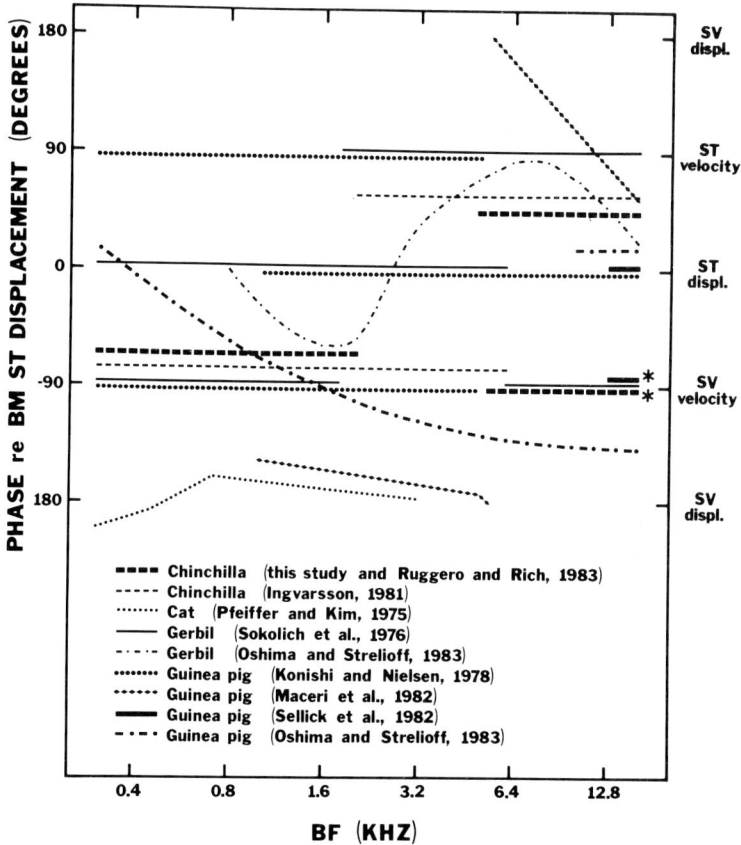

Fig. 3. Derived phases of IHC depolarization relative to BM displacement toward scala tympani. Results are from 10 single-unit studies in the auditory nerve of four species (chinchilla, cat, gerbil and guinea-pig.) The original neural data have been corrected for synaptic/neural and BM travel-time delays as necessary [except for the gerbil data of Oshima and Strelioff (1983) for which no adequate correction data are available] using appropriate click latencies; cat rarefaction latencies were derived from condensation click data of Kim and Molnar (1979) and guinea-pig latencies were taken from results of Evans (1972). 'This study' refers to Ruggero and Rich (1987). The asterisks next to the lines representing the latter results and those of Sellick *et al.* (1982a) indicate that, in both cases, the response phases of high-CF neurons shift by some 90–130 deg at high stimulus intensities.

[Reproduced, with permission, from Fig. 15 of Ruggero and Rich (1987).]

by showing that its responses to very-low-frequency stimuli (100 Hz or less) can be synchronous with BM displacement or motion toward scala tympani. Figure 2 illustrates the stimulation of an auditory-nerve fiber by trapezoidal displacements applied to the round window of a guinea-pig cochlea with a plugged helicotrema (Konishi and Nielsen, 1978). The responses of a fiber with high characteristic frequency (CF; 17.7 kHz) are displayed as arrays of dots, one for each action potential, with a superimposed tracing indicating the endocochlear potential. As is evident from the steady elevation of the endocochlear potential (top of Fig. 2), which indicates hyperpolarization of OHCs (Tasaki *et al.*, 1954), this fiber was tonically excited by a constant displacement of the BM toward scala tympani. Responses such as depicted in Fig. 2 were typical for an overwhelming majority of high-CF auditory-nerve fibers (Konishi and Nielsen, 1978).

Following the studies of Konishi and Nielsen (1973, 1978) and Zwislocki and Sokolich (1973), many other investigations have measured the phases of auditory-nerve fiber responses to low-frequency stimuli. The results of these investigations, summarized in Fig. 3, have often been perplexingly inconsistent. On first examination of Fig. 3, little agreement seems to exist among the various sets of data: at any given CF, neural response phases relative to BM displacement are found in every quadrant. Nevertheless, certain generalizations are possible (Ruggero and Rich, 1987): (i) the phases of responses to very-low-frequency stimuli — for which BM response phases should not change as a function of cochlear location — are often reported to vary systematically with fiber CF; (ii) at near-threshold stimulus levels, high-CF fibers respond roughly synchronously with peak BM displacement or velocity toward scala tympani; (iii) for low-CF fibers, excitation occurs roughly in phase with BM peak velocity toward scala vestibuli. Thus, while the response phases of low-CF auditory-nerve fibers may be consistent with those of apical IHCs, the near-threshold responses of high-CF fibers apparently lead IHC depolarization by at least 90 deg.

Peak Splitting and Intensity-Dependent Phase Shifts in Auditory-Nerve Responses to Low-Frequency Tones

When stimulated by intense low-frequency tones, auditory-nerve fibers can be excited at two distinct phases within each stimulus period (Kiang and Moxon, 1972). This phenomenon, termed 'peak splitting', resembles what might result from a second harmonic component present in the stimulus but is actually intrinsic to the cochlea (Kiang and Moxon, 1972; Sellick *et al.*, 1982a; Gifford and Guinan, 1983; Ruggero

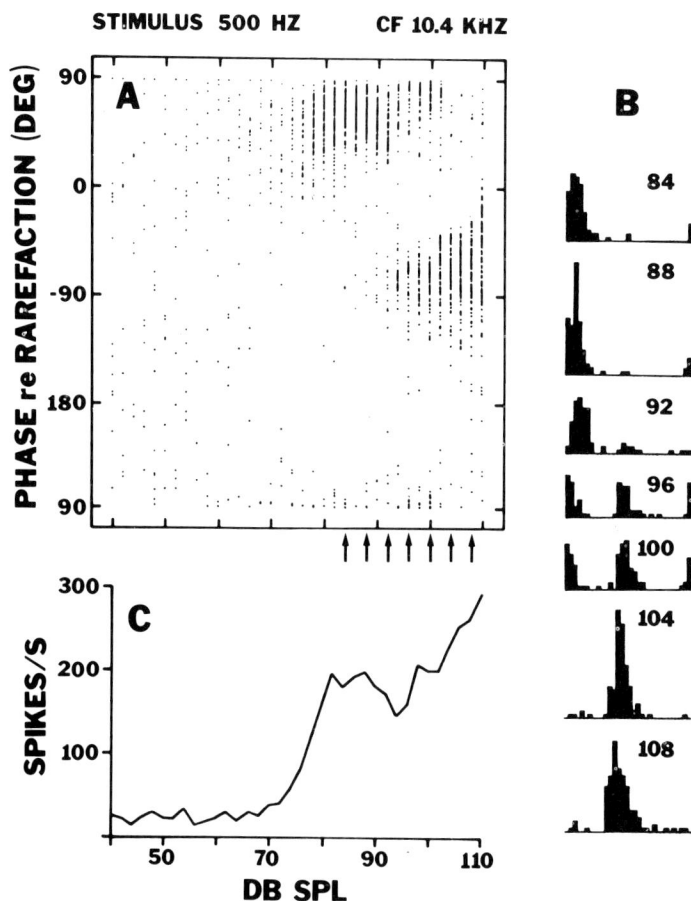

Fig. 4. Intensity dependence in the responses of an auditory-nerve fiber to 500 Hz tones. (A) Dot display showing the response phase of each action potential relative to rarefaction at the eardrum (ordinate) vs. sound level (abscissa). (B) Period histograms constructed for responses at selected stimulus intensities, indicated by arrows under (A). (C) Rate–intensity function for the data of (A). Note the bimodality of response phases at 94–102 dB SPL (A and B) and the rate notch at the same stimulus levels (C). [Reproduced, with permission, from Fig. 1 of Ruggero and Rich (1989).]

and Rich, 1983, 1989; Liberman and Kiang, 1984; Kiang, 1990). For a given stimulus frequency, peak splitting is observable within a narrow range of stimulus intensities, marking an abrupt transition of excitation phase, and it is sometimes accompanied by a notch in the rate–intensity function (Fig. 4). The rate notch is seldom detectable, but peak splitting

CF 800-1600 HZ

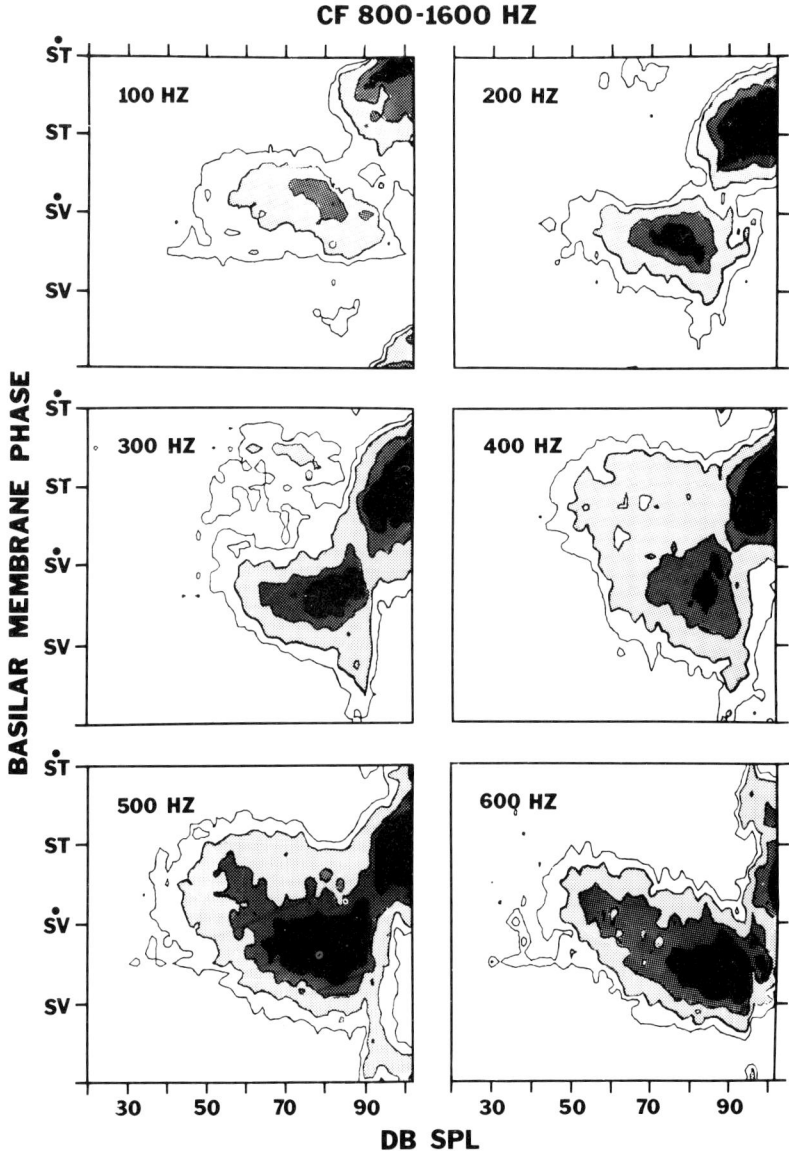

Fig. 5. Intensity dependence of response phases of apical auditory-nerve fibers. The contour plots represent averaged phases, relative to BM vibration, for responses to tones with frequencies 100–600 Hz. In each plot the abscissa indicates stimulus intensity and the ordinate shows the phase of BM motion: ST, maximal displacement toward scala tympani; ṠT, maximal velocity toward scala tympani; SV, maximal displacement toward scala vestibuli; ṠV, maximal velocity toward scala vestibuli. Contour lines are drawn for firing rates of 100, 150, 250 and 350 spikes/sec. The black areas represent rates higher than 350 spikes/s, the darkest gray area represents rates between 250 and 350 spikes/sec, and so on. The contour plot for each frequency represents responses for many neurons from several chinchillas. 100 Hz: 57 neurons (from 10 chinchillas); 200 Hz: 51 (12); 300 Hz: 41 (11); 400 Hz: 44 (10); 500 Hz: 33 (8); 600 Hz: 18 (10). [Reproduced, with permission, from Fig. 2 of Ruggero and Rich (1989).]

is quite common and large phase shifts (ranging in magnitude between 90 and 180 deg) exist in the responses to low-frequency tones of the vast majority of chinchilla auditory–nerve fibers (Ruggero and Rich, 1989). A convincing demonstration of the generality of the phenomenon in the chinchilla is obtained by pooling together the responses of neurons recorded from many cochleae (Fig. 5). In spite of uncertainties in the computation of neural-re-basilar-membrane response phases (including possible errors in calibrating the acoustic stimuli, in measurements of BM responses and in estimating neural/synaptic delays), Fig. 5 makes it clear that most auditory-nerve fibers with CFs between 800 and 1600 Hz undergo a large phase shift (approaching 180 deg) at about 90 dB SPL. Similar phase shifts can be demonstrated in all CF ranges in the auditory nerves of cat and chinchilla (Gifford and Guinan, 1983; Ruggero and Rich, 1989) and in spiral-ganglion cells at the base of the guinea-pig cochlea (Sellick *et al.*, 1982a), and may well be a general feature of mammalian auditory-nerve function.

Absence of Correlates of Intensity-Dependent Phase Shifts in Basilar-Membrane or Outer Hair Cell Responses to Low-Frequency Tones

It is likely that neither peak splitting nor abrupt phase shifts originate in the vibration of the BM:BM responses to low-frequency tones are devoid of harmonic distortion and grow linearly with stimulus intensity, and what intensity-dependent phase shifts occur are confined to near-CF frequencies (Rhode, 1971; Sellick *et al.*, 1982b; Ruggero *et al.*, 1992; reviewed by Ruggero, 1992b). The one study that directly addressed this issue came to the conclusion that, at least in the basal region of the chinchilla cochlea, stimuli that lead to robust and almost universal intensity-dependent phase shifts in auditory-nerve responses elicit BM vibrations whose phases do not vary with stimulus intensity (Ruggero *et al.*, 1991). Thus, the origin of the abrupt phase shifts must be sought in signal transformations that take place beyond the basilar membrane, presumably in the organ of Corti.

If present in receptor potentials, peak splitting should manifest itself as a large second harmonic, and the abrupt phase shifts of neural excitation should appear as commensurate shifts in the phases of the fundamental frequency component. In fact, neither feature is present in responses to low-frequency tones recorded intracellularly from OHCs or extracellularly from the organ of Corti (Cody and Mountain, 1989; Dallos and Cheatham, 1989). Intensity-dependent phase reversals do occur in responses of OHCs to tones with frequency near CF, but not in responses to tones with frequency well below CF (Zwislocki and Smith, 1989; Kössl and Russell, 1992). This contrasts with the abrupt phase shifts in auditory-nerve responses, which are almost universally present for stimulation with low-frequency tones, regardless of CF (Ruggero and Rich, 1989).

Possible Correlates of Intensity-Dependent Phase Shifts in Inner Hair Cell Responses to Low-Frequency Tones

There is stronger evidence that counterparts of peak splitting or intensity-dependent phase shifts are present in the receptor potentials of IHCs. Phase shifts, albeit varying widely in magnitude (between 12 and 210 deg), can be measured routinely in basal IHCs (Cody and Mountain, 1989). Among apical IHCs, however, 'the phenomenon is unusual' (Dallos and Cheatham, 1989). The responses of IHCs to intense, low-frequency tones may exhibit large second-harmonic components, sometimes creating frequency doubling reminiscent of neural peak splitting. Again, this phenomenon is relatively common in basal IHCs (Cody and Mountain, 1989) but less usual among apical IHCs (Dallos and Cheatham, 1989). Current injection experiments in basal IHCs show that frequency doubling in receptor potentials is accompanied by parallel resistance changes (Cody and Mountain, 1989). This suggests that the second-harmonic distortion is present at the input of mechanoelectrical transduction, presumably as the driving force for stereocilia motion.

Can the Low-Frequency Response Phase of Inner Hair Cells and Auditory-Nerve Fibers be Reconciled?

Figure 6 summarizes what is known about the timing of mammalian auditory-nerve excitation, relative to BM displacement, in response to low-frequency stimulation. Auditory-nerve fibers tend to be excited either in phase with peak BM velocity toward scala vestibuli or peak BM displacement/velocity toward scala tympani. Which of these preferred phases is dominant depends on CF (or, equivalently, longitudinal cochlear location) and on stimulus intensity. At any particular stimulus intensity, the neural-re-basilar membrane phase nearly reverses in the region of the cochlea with CFs 3–6 kHz. At any particular CF region, the neural-re-basilar membrane phase also nearly reverses at intensities of about 90 dB SPL. IHC depolarization appears to be at odds with auditory-nerve excitation in at least two respects, namely: (i) at the

VIII–N. EXCITATION re BASILAR MEMBRANE

Stimulus Frequencies ≪ CF

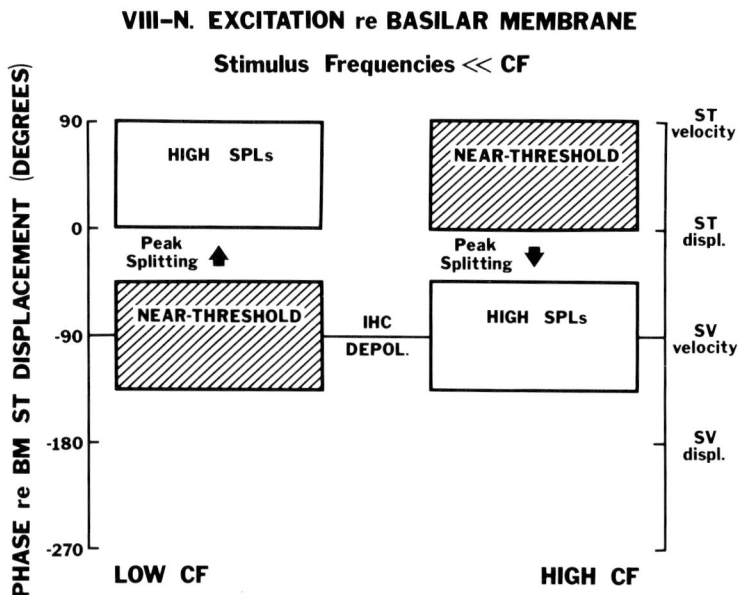

Fig. 6. Schema depicting the proposed excitatory response of IHCs under low-frequency stimulation (i.e. stimulus frequencies lower than CF/2). The rectangles indicate IHC phases derived from auditory-nerve responses. Although based primarily on chinchilla recordings (Ruggero and Rich, 1987, 1989), the phases indicated by the rectangles are probably representative of results in several other species. The solid horizontal line indicates the low-frequency phases usually obtained from intracellular recordings from IHCs. [Reproduced, with permission, from Fig. 2.14 of Ruggero (1992a).]

cochlear base, neural responses to near-threshold, low-frequency tones phase lead IHC depolarization by 90–180 deg; (ii) at the apical half of the cochlea, IHC counterparts of intensity-dependent phase reversals are unusual (Dallos and Cheatham, 1989).

What is the origin of the paradoxical phase mismatch at the base of the cochlea? One possible explanation arises from the observation that, at very low frequencies, the magnitude of the extracellular microphonic can exceed the intracellular receptor potentials of IHCs and dominate the transmembrane voltage (Russell and Sellick, 1983). Under such circumstances, when the microphonic within the organ of Corti is most negative (synchronous with maximal BM displacement toward scala tympani) the transmembrane potential would be at a minimum, leading to transmitter release. This explanation appears unlikely because the electroanatomy of the organ of Corti probably confines the influence of cochlear microphonics upon IHC potentials to the case of intense stimulation at very low frequencies (Dallos, 1983). In fact, the phase lead of neural excitation relative to IHC depolarization extends, at least in the chinchilla, to frequencies as high as several hundred hertz (Figure 3; Ruggero and Rich, 1983, 1987; Ruggero *et al.*, 1986). At these stimulus frequencies, at moderate stimulus levels, the extracellular microphonic is much smaller than the transmembrane receptor potential (Russell and Sellick, 1983).

Three other explanations may be advanced for the phase mismatches between the responses of auditory-nerve fibers and IHCs. (i) The auditory-nerve and IHC data sets may not be strictly comparable, due to either dissimilar stimulation conditions (e.g. different stimulus intensities or frequencies) or genuine species differences (but compare the IHC data of Fig. 1A and the spiral-ganglion data of Sellick *et al.*, 1982a, in Fig. 3, both from the basal region of the guinea pig cochlea). (ii) The determinations of response phases relative to BM motion are incorrect (e.g. due to a hypothetical mismatch between cochlear microphonics and BM displacement). (iii) The recordings from IHCs are flawed (e.g. because the microelectrode interferes with cochlear micromechanics; Zwislocki, 1985).

What Deflects the Stereocilia of Inner Hair Cells?

The nature of the forces that push or pull on the stereocilia of IHCs must reflect the anatomical relation between the stereocilia and the tectorial membrane. One possibility, consistent with the CF dependence of auditory-nerve response phases (Fig. 6), is that the stereocilia of IHCs contact the tectorial membrane only at the cochlear base, where their hair bundles are much longer than those of OHCs (Lim, 1980, 1986). Another possibility is that the IHCs and the tectorial membrane remain unattached throughout the cochlea. In this case, a CF dependence of neural response phases would still be possible if the relative motions of the reticular lamina and the tectorial membrane were to vary as a function of cochlear location (Freeman and Weiss, 1990). Yet a third possibility is that the stereocilia of apical IHCs contact the tectorial membrane only at high stimulus intensities, thereby providing a basis for the intensity-dependent phase reversal. Deciding which of these possibilities, if any, is correct will necessitate much new data and, probably, some technical breakthroughs.

Acknowledgements

This review was written with support of NIH Grants DC-00110 and

DC-00419. I thank M. A. Cheatham, P. Dallos and N. C. Rich for their helpful comments.

References

Békésy, G. von (1951). Microphonics produced by touching the cochlear partition with a vibrating electrode. *J. Acoust. Soc. Am.* **23:** 29–35.

Billone, M. and Raynor, S. (1973). Transmission of radial shear forces to cochlear hair cells. *J. Acoust. Soc. Am.* **54:** 1143–1156.

Cody, A. R. and Mountain, D. C. (1989). Low-frequency responses of inner-hair cells: evidence for a mechanical origin of peak splitting. *Hearing Res.* **41:** 89–100.

Dallos, P. (1983). Some electrical circuit properties of the organ of Corti. I. Analysis without reactive elements. *Hearing Res.* **12:** 89–119.

Dallos, P. (1984). Some electrical circuit properties of the organ of Corti. II. Analysis including reactive elements. *Hearing Res.* **14:** 281–291.

Dallos, P. and Cheatham, M. A. (1976). Production of cochlear potentials by inner and outer hair cells. *J. Acoust. Soc. Am.* **60:** 510–512.

Dallos, P. and Cheatham, M. A. (1989). Nonlinearities in cochlear receptor potentials and their origins. *J. Acoust. Soc. Am.* **86:** 1790–1796.

Dallos, P. and Santos-Sacchi, J. (1983). AC receptor potentials from hair cells in the low-frequency region of the guinea pig cochlea. In *Mechanisms of Hearing* (W. R. Webster and L. Aitkin, eds), pp. 11–16. Monash University Press, Clayton, Victoria, Australia.

Dallos, P., Billone, M. C., Durrant, J. D., Wang, C.-Y. and Raynor, S. (1972). Cochlear inner and outer hair cells: functional differences. *Science* **177:** 356–358.

Davis, H. (1958). Transmission and transduction in the cochlea. *Laryngoscope* **68:** 359–382.

Derbyshire, A. J. and Davis, H. (1935). The action potentials of the auditory nerve. *Am. J. Physiol.* **113:** 476–504.

Evans, E. F. (1972). The frequency response and other properties of single fibers in the guinea pig cochlear nerve. *J. Physiol.* **226:** 263–287.

Flock, Å. (1971). Sensory transduction in hair cells. In *Handbook of Sensory Physiology*, Vol. 1: *Principles of Receptor Physiology* (W. R. Lowenstein, ed.), pp. 396–441. Springer-Verlag, Berlin.

Freeman, D. M. and Weiss, T. F. (1990). Hydrodynamic forces on hair bundles at low frequencies. *Hearing Res.* **48:** 17–30.

Gifford, M. L. and Guinan, Jr., J. J. (1983). Effects of crossed olivocochlear-bundle stimulation on cat auditory nerve fiber responses to tones. *J. Acoust. Soc. Am.* **74:** 115–123.

Goldstein, M. H. (1968). The auditory periphery. In *Medical Physiology* (V. B. Mountcastle, ed.), pp. 1465–1498. Mosby, St Louis, MO.

Hudspeth, A. J. and Corey, D. P. (1977). Sensitivity, polarity and conductance change in the response of vertebrate hair cells to controlled mechanical stimuli. *Proc. Natl. Acad. Sci. USA* **76:** 2407–2411.

Ingvarsson, K. (1981). Dynamics of cochlear hair cells as inferred from cochlear potentials and responses from auditory nerve fibers. Ph. D. thesis, Northwestern University, Evanston, IL.

Kiang, N. Y. S. (1990). Curious oddments of auditory-nerve studies. *Hearing Res.* **49:** 1–16.

Kiang, N. Y. S. and Moxon, E. C. (1972). Physiological considerations in artificial stimulation of the inner ear. *Ann. Otol. Rhinol. Laryngol.* **81:** 714–730.

Kim, D. O. and Molnar, C. E. (1979). A population study of cochlear nerve fibers: comparison of spatial distributions of average-rate and phase-locking measures of responses to single tones. *J. Neurophysiol.* **42:** 16–30.

Konishi, T. and Nielsen, D. W. (1973). Temporal relationship between motion of the basilar membrane and initiation of nerve impulses in the auditory nerve fibers. *J. Acoust. Soc. Am.* **53:** 325.

Konishi, T. and Nielsen, D. W. (1978). The temporal relationship between basilar membrane motion and nerve impulse initiation in auditory nerve fibers of guinea pigs. *Jap. J. Physiol.* **28:** 291–307.

Kössl, M. and Russell, I. J. (1992). The phase and magnitude of hair cell receptor potentials and frequency tuning in the guinea pig cochlea. *J. Neurosci.* **12:** 1575–1586.

Liberman, M. C. and Kiang, N. Y. S. (1984). Single-neuron labeling and chronic cochlear pathology. IV. Stereocilia damage and alterations in rate- and phase-level functions. *Hearing Res.* **16:** 75–90.

Lim, D. J. (1980). Cochlear anatomy related to cochlear micromechanics. A review. *J. Acoust. Soc. Am.* **67:** 1686–1695.

Lim, D. J. (1986). Functional structure of the organ of Corti: a review. *Hearing Res.* **22:** 117–146.

Maceri, D. R., Sokolich, W. G. and Strelioff, D. (1982). A reinvestigation of the response phase of auditory nerve fibers in guinea pigs. *Abstr. Assoc. Res. Otolaryngol. Midwinter Meeting* Vol. 5, p. 10.

Nuttall, A. L., Brown, M. C., Masta, R. I. and Lawrence, M. (1981). Inner hair cell responses to the velocity of basilar membrane motion in the guinea pig. *Brain Res.* **211:** 171–174.

Oshima, W. and Strelioff, D. (1983). Responses of gerbil and guinea pig auditory nerve fibers to low-frequency sinusoids. *Hearing Res.* **12:** 167–184.

Pfeiffer, R. R. and Kim, D. O. (1975). Cochlear nerve fiber responses: distribution along the cochlear partition. *J. Acoust. Soc. Am.* **58:** 867–869; erratum: *J. Acoust. Soc. Am.* **60:** 966 (1976).

Rhode, W. S. (1971). Observations of the vibration of the basilar membrane in squirrel monkeys using the Mössbauer technique. *J. Acoust. Soc. Am.* **49:** 1218–1231.

Ruggero, M. A. (1992a). Physiology and coding of sound in the auditory nerve. In *The Mammalian Auditory Pathway: Neurophysiology* (A. N. Popper and R. R. Fay, eds), pp. 34–93. Springer-Verlag, New York.

Ruggero, M. A. (1992b). Responses to sound of the basilar membrane of the mammalian cochlea. *Curr. Opin. Neurobiol.* **2:** 449–456.

Ruggero, M. A. and Rich, N. C. (1983). Chinchilla auditory-nerve responses to low-frequency tones. *J. Acoust. Soc. Am.* **73:** 2096–2108.

Ruggero, M. A. and Rich, N. C. (1987). Timing of spike initiation in cochlear afferents: dependence on site of innervation. *J. Neurophysiol.* **58:** 379–403.

Ruggero, M. A. and Rich, N. C. (1989). 'Peak splitting': intensity effects in cochlear afferent responses to low frequency tones. In *Cochlear Mechanisms – Structure, Function and Models* (J. P. Wilson and D. T. Kemp, eds), pp. 259–266. Plenum, New York.

Ruggero, M. A., Rich, N. C. and Robles, L. (1991). Comparison of cochlear-nerve and basilar-membrane responses to low-frequency tones: absence of macromechanical basis for 'peak splitting'. *Abstr. Assoc. Res. Otolaryngol. Midwinter Meeting* Vol. 14, p. 78.

Ruggero, M. A., Robles, L. and Rich, N. C. (1986). Basilar membrane mechanics at the base of the chinchilla cochlea. II. Responses to low-frequency tones and relationship to microphonics and spike initiation in the VIII-nerve. *J. Acoust. Soc. Am.* **80:** 1375–1383.

Ruggero, M. A., Rich, N. C. and Recio, A. (1992). Basilar membrane responses to clicks. In *Auditory Physiology and Perception* (Y. Cazals, L. Demany and K. Horner, eds), pp. 85–91. Pergamon Press, London.

Russell, I. J. and Sellick, P. M. (1983). Low-frequency characteristics of intracellularly recorded receptor potentials in guinea-pig cochlear hair cells. *J. Physiol.* **338:** 179–206.

Russell, I. J., Richardson, G. P. and Cody, A. R. (1986). Mechanosensitivity of mammalian auditory hair cells *in vitro. Nature* **321:** 517–519.

Sellick, P. M., Patuzzi, R. and Johnstone, B. M. (1982a). Modulation of responses of spiral ganglion cells in the guinea pig cochlea by low frequency sounds. *Hearing Res.* **7:** 199–221.

Sellick, P. M., Patuzzi, R. and Johnstone, B. M. (1982b). Measurement of basilar membrane motion in the guinea pig using the Mössbauer technique. *J. Acoust. Soc. Am.* **72:** 131–141.

Sokolich, W. G., Hamernick, R. P., Zwislocki, J. J. and Schmiedt, R. A. (1976). Inferred response polarities of cochlear hair cells. *J. Acoust. Soc. Am.* **59:** 963–974.

Tasaki, I., Davis, H. and Eldredge, H. (1954). Exploration of cochlear potentials in guinea pigs with a microelectrode. *J. Acoust. Soc. Am.* **26:** 765–773.

Wilson, J. P. and Johnstone, J. R. (1975). Basilar membrane and middle-ear vibration in guinea pig measured by capacitive probe. *J. Acoust. Soc. Am.* **57:** 705–723.

Zwislocki, J. J. (1985). Cochlear function – an analysis. *Acta Otolaryngol.* **100:** 201–209.

Zwislocki, J. J. and Smith, R. L. (1989). Phase reversal in OHC response at high sound intensities. In *Cochlear Mechanisms – Structure, Function and Models* (J. P. Wilson and D. T. Kemp, eds), pp. 163–168. Plenum, New York.

Zwislocki, J. J. and Sokolich, W. G. (1973). Velocity and displacement responses in auditory-nerve fibers. *Science* **182:** 64–66.

The Pharmacology of the Outer Hair Cell Motor

J. F. ASHMORE, F. MAMMANO, J. E. GALE AND M. J. TUNSTALL

Department of Physiology, School of Medical Sciences, University of Bristol, Bristol BS8 1TD, UK

Introduction

One of the more exciting developments in the study of cochlear mechanisms over the past decade has been the realization that the properties of outer hair cells (OHCs) may be able to account for the majority of the features of sensitivity and tuning in the mammalian cochlea. In turn, the critical property of OHCs, the ability of the cells to generate forces rapidly and so enhance cochlear mechanics, may itself be determined by the biophysical properties and characteristics of an electrotransducer or 'motor molecule' packed in a dense array on the basolateral membrane of the cell, (e.g. Ashmore, 1992, 1994; Dallos *et al.*, 1993).

In other neurobiological systems, an essential part of the programme of identifying and characterizing any biological membrane is finding how its properties are altered by external agents. Such agents can then be used as probes to characterize the molecules embedded in the membrane. To this end we have identified agents which specifically modify the properties of the OHC 'motor' and which give clues about the nature of the protein and how it interacts with the hair cell membrane.

Several agents have been identified as inhibitors of fast force generation in OHCs, including polyvalent cations and ions of the lanthanide series (Flock *et al.*, 1986; Santos-Sacchi, 1991; Ashmore, 1994). Surprisingly, agents that interfere with cell motility in other systems do not have an effect (Holley and Ashmore, 1988), and all the evidence points to the basolateral membrane, where the membrane and underlying cytoskeleton meet, as the site of OHC motility. Nevertheless, there are mechanical stimuli including osmotic stresses (producing compression and distension of the cell membrane) which affect the motor activity. It seems most likely that these latter inhibitory effects arise from a strong thermodynamic coupling between the motor molecule and its mechanical environment (Iwasa, 1993; Gale and Ashmore, 1994).

Amongst the organic compounds which block the OHC force generator are the salicylates, present in the body after aspirin ingestion. Aspirins have been known for a long time to interact with the peripheral auditory system where the effects include reversible elevation of thresholds (Myers and Bernstein, 1965; Douek *et al.*, 1986) reduction of otoacoustic emissions (McFadden and Plattsmier, 1984) and tinnitus. Various explanations have been advanced for these actions, most by invoking an action of salicylates on the OHCs. Nevertheless, because of the delicate mechanical organization of the cochlea, such drug interventions could produce a variety of effects on other cochlear subsystems (Stypulkowski, 1990). Sites which arise from alteration of the cochlear circulation have also received attention (Didier *et al.*, 1993).

The purpose of the present work is to pinpoint the mode of action of salicylates and related compounds using *in vitro* systems of (i) the isolated cochlea and (ii) the OHCs. It will be shown how these preparations allow direct functional monitoring of the motor molecule.

Methods

All experiments used preparations derived from adult albino guinea-pigs (200–300 g). The isolated temporal bone preparation (Ulfendahl *et al.*, 1989) was used to drive assemblies of hair cells *in situ* by electrical current. The cochlea was opened to allow access via scala media to the cochlear partition in the third and fourth turns (Mammano and Ashmore, 1993). This approach allowed access to the reticular lamina and the upper surface of the basilar membrane while opening the feedback loop involving the forward mechanotransducer. Once the tectorial membrane had been peeled away, a 5 μm diameter glass stimulating electrode was placed over OHCs under visual control through a ×40 WI objective. Intracellular microelectrodes were also positioned under visual control.

Motion of the basilar membrane and the reticular lamina were measured using a Michelson interferometer driven from a stabilized laser source to measure the motion along the optical axis of 10 μm diameter silvered glass beads. The interferometer was designed so that it could be kept in quadrature under software control. The cochlea was perfused with an oxygenated solution with the composition of perilymph (in mM): Na, 147; K, 2.5; Ca, 1; Mg, 2; Cl, 140; phosphate buffered to pH 7.4) and kept iso-osmotic (320 mOsm) by addition of D-glucose where appropriate.

In the second system, isolated cells were studied using conventional patch recording methods (Housley and Ashmore, 1992). They were recorded in a chamber mounted on an inverted microscope. Cells were recorded in a small (100 μl) volume bath with a solution exchange time of 20 sec. For pH_i recording, cells were loaded with 50 μM BCECF Dextran (MW 10,000) via the patch pipette and the fluorescence measured with a transputer-based imaging system (Kolston and Ashmore, 1992). The pH_i values were calibrated using the measured change of fluorescence when 5 mM solutions of a weak acid and a weak base were applied in sequence (Buckler *et al.*, 1992). This null-point method was preferred over the more conventional use of nigericin.

Cell membrane capacitance was measured using sinusoidal voltage commands applied to the cell during whole cell recording (Fidler and Fernandez, 1989). pipettes were coated with ski wax to reduce capacitative loss and this proved essential to maximize the sensitivity of the capacitance measurement. The capacitance measuring method used a software implementation of a lock-in amplifier. Phase and amplitude changes in the measured currents were detected when a 10 mV amplitude sinusoidal voltage command was applied at a fixed frequency, typically at 1.5 kHz (Ashmore *et al.*, 1994). This technique allowed the membrane capacitance to be measured at approximately 10 samples/ sec for on-line computations and allowed the full range of the voltage-dependent capacitance to be measured in less than 2 sec. Experiments in both systems were carried out at 24–30 deg C.

Results

Effects of salicylate on tuning in the whole organ of Corti

The normal pattern of motion of the basilar membrane in response to current passed across the partition was a damped oscillation of amplitude up to 5 nm at a characteristic frequency that corresponded to the recorded site of the basilar membrane. The amplitude of the reticular lamina motion was 4–5 times larger, but in antiphase (Mammano and Ashmore, 1993). The motion showed no sign of saturation for currents less than 200 μA. The membrane moved towards scala media when current flowed into a pipette placed above the organ of Corti. The effect of transcochlear current was also measured using

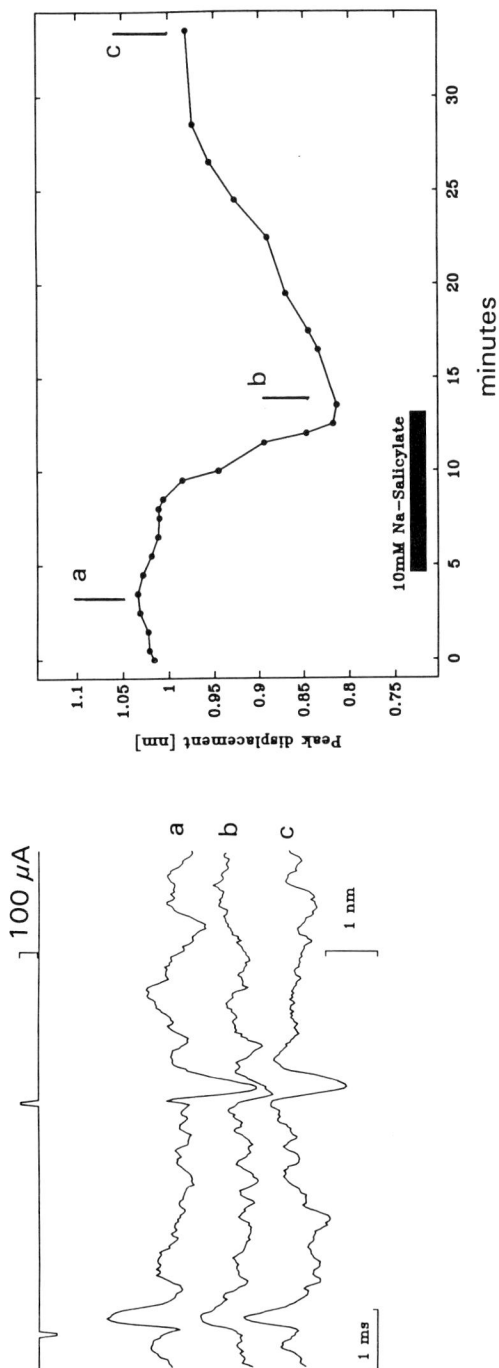

Fig. 1. Effect of salicylate superfusion on the motion of the cochlear partition produced by transcochlear electrical stimulation. The response of the basilar membrane to 50 μsec current pulse represents current flowing into scala media, producing a basilar membrane motion towards scala media. The amplitude of the motion of the cochlear partition was reduced by 20% on superfusion with 10 mM salicylate. The effect was reversible with short (min) duration applications. Applications for 10 min produced irreversible changes

intracellular microelectrodes placed in the OHCs under visual control and it was found that this direction of current caused a depolarizing change in the potential across the cells' basolateral membrane. Conversely current passing from scala media to scala tympani hyperpolarized the cells. The data are consistent with the interpretation that OHCs generate sufficient force to distort the organ of Corti.

When 10 mM salicylate was added to the bath, a reduction in the observed amplitude of the motions of both basilar membrane and reticular lamina occurred (Fig. 1). The reduction was 20–50% but never complete probably because the perfusion was limited by the reticular lamina boundary. The effect was reversible when applications lasted less than 10 min. Such results, however, can be most economically explained if salicylate acted on the the OHCs alone. The current flow around the cells is complex and it remains possible that a secondary effect of salicylate could arise from an action on conductances of other cells of the partition. This seems unlikely as in recordings from isolated cells of the partition, 5 mM salicylate was found to reduce membrane conductance and would therefore have lead to an expected increase in OHC polarization.

Effects of salicylate on pH_i

Salicylate is a weak acid (pK_a = 2.9) and, as salicylic acid, readily permeates cell membranes. The resulting cell acidification occurs rapidly as the acid dissociates in the cytoplasm and has a variety of effects on the cell metabolism. Effects of incubation of OHCs in salicylate have been described which include disruption of the lateral cisternal system (Dieler *et al.*, 1991), and these effects may well be a consequence of cell acidification.

To investigate the effects on pH_i, the fluorescent dye BCECF (either introduced through the patch pipette or as the AM ester) was used on isolated cells. Salicylate 5 mM produced a fall in pH_i of 0.5–0.8 units. The acidification recovered with a pH_i overshoot. This is characteristic of pH_i regulation processes, involving Na:H and bicarbonate exchangers, that have been seen in other systems.

As described previously (Ashmore, 1990; Dieler *et al.*, 1991), salicylates reduced cell motility. It was found that the effects of short (100–300 sec) bath perfusions of 5 mM salicylate were reversible under the experimental conditions used. Outward currents in the cell were reduced by 15%. There was a small decrease in the length of the cell during the application of the drug when the cells were not voltage-clamped under whole cell recording conditions. These length changes can be

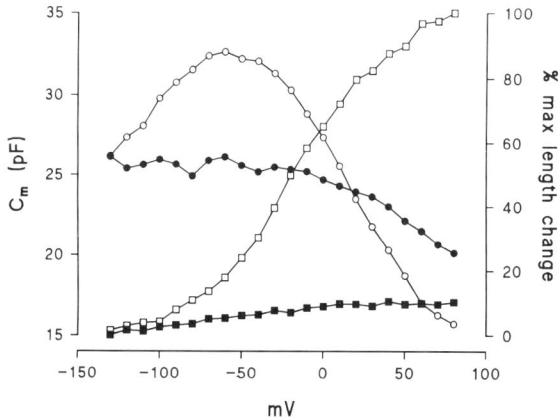

Fig. 2. Effect of salicylate on rapid changes of length. Membrane capacitance (circles) and normalized cell length changes (squares) measured simultaneously before (open symbols) and during (closed symbols) 5 mM salicylate application. The cell length changes, measured with a photodiode, were reduced by over 80%. Cell held at −50 mV and stepped to the potentials shown.

most simply be explained by salicylate loading of the cell and the consequent osmotic imbalance (see Discussion).

Effects of salicylate on gating charge movement

Salicylate reduced the gating charge movement in hair cells (Ashmore, 1990). This charge, $q_T(V)$, is thought to arise from the conformational change in the motor protein (Santos-Sacchi, 1991; Ashmore, 1992). Since the voltage-dependent membrane capacitance $C_{nl}(V)$ is related to charge movement by $C_{nl} = dq_T/dV$, these measurements were carried out using a lock-in amplifier to measure membrane capacitance directly but using a software implementation. This method gave a fast on-line measurement of capacitance. Bath application of 5 mM salicylate reduced the voltage-dependent capacitance and motility almost completely (Figs 2 and 3).

The reduction in cell motility was not simply due to the effect of pH_i. Before the salicylate was applied, 10 mM butyrate produced a comparable change in pH_i but little effect on the cell capacitance. During application of such a weak acid there was little detectable shift in the position of the maximum of the capacitance, C_{nl}. The simplest explanation is that the salicylate reduces activity of the motor independently of cell pH_i and without altering the electric field in which the motor operates, i.e. salicylate acts in a voltage-independent manner.

Dose–response curve for the action of salicylate

Continuous measurement of voltage-dependent cell capacitance also gives a rapid method of measuring salicylate dose response. The membrane capacitance was measured while the salicylate was slowly perfused on to the cell held at –50 mV. The application was carried out sufficiently slowly to ensure that the action of the drug was in near equilibrium with the cell and not diffusion limited. To monitor the concentration of salicylate in the bath, a fluorescent probe was also added (Fig. 4) and allowed the concentration to be determined directly. The results of these experiments show that half-maximal inhibition of C_{nl} occured with K_m = 3.9 mM and with Hill coefficient, n = 3.45. Holding the cell at a range of holding potentials from –100 mV to +50 mV gave identical values for K_m and n. This also supports the idea that salicylate acts in a voltage-independent manner.

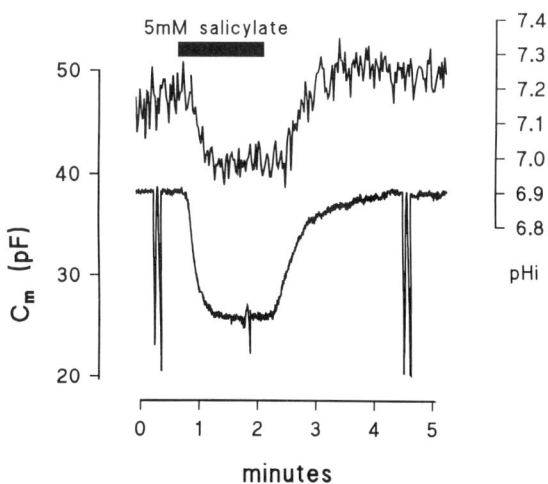

Fig. 3. Effect of 5 mM salicylate on the voltage-dependent capacitance of an OHC during simultaneous pH measurement. Lower trace, membrane capacitance measured under whole-cell voltage-clamp as described in text. Upper trace, change in pH_i. 50 μM BCECF in the patch pipette. The transients in the capacitance traces arise when voltage command ramps were used to measure membrane capacitance.

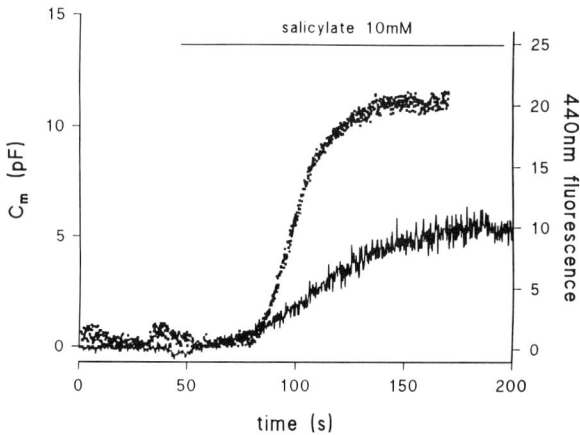

Fig. 4. Salicylate dose–response curve. Simultaneous recording of cell capacitance (dots) and fluorescence (continuous line). FITC dextran (10 μM; MW 10,000) was added along with salicylate solution and the fluorescence at 440 nm measured near the cell. The half-maximal suppression of capacitance was measured at –50 mV. Note that the fluorescence began to rise before the change of cell capacitance, indicating a co-operative action of salicylate.

Discussion

The experiments on the effects of salicylate provide evidence that this agent selectively inhibits the OHC force generator in the organ of Corti. It can be further shown that the OHC motor can be inhibited by salicylate and other amphiphilic anions with similar aromatic ring structures. The pharmacokinetics and voltage independence is consistent with a scheme where salicylate enters the membrane in the unionized form (i.e. as salicylic acid) and then acts on the hair cell motor directly.

Models for the motor molecule

The value of the Hill coefficient (3.45) is consistent with the 'motor' being able to bind three or four salicylates. The simplest model (model 1; Fig. 5) is one where each motor molecule (M) has 4 binding sites (S) and occupancy of all these sites by salicylate inhibits the conformation change, i.e.

$$4S + M \leftrightarrow (4S.M)^*$$

Model 1   Model 2   Model 3

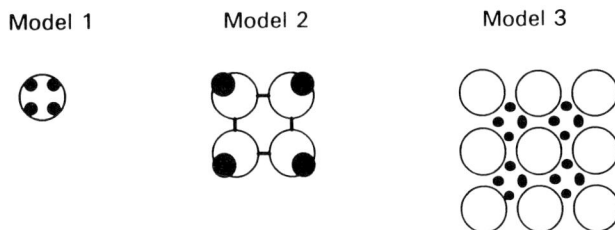

Fig. 5. Models for the OHC motor. Three models discussed in the text. Solid symbols represent salicylate molecules, open circles represent the motor protein. In model 3 each 'motor' itself may be composed of subunits.

where * signifies an inactive (i.e. voltage-independent) configuration of the motor.

There have been suggestions that the motor is a tetramer (Kalinec *et al.*, 1992). Model 2, therefore, is one where each component motor molecule binds a single salicylate molecule:

$$S + M \leftrightarrow (S.M).$$

Only when all four subunits of the tetramer are bound will the motor be inhibited. There are few data that distinguish between models 1 and 2 since the nature and identity of the subunits has yet to be determined.

Nevertheless, the high value obtained for the binding constant, K_m, is not readily compatible with a specific binding to a protein (e.g. the binding of aspirin to cyclooxygenase has a binding coefficient measured in the micromolar range). The value of K_m therefore suggests that salicylate action may involve the membrane in which the motor is embedded.

Model 3 is one where the agonist inserts into the membrane. This would be expected of an amphiphilic compound such as salicylic acid. The motor particle appears as a 12 nm diameter particle in freeze etch images (Kalinec *et al.*, 1992). If the motor is tightly packed, it is possible to imagine that salicylate inserts into the lipid environment between the motor particles. It should be noted that if salicylate were in thermodynamic equilibrium with the cell membrane, with the lipid offering no diffusion boundary, a concentration of 3.9 mM (K_m) corresponds to about four molecules absorbed per 12 nm × 12 nm of membrane. This argument suggests that if we identify the particle observed in freeze replicas with the motor, inhibition occurs when salicylate inserts regularly into the motor lattice and only in this configuration is charge movement prevented. It suggests that lipophilic group structure may be critical for the action of this class of agent.

The significance of the anionic side group of the molecule is not so clear. The absence of any potential shift in the binding (Fig. 2) suggests that screening of the motor voltage-sensor due to the negative charge on the carboxyl group cannot be strong. It should be noted that surface charge effects introduced by salicylate's negative charge has been proposed as a mechanism to explain its action on bilayer channels (McLaughlin, 1973). The voltage independence of salicylate inhibition is in marked contrast to the effects of positively charged cations such as gadolinium. In that case both hair cell motility is inhibited and the potential dependence of gating charge (and hence membrane capacitance) shifts along the voltage axis (Santos-Sacchi, 1991).

A further difference between amphiphilic anions and polyvalent cations is that the latter have an effect on the gating currents at much lower concentrations (i.e. 10–500 μM; Santos-Sacchi, 1991; Gale, unpublished results). The most likely source of these differences is that such ions may be selectively adsorbed onto the membrane surface and produce strong charge screening effects; such ions may also bind to the motor protein itself.

When the motor is inhibited it is reasonable to ask whether it becomes 'fixed' in a particular configuration. Gating charge movement models link the gating charges to conformational changes of the motor protein (Santos-Sacchi, 1991; Ashmore, 1992; Iwasa, 1993). Such models also suggest that, in order to explain the length changes in OHCs, the motor molecule occupies the largest area in the plane of membrane at negative membrane potential. The present data suggest that the cells should become longer in the presence of salicylate if the motor is 'frozen out' into a high-area configuration, equivalent to the hyperpolarized state. Experimentally the effect is complicated since the cell loads with salicylate, thereby causing the cell to swell. In addition, the amphilic nature of salicylic acid may also alter the interaction between the motor proteins. Nevertheless, it should be noted that the resulting changes in the mechanical properties of the cells and in the basolateral membrane will have significant effects on the macroscopic properties of the cochlear partition.

Acknowledgements

This work was supported by the Medical Research Council, the Wellcome Trust, the EC (Esprit SSS 6961) and the Hearing Research

Trust. We thank Dr Paul Kolston for critical assistance with these experiments.

References

Ashmore, J. F. (1990). Effect of salicylate on a rapid charge movement in outer hair cells isolated from the guinea-pig cochlea. *J. Physiol.* **412**: 46P.

Ashmore, J. F. (1992). Mammalian hearing and the cellular mechanisms of the cochlear amplifier. In *Sensory Transduction* (S. Roper and D. P. Corey, eds), pp. 396–412. Rockefeller University Press, New York.

Ashmore, J. F (1994). The cellular machinery of the cochlea. *Exp. Physiol.* **79**: 113–134.

Ashmore, J. F., Gale, J. E. and Tunstall, M. J. (1994). Capacitance measurements using a CED1401 interface. *J. Physiol.* **476**: 7P.

Buckler, K. I., Vaughan-Jones,R. D., Peers, C. and Nye, P. C. G. (1991). Intracellular pH and its regulation in isolated type I carotid body cells of the neonatal rat. *J. Physiol.* **436**: 107–130.

Carlyon, R. P. and Butt, M. (1993). Effects of aspirin on human auditory filters. *Hearing Res.* **66**: 233–244.

Dallos, P., Hallworth, R. and Evans, B. N. (1993). Theory of electrically driven shape changes of cochlear outer hair cells. *J. Neurophysiol.* **70**: 290–323.

Didier, A., Miller, J. M. and Nuttall, A. L. (1993). The vascular component of sodium salicylate ototoxicity in the guinea pig. *Hearing Res.* **69**: 199–206.

Dieler, R., Shehata, W. E. and Brownell, W. E. (1990). Concomitant salicylate-induced alterations of outer hair cell subsurface cisternae and electromotility. *J. Neurocytol.* **20**: 637–653.

Douek, E. E., Dodson, H. C. and Bannister, L. H. (1983). The effects of salicylate on the cochlea of the guinea-pig. *J. Laryngol. Otol.* **93**: 793–799.

Fidler, N. and Fernandez, J. M. (1989). Phase-tracking: an improved detection technique for cell membrane capacitance measurements. *Biophys. J*, **56**: 1153–1162.

Flock, Å., Flock, B. and Ulfendahl, M. (1986). Mechanism of movement in outer hair cells and a possible structural basis. *Arch Otolaryngol.* **243**: 83–90.

Gale, J. E. and Ashmore, J. F. (1994). Charge displacement induced by rapid stretch in the basolateral membrane of the guinea pig outer hair cell. *Proc. R. Soc. Lond. B* **255**: 243–249.

Holley, M. C. and Ashmore, J. F. (1988). On the mechanism of a high frequency force generator in outer hair cells isolated from the guinea pig cochlea. *Proc. R. Soc. Lond. B* **232**: 413–429.

Housley, G. D. and Ashmore, J. F. (1992). Ionic currents of outer hair cells isolated from the guinea-pig cochlea. *J. Physiol.* **448**: 73–98.

Iwasa K. H. (1993). Effect of stress on membrane capacitance of the auditory outer hair cell. *Biophys. J.* **65**: 492–498.

Kalinec, F., Holley, M. C., Iwasa, K. H., Lim, D. J. and Kachar, B. (1992). A membrane-based force generation mechanism in auditory sensory cells. *Proc. Natl. Acad. Sci. USA* **89**: 8671–8675.

Kolston, P. J. and Ashmore, J. F. (1992). Using transputers to measure calcium changes in isolated cochlear cells. *J. Physiol.* **452**: 170P.

Mammano, F. and Ashmore, J. F. (1993). Reverse transduction measured in the isolated cochlea by laser Michelson interferometry. *Nature* **365**: 838–841.

McLaughlin, S. (1973). Salicylates and phospholipid bilayers. *Nature* **243**: 234–236.

McFadden, D. and Plattsmier, H. S. (1984). Aspirin abolishes spontaneous otoacoustic emissions. *J. Acoust. Soc. Am.* **76**: 443–448.

Santos-Sacchi, J. (1991). Reversible inhibition of voltage-dependent outer hair cell motility and capacitance. *J. Neurosci.* **11**: 3096–3110.

Shehata, W. E., Brownell, W. E. and Dieler, R. (1991). Effects of salicylate on shape, electromotility and membrane characteristics of isolated outer hair cells from guinea-pig cochlea. *Acta Otolaryngol.* **111**: 707–718.

Stypulkowski, P. H. (1990). The mechanisms of salicylate ototoxicity. *Hearing Res.*, **46**: 113–146.

Ulfendahl, M., Flock, Å. and Khanna, S. M. (1989). A temporal bone preparation for the study of cochlear micromechanics at the cellular level. *Hearing Res.* **40**: 55–64.

Cochlear Plasticity: Effects of Sound Overexposure on Evoked Otoacoustic Emissions

BRENDA L. LONSBURY-MARTIN, GLEN K. MARTIN AND
MARTIN L. WHITEHEAD

*Department of Otolaryngology, University of Miami Ear Institute,
Miami, FL 33101, USA*

Introduction

It is well accepted that acoustic overstimulation causes irreversible hearing loss. This knowledge is based on historical observations of poor hearing in workers who toiled in noisy environments, and on the results of many years of experimental work in animal models using deliberate exposures to excessive sound. Along with such irrevocable outcomes, other observances support the notion that a human ear possesses innate properties that make it either more or less vulnerable to the potentially damaging effects of loud sounds. The concepts of susceptibility and resistance are grounded in natural observations that in occupations involving exposure to intense noise, the ears of different workers are unique in terms of the amount of job-related hearing loss they eventually exhibit.

Deliberate exposure of various animal models to discrete episodes of excessive sound has been a long-used strategy in investigating the fundamental nature of the processes that cause noise-induced hearing loss (NIHL). A tactic typical of early studies was to subject experimental animals to a single exposure of a continuous, intense sound in order to document the functional and anatomical effects of overstimulation. However, more recent studies have used either chronic or repeated episodes of overexposure to more moderate sounds to mimic more reasonably the acoustic conditions typically encountered by humans.

The findings of many of these latter studies have reinforced the practically based view that a given ear is more or less vulnerable to the potentially harmful effects of noisy sounds, and that this ability can be modified somewhat by experience.

In a series of studies on the ability of prior acoustic exposure to alter susceptibility to subsequent overstimulation, intermittent re-exposures of chinchillas to a moderate, low-frequency octave band of noise (OBN) were shown to produce considerable recovery of the initially observed deficits in behavioral thresholds and cochlear-nerve fiber activity (Clark *et al.*, 1987; Sinex *et al.*, 1987). Other investigators using a similar OBN exposure demonstrated the ability of the auditory-brainstem response (ABR) in chinchillas to recover from the initial detrimental effects of overstimulation during repeated exposures to the same noise (Subramaniam *et al.*, 1991a,b; Henderson *et al.*, 1992). Finally, by exposing guinea-pigs to a non-damaging level of a low-frequency tonal stimulation, Canlon *et al.* (1988) showed an increased resistance of ABRs to a subsequent exposure to the identical stimulus at higher levels that were known to be harmful. Together, the experimental findings in animal models provide basic evidence that the peripheral auditory apparatus is capable of developing activity-dependent processes that make it less susceptible to injury from overuse.

Evoked otoacoustic emissions (EOAEs), discovered over 15 years ago (Kemp, 1978), are assumed to reflect the natural vibromechanical response of outer hair cells (OHCs) to incoming sounds (Brownell, 1990). The results of early animal research on EOAEs using experimental methods known to adversely influence cochlear function (e.g. acoustic overstimulation, ototoxic-drug dosing, depletion of oxygen supplies; see review in Whitehead *et al.*, 1994) suggested that emitted responses would be useful as objective, non-invasive indicators of the status of OHC function. It is well established from histopathologic studies of both human temporal bones and intentionally sound-exposed animal cochleae that OHCs are preferentially damaged by excessive sounds during the initial stages of NIHL. Thus, the EOAE testing of the micromechanically based activity of OHCs is ideally suited for investigating the processes underlying the basic nature of the ear's response to overstimulation. Of the different subclasses of EOAEs that have been distinguished on the basis of the type of auditory stimulus needed to elicit them (see review in Probst *et al.*, 1991), distortion-product otoacoustic emissions (DPOAEs) measured in animals have the benefits of being robust, rapid to measure, straightforward to interpret, and frequency specific for stimuli as high as, at least, ~20 kHz (e.g. Lonsbury-Martin *et al.*, 1987; Martin *et al.*, 1987).

Such features would clearly be useful in examining the cochlea's ability to develop processes that protect it from sound overexposure.

In our own laboratory, we have been conducting a series of studies on noise-induced changes in ear function using DPOAEs in both experimental animal and human models. By using DPOAEs to evaluate the condition of OHC activity in sound-exposed ears, we reasoned that the processes underlying use-dependent changes in cochlear function could be examined in more detail than has been possible with the physiologic methods used in the earlier, pertinent studies that empha-sized the neural consequences of overstimulation. The principal goals of these ongoing experimentally based studies are to specify: (i) the ability of an ear to resist the damaging effects of repeated exposures to moderately intense sounds; (ii) the properties of DPOAEs that predict a given ear's ability to be more or less susceptible to potentially injurious noises; and (iii) the ability of DPOAEs to detect the initial damage processes of a developing noise-based impairment. The purpose of this report is to review our findings, to date, and to identify future directions of EOAE research that can optimize our understanding of the cochlea's ability to adjust to overuse.

Methods

Rabbits and humans were studied using generally similar procedures. First, $2f_1 - f_2$ DPOAEs were obtained from both types of subjects while awake, within the confines of a sound-treated room, although a few of the rabbit experiments were performed under general anesthesia. Second, DPOAEs were examined either as a series of response/growth or input/output (I/O) functions, which describe emission levels at discrete frequencies as a function of systematic increases in stimulus level, or as frequency functions (i.e. 'audiograms') that plot DPOAE levels in response to constant-level primaries (rabbits: 45 dB SPL; humans: 75 dB SPL) as a function of stimulus frequency. In most cases, both measures of DPOAE activity were elicited with equilevel primaries, i.e. $L_1=L_2$. Third, both these measurement forms were interpreted by assuming that the DPOAE-generation site is best represented as the geometric-mean (GM) frequency [i.e. $(f_1 \times f_2)^{0.5}$] of the eliciting f_1 and f_2 primary tones (Martin *et al.*, 1987).

Subjects

A number of 3 month old, female rabbits, as detailed below, were obtained from commercial suppliers. All but one animal was pigmented. Human subjects (mean = 30 ± 7 years) were recruited as normal hearing individuals (n = 7 males, n = 7 females), according to the informed-consent guidelines established by our institute's internal review board for human experimentation.

DPOAE testing

The instrumentation used for acquiring rabbit and human DPOAEs has been described in great detail in previous reports [rabbit: Lonsbury-Martin *et al.* (1987), Whitehead *et al.* (1992); human: Lonsbury-Martin *et al.* (1990), Whitehead *et al.* (1993)]. Briefly, for the generation of the f_1 and f_2 primary tones, and the sampling and processing of the emitted response, a PC system either controlled separate frequency synthesizing or spectral-averaging instruments, respectively, or instructed an on-board digital signal-processor to serve these functions. Two types of stimulus transducer/microphone systems incorporated into unique acoustic specula were used to produce and measure DPOAEs. Specifically, the f_1 and f_2 primaries were generated by dedicated earspeakers (rabbit: Beyer DT-48; human: Etymotic Research ER-2), and mixed acoustically in the ear canal, whereas the ear-canal signal was measured in the rabbit with a 1/2 in. condenser microphone (Bruel & Kjaer 4166) fitted with a probe tube, and in the human with a subminiature microphone (Etymotic Research ER 10B) assembly that was specifically designed for recording human emissions. The DPOAEs were determined to be present if the level of the time average of a specified number of samples of the emitted response at the $2f_1 - f_2$ frequency was ≥3 dB above the noise floor measured at ±5 frequency bins distributed around that frequency (i.e. ±50 Hz).

Behavioral testing

Human subjects underwent standard audiometric testing to document pure-tone sensitivity, speech understanding and middle-ear function. In some experiments, the behavioral hearing of rabbits was tested using an established procedure based on the classical conditioning of the eye's nictitating membrane (Martin *et al.*, 1980, 1989).

Experimental design

Rabbit studies. The rabbit experiments were designed to take advantage of the OHC-testing ability of DPOAEs by using this measure to detail the response of this class of hearing receptor to overstimulation. The primary goal of one study (Franklin *et al.*, 1991) was to establish whether the rabbit ear (n = 3 from separate pigmented animals, and n = 1 from an albino rabbit) was capable of developing a use-dependent plasticity over time in the form of an increased resistance to the detrimental effects of repeated exposure to a moderately intense noise, similar to the type of invulnerability reported previously for chinchillas and guinea-pigs. Toward this end, behavioral tests of pure-tone sensitivity and DPOAE testing were performed daily over a 2 hr period to depict the development of NIHL over time caused by an essentially continuous exposure (i.e. 22 hr/day) to a 95 dB SPL OBN centered at 1 kHz. Behavioral thresholds were tested at 11 frequencies distributed at half-octave intervals within the optimal region of rabbit hearing (i.e. 0.35–22.6 kHz). The DPOAEs were obtained as frequency functions by describing emission levels evoked by 45 dB SPL primaries as a function the GM frequency between 1 and 17 kHz, and as I/O functions by plotting emission level as a function a systematic 5 dB decreases in primary-tone levels, from 75 dB SPL to the noise floor, at nine frequencies distributed at half-octave intervals along the frequency function.

The criterion measure was the number of exposure days required to reduce DPOAEs between 1 and 4 kHz to noise-floor levels, i.e. by ~20–30 dB. Following recovery of DPOAEs to baseline levels, the same animals were re-exposed to the identical OBN to establish if the initial exposure induced a protective effect that could be documented in the form of an increase in the number of exposure days needed to cause a similar decrease in low-frequency DPOAEs. These overstimulation sessions continued at a rate of about once every three weeks until a permanent reduction of $\geq$10 dB was documented for any DPOAE between 2 and 10 kHz.

In a second set of experiments (Mensh *et al.*, 1993a), reversible alterations in DPOAE levels caused by brief exposures to moderately intense tones were used to investigate if the use-dependent effects caused by chronic overstimulation could be established over a timeframe that was amenable to more analytical experiments aimed at uncovering the nature of the resistance process. Toward this goal, the ability of brief, repeated exposures to a 100 dB SPL pure tone at 1.7 kHz (i.e. half-octave below f_2) to induce resistance in 1.5 kHz DPOAEs elicited by 50 dB SPL primaries (f_1 = 2, f_2 = 2.5 kHz) was examined in 12 anesthetized rabbits. Two brief tonal-overstimulation episodes, 40 min apart, comprised an exposure session, which was repeated on three distinct occasions, each separated from the others by several days. Each exposure period consisted of about 100 sec of overstimulation in the form of either repetitive 10 sec tonal bursts, with 55 sec 'recovery' interburst intervals, or 60 sec of continuous exposure to the moderately intense tonal stimulus. The rate of the exposure-induced decrement in DPOAE level during the overstimulation (i.e. the susceptibility index) was used as a quantitative metric of the ear's vulnerability to over-exposure.

In a final rabbit study on eight animals (Mensh *et al.*, 1993b), the rate of decrement in DPOAE level during tonal overexposure was used as an index to predict that ear's response to overexposure by the same moder-ately intense OBN that was utilized in the 'resistance' experiments described above. Toward this end, the identical brief, tonal exposures administered to the anesthetized animals were used to establish the vulnerability index, after which the rabbits were exposed while awake to the OBN, until the criterion amount of reduction in low-frequency DPOAEs occurred.

Human studies. The primary purpose of the human experiments was to test the ability of DPOAEs to detect the transient effects of exposure to brief, intense pure tones on OHC function. Toward this goal, a paradigm was devised that incorporated four distinct combinations of L_1 and L_2 values, so that the benefits of offsetting and/or lowering primary-tone levels could be determined with respect to specifying the sound-induced reversible changes in DPOAE levels. The specific protocols consisted of primaries of equilevel ($L_1 = L_2 = 60$ or 55 dB SPL), or off-set levels, with $L_1 = 60$ or 55 dB and $L_2 = 35$ or 30 dB, respectively. Following baseline measures of DPOAEs in response to each protocol for one ear from each normal-hearing subject, the ears were exposed to a 3 min, 105 dB SPL pure at 2.8 kHz, i.e. half-octave below the GM frequency of 4 kHz ($f_1 = 3.6$ and $f_2 = 4.4$ kHz), under closed-sound conditions. In four subjects, the actual shifts in hearing sensitivity induced by identical overstimulations were also measured by connecting the DPOAE probe to the clinical audiometer.

Results

An early observation in the chronic overstimulation studies in rabbits was that during both the repeated development and recovery stages of the exposure paradigm, DPOAEs accurately tracked the frequency

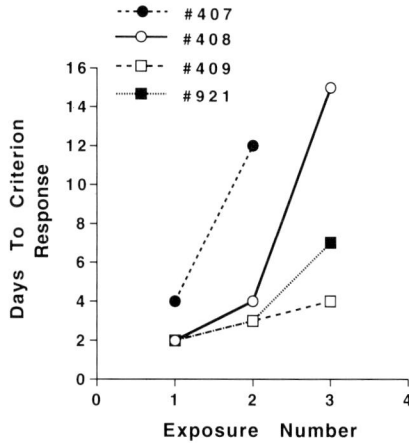

Fig. 1. Summary of findings indicating a progressive increase in resistance to the effects of a sound-induced reduction in DPOAE levels with repeated OBN exposure. The line plot illustrates details of the rate at which each rabbit developed this increased resistance over the repeated noise-exposure sessions.

pattern described by the behaviorally determined threshold shifts, thus, ensuring that the type of NIHL studied here was cochlear-based. The major finding illustrated in Fig. 1 was that the cumulative OBN exposures resulted in a reduced susceptibility to NIHL in that, with additional doses of the exposure/recovery regimen, an increasingly greater number of days of overstimulation was required for rabbits to attain the criterion reduction in DPOAE levels.

The principal results of the follow-up short-term, tonal-exposure studies were that, rather than developing a resistance to the DPOAE-reducing effects of repeated overexposure, the ears became more susceptible to exposures delivered 40 min after the first exposure of a session, than they were initially. Further results showed, however, that susceptibility returned to its original baseline level during the 2–3 day intersession interval. However, a clear finding was that each rabbit exhibited a unique rate of decrement in DPOAE level during the exposure period, even though emission levels elicited by the suprathreshold primaries and detection thresholds determined from I/Os were similar between animals.

The primary outcome of the experiments using the susceptibility index describing the DPOAE-reducing effects of brief tonal over-exposures showed a relation between the slope of the short-term loss function and the number of chronic OBN-days required to reach the criterion-loss response. Thus, rabbits showing slow rates of tonal-induced

Active hearing

A   R-157L

B   R-157L

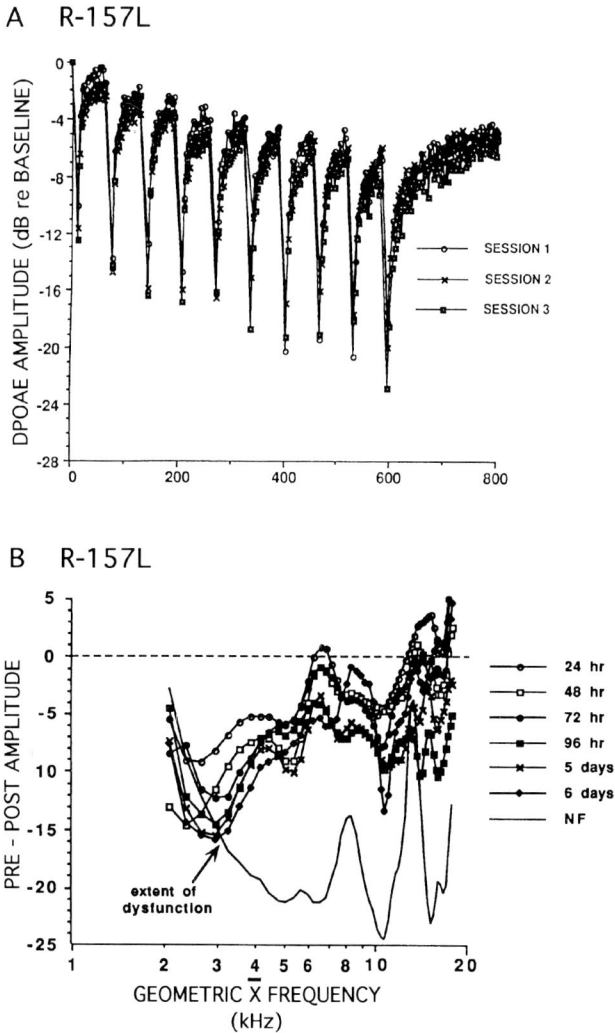

Fig. 2. Example of the rate of reversible tonal-induced reductions in DPOAE levels and the subsequent frequency/level effects of chronic OBN exposure on the emitted response. (A) Repeatability of the initial episode of the short-term exposure protocol for the left ear of rabbit 157, for three separate sessions. Note that the rate of decrement over the period consisting of ten 10 sec exposure bursts is reasonably constant across the exposure sessions. (B) The effects of the chronic OBN exposure for the rabbit above, which displayed relatively shallow slopes for the tonal-induced DPOAE-loss function, thus suggesting 'resistant' ears. Note that the animal endured the chronic noise for 6 days, without ever achieving the criterion reduction in DPOAE levels over the required frequency region.

reductions in 1.5 kHz DPOAEs were more resistant to the effects of chronic noise exposure than were animals exhibiting rapid reduction rates. In Fig. 2, an example is presented of the constancy of the acute-exposure effects for an animal that exhibited a relatively slow decrement in DPOAE levels (Fig. 2A), and the subsequent effects of almost a week of overexposure to the OBN on DPOAEs elicited by low-level primaries (Fig. 2B).

Finally, in the human studies of the ability of equal vs. offset level primaries to detect reversible cochlear dysfunction, the major finding was that stimulus protocols with unequal primaries demonstrated an increased sensitivity to the effects of tonal overstimulation. In particular, the protocol with the lowest-level offset primaries (i.e. $L_1 = 55$, $L_2 = 30$) was most sensitive to the overexposure effects. Additionally, when DPOAE recovery-time courses were compared to the behaviorally measured temporary threshold shifts (TTSs) determined for the same ears, the reasonable similarity between the two measures suggested that emitted responses were as sensitive to threshold elevations as was routine pure-tone audiometry. An example of the fair correspondence between the two response measures is shown in Fig. 3 for an individual in which the tonal overstimulation was relatively effective in producing a detectable TTS. Further analysis of the recovery

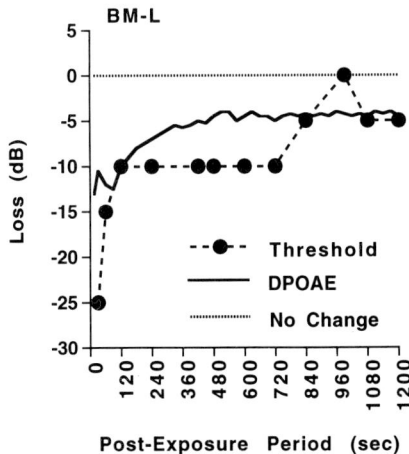

Fig. 3. Example from a 48 year old female subject of loss functions (i.e. post-exposure minus pre-exposure data) showing behavioral threshold shifts (solid circles) measured at the indicated post-exposure times compared to DPOAE levels (bold line), which were smoothed by a sliding three-point transformation process. The DPOAE data were obtained using the experimental protocol consisting of primary tones set to $L_1 = 55$ and $L_2 = 30$ dB SPL.

process was achieved by fitting logarithmic functions to each subject's data. Inspection of the resulting equations, which were extracted by a least-squares fit analysis, revealed that the initial amount of DPOAE loss (y-intercept), and the rate of recovery (slope) educed from the log functions predicted the degree of vulnerability of the ear to over-exposure. Thus, these derived parameters appeared to correspond quite accurately to the vulnerability of the related behavioral hearing to overstimulation.

Discussion

The findings in rabbits that the ear is capable of exhibiting an activity-based plasticity in response to repeated chronic exposures to moderate sounds support similar results discovered by earlier animal experiments using more centrally based measures of auditory-system function. In the present work, the use-dependent changes in DPOAEs suggest the action of a dynamic mechanism that appears to influence noise-damaged OHC elements so that behavioral and physiological recovery from overstimulation occurs in the form of an increasing resistance to the effects of repeated overexposure. The DPOAE results also support the notion that an optimized set of exposure-stimulus variables can be identified in order to 'train' OHCs to be less vulnerable to the damaging effects of overactivation.

The short-term tonal-exposure experiments in rabbits uncovered a transitory increased susceptibility of OHCs to overstimulation, and thus failed to replicate the earlier observations in more chronically exposed animals of an increased resistance. However, the documented increase in overall vulnerability was not lasting, because the identical exposures repeated a few days later resulted in a similar rate of overstimulation-induced reduction in DPOAEs. The finding that this short-term index of susceptibility to brief tonal exposures could be used to predict the effects of a later chronic-noise exposure on a rabbit's DPOAE-frequency function suggested that deliberately induced reversible changes in the dynamic features of emission activity might provide a fair test of the vulnerability of the ear to the harmful effects of more lengthy exposures to moderately intense sounds.

Finally, the brief tonal-exposure experiments in humans revealed several findings that further the study of NIHL. First, stimulus protocols that incorporate primary tones with moderately low and unequal levels appear to detect best the behavioral effects of reversible, tonal overstimulation. In addition, several parameters derived from the logarithmic functions that best describe the DPOAE recovery process in the form of the *y*-intercept and slope were identified as potential measures of the vulnerability of the human ear to overexposure in that these variables appeared to predict the susceptibility of behavioral hearing to the same overexposure.

Together, the results of the experimental studies in animals and humans using DPOAEs to effectively register the reaction of OHCs to overstimulation suggest that these emitted responses are useful for analyzing use-dependent changes in cochlear activity. Additionally, the findings imply that measures of DPOAE activity would be promising in industrial hearing-conservation programs as potential indicators of the onset of noise damage. Finally, emitted responses may provide the opportunity to predict which ears are most resistant to overuse, and to develop acoustic 'training' environments that have the potential to make OHCs more resilient to the potentially harmful effects of overstimulation.

Acknowledgements

Portions of this work were supported by grants from the Deafness Research Foundation, and the Public Health Service (DC00613, DC01668, ES03500). The authors thank Barden B. Stagner, Marcy J. McCoy, and Mayte T. Ruiz for assisting in data collection and processing, and the preparation of illustrations.

References

Brownell, W. E. (1990). Outer hair cell electromotility and otoacoustic emissions. *Ear Hear.* **11:** 82–92.

Canlon, B., Borg, E. and Flock, A. (1988). Protection against noise trauma by pre-exposure to a low level acoustic stimulus. *Hearing Res.* **34:** 197–200.

Clark, W. W., Bohne, B. A. and Boettcher, F. A. (1987). Effect of periodic rest on hearing loss and cochlear damage following exposure to noise. *J. Acoust. Soc. Am.* **82:** 1253–1264.

Franklin, D. J., Lonsbury-Martin, B. L., Stagner, B. B. and Martin, G. K. (1991). Altered susceptibility of $2f_1–f_2$ acoustic-distortion products to the effects of repeated noise exposure in rabbits. *Hearing Res.* **53:** 185–208.

Henderson, D., Campo, P., Subramaniam, M. and Fiorino, F. (1992). Development of resistance to noise. In *Noise-Induced Hearing Loss* (A. L.

Dancer, D. Henderson, R. J. Salvi and R.P. Hamernik, eds), pp. 476–488. Mosby Year Book, St Louis.

Kemp, D. T. (1978). Stimulated acoustic emissions from within the human auditory system. *J. Acoust. Soc. Am.* **64:** 1386–1391.

Lonsbury-Martin, B. L., Martin, G. K., Probst, R. and Coats, A. C. (1987). Acoustic distortion products in rabbits. I. Basic features and physiological vulnerability. *Hearing Res.* **28:** 173–189.

Lonsbury-Martin, B. L., Harris, F. P., Hawkins, M. D., Stagner, B. B. and Martin, G. K. (1990). Distortion-product emissions in humans: I. Basic properties in normally hearing subjects. *Ann. Otol. Rhinol. Laryngol.* **99:** Suppl. 47, 3–13.

Martin, G. K., Lonsbury-Martin, B. L. and Kimm, J. (1980). A rabbit preparation for neurobehavioral auditory research. *Hearing Res.* **2:** 65–78.

Martin, G. K., Probst, R., Scheinin, S. A., Coats, A. C. and Lonsbury-Martin, B. L. (1987). Acoustic distortion products in rabbits. II. Sites of origin revealed by suppression and pure-tone exposures. *Hearing Res.* **28:** 191–208.

Martin, G. K., Stagner, B. B., Coats, A. C. and Lonsbury-Martin, B. L. (1989). Endolymphatic hydrops in rabbits: Behavioral thresholds, acoustic distortion products, and cochlear pathology. In *Meniere's Disease: Pathogenesis, Pathophysiology, Diagnosis and Treatment* (J. B.Nadol, ed.), pp. 205–219. Kugler and Ghedini, Berkeley, CA.

Martin, G. K., Ohlms, L.A,. Franklin, D. J., Harris, F. P. and Lonsbury-Martin, B. L. (1990). Distortion-product emissions in humans: III. Influence of sensorineural hearing loss. *Ann. Otol. Rhinol. Laryngol.* **99:** Suppl. 147, 29–44.

Mensh, B. D., Lonsbury-Martin, B. L. and Martin, G. K. (1993a). Distortion-product otoacoustic emissions in rabbit: II. Prediction of chronic-noise effects by brief pure-tone exposures. *Hearing Res.* **70:** 65–72.

Mensh, B. D., Patterson, M. C., Whitehead, M. L., Lonsbury-Martin, B. L. and Martin, G. K. (1993a). Distortion-product otoacoustic emissions in rabbit: I. Altered susceptibility to repeated pure-tone exposures. *Hearing Res.* **70:** 50–64.

Probst, R., Lonsbury-Martin, B. L. and Martin, G. K. (1991). A review of otoacoustic emissions. *J. Acoust. Soc. Am.* **89:** 2027–2067.

Sinex, D. G., Clark, W. W. and Bohne, B. A. (1987). Effects of periodic rest on physiological measures of auditory sensitivity following exposure to noise. *J. Acoust. Soc. Am.* **82:** 1265–1273.

Subramaniam, M., Campo, P. and Henderson, D. (1991a). The effect of exposure level on the development of progressive resistance to noise. *Hearing Res.* **52:** 181–188.

Subramaniam, M., Campo, P. and Henderson, D. (1991b). Development of resistance to hearing loss from high frequency noise. *Hearing Res.* **56:** 65–68.

Sutton, L. A., Lonsbury-Martin, B. L., Martin, G. K. and Whitehead, M. L. (1994). Sensitivity of distortion-product otoacoustic emissions in humans to tonal over-exposure: time course of recovery and effects of lowering L_2. *Hearing Res.* **75:** 161–174.

Whitehead, M. L., Lonsbury-Martin, B. L. and Martin, G. K. (1992). Evidence for two discrete sources of $2f_1$-f_2 distortion-product otoacoustic emission in rabbit: I. Differential dependence on stimulus parameters. *J. Acoust. Soc. Am.* **91:** 1587–1607.

Whitehead, M. L., Lonsbury-Martin, B. L. and Martin, G. K. (1993). The influence of noise on the measured amplitudes of distortion-product otoacoustic emissions. *J. Speech Hearing Res.* **36:** 1097–1102.

Whitehead, M. L., Lonsbury-Martin, B. L., Martin, G. K. and McCoy, M. J. (1995). Otoacoustic emissions: animal models and clinical observations. In *Clinical Aspects of Hearing* (T. R. Van De Water, A. H. Popper and R. R. Fay, eds), in press. Springer-Verlag, New York.

Comparative Aspects of Interactive Communication

PETER M. NARINS

Department of Physiological Science, University of California, Los Angeles, CA 90095, USA

Introduction

In this paper I shall try to explore the way in which several groups of animals utilize their acoustic habitat, and to describe, when known, the specific adaptations of the auditory system which have evolved to communicate biologically significant information under natural (noisy) conditions. Specifically, I hope to stimulate the study of animals other than the standard mammalian preps by stressing that the natural acoustic behavior of an animal often leads to insights concerning the underlying physiological mechanisms. This neuroethological approach has yielded a great deal of information about bat echolocation (Neuweiller, 1984; Pollak and Casseday, 1989; Simmons and Dear, 1991; Suga, 1989), sound localization by owls (Knudsen, 1983; Konishi, 1983; Moiseff and Konishi, 1981), electrolocation and communication in weakly electric fishes (Bullock, 1973; Heiligenberg and Rose, 1985; Hopkins, 1988), insect sound reception (Huber and Thorsen, 1985; Pires and Hoy, 1992), song learning in birds (Konishi, 1985; Marler and Sherman, 1983) and frog acoustic communication (Capranica, 1965; Feng *et al.*, 1990; Schwartz and Megela Simmons, 1990; Walkowiak, 1992).

Behavioral Strategies for Communicating in Noise

What features of a vocal communication system are selected for in a highly noisy environment? Specifically, what options have evolved in such systems to enhance detection of certain sounds and reject unwanted sounds? Several possibilities suggest themselves: an individual may produce call energy in a species-specific frequency band. This 'spectral separation' strategy implies that a user of the acoustic space possesses adequate neural selectivity to reject or significantly attenuate a heterospecific call. Additionally, a species may restrict calling to specific times of the day or may simply avoid acoustic overlap with its nearest neighbors on a minute, second or even millisecond time scale (temporal separation).

Some animals adopt another strategy: they are able to exploit alternate, low-noise channels through which they interact with their neighbors. This appears to be the unusual tack taken by the white-lipped frog of Puerto Rico, *Leptodactylus albilabris*. This is a ground-dwelling, nocturnally active amphibian found in the marshes, ditches and along mountain streams throughout much of the island of Puerto Rico. Males of *L. albilabris* produce two distinct vocalizations; the species' advertisement call ('chirps') and the species' aggressive call ('chuckles'), emitted during male-male acoustic interactions (Lewis and

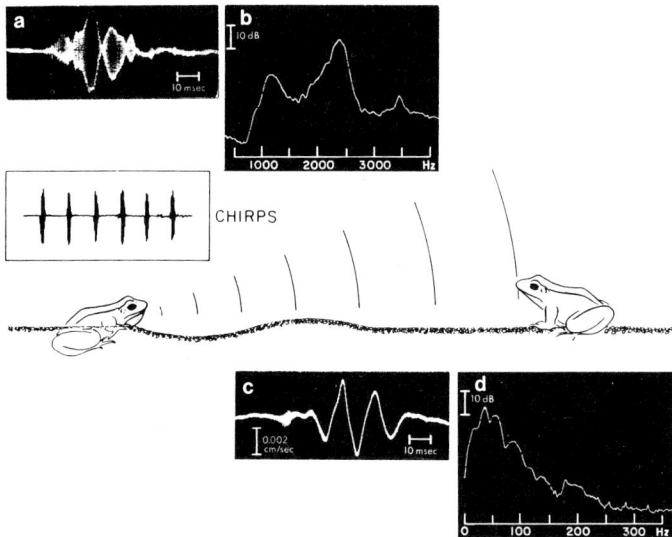

Fig. 1. Communication signals of the white-lipped frog, *Leptodactylus albilabris*, of Puerto Rico. An adult male (left) is partially buried in the muddy substrate while calling. During a 'chirp' vocalization, his vocal sac expands rapidly and within 20 msec of the onset of vocalization strikes the substrate forcefully, generating a vertically polarized surface (Rayleigh) wave which propagates outward in all directions at a velocity of ~100 m/sec. (A) and (B) show the waveform and amplitude spectrum of a single airborne chirp; (C) and (D) show the waveform and amplitude spectrum of the seismic signal.

Narins 1985; Lopez *et al.*, 1988). During call production, a male's vocal pouch expands explosively, striking the substrate impulsively as it inflates. Using an array of three orthogonal geophones in the field, Lewis and Narins (1985) showed that the resulting 'thump' generates a vertically polarized surface (Rayleigh) wave. This wave propagates in the muddy soil outward in all directions from the source at a velocity of roughly 100 m/sec, and with a peak acceleration of 0.002 g at a distance of 1 m. Spectral analysis of the thump reveals that the peak energy in this signal falls at 50 Hz (Fig. 1).

To test if the propagating thumps were being detected by the white-lipped frogs, Lewis and his colleagues used the solenoid from an electric typewriter to construct an electromechanical 'thumper'. By triggering the thumper from a tape recording of the male's call, we were able to simulate the thump rate and pattern of a calling white-lipped frog. Males within 3 m of our artificial frog consistently entrained their calls to its pattern, suggesting the males can detect seismic signals transmitted through the soil and alter their calling behavior in choruses in response to them. Ongoing experiments in Puerto Rico are attempting to determine if males respond differentially to the acoustic and seismic components of the frog's call.

Neurophysiological Correlates of Seismic Behavior

In a correlative study, we found an extremely sensitive population of fibers originating in the sacculus in the inner ear of *L. albilabris* which responds vigorously when the frog is subjected to whole-body vibrations at frequencies below 160 Hz (Narins and Lewis, 1984). The most sensitive fibers in this species show clear stimulus-evoked modulations of their resting discharge rates in response to sinusoidal seismic stimuli with peak accelerations on the order of 0.001 cm/sec^2 (~0.000001 g). Single vibration-sensitive fibers in the white-lipped frog saturate at whole animal displacements of 10 Å peak-to-peak. Assuming a conservative 20 dB dynamic range for the fibers, the *in vivo* frog saccule and the mammalian cochlea exhibit roughly equal sensitivites to displacement.

Recent studies of the white-lipped frog in our laboratory have revealed *two* populations of nerve fibers, one originating in the amphibian papilla (low-frequency airborne sound sensor) and one from the sacculus (sensor with exceptional sensitivity to whole-body vibrations) which respond to *both* airborne sound and whole-body vibrations (Christensen-Dalsgaard and Narins, 1993). We recorded from the VIIIth nerve near its entry to the brain-stem and analyzed responses to both sound and vibration stimuli for 90 fibers from two species. The fibers were classified as amphibian papilla (AP), basilar papilla (BP), saccular or vestibular based on their location in the nerve. Only AP and saccular fibers responded to vibrations. The AP-fibers responded to vibrations from 0.01 cm/sec^2 and to sound from 40 dB SPL by increasing their spike rate. The characteristic frequencies (CFs) for these bimodal fibers ranged from 60–900 Hz, and only fibers with CFs below 500 Hz responded to vibrations. The fibers had identical CFs for sound and vibration. The saccular fibers had CFs ranging from 10 to 80 Hz with 22 fibers having CFs at 40–50 Hz. The bimodal fibers from this organ responded to sound from 70 dB SPL and to vibrations from 0.01 cm/sec^2 (Fig. 2).

In other words, sound and vibration stimuli appear to converge onto the same cells in two of the inner ear sensors in the frog. Upon reflection, this is not surprising given the striking similarity of the anatomy and physiology of the sensory cells of the auditory and vestibular systems. Clearly in this case, we might adopt the view that auditory and vibratory signals differ principally in the pathway they take *en route* to the inner ear sensors.

Future Neuroethological Studies

The Cape Mole-rat, Georychus capensis

It is known that many terrestrial animals are sensitive to low-level substrate vibrations (Christensen-Dalsgaard and Jorgensen, 1988; Hartline, 1971; Koyama *et al.*, 1982; Narins, 1990; Narins and Lewis, 1984; Ross and Smith, 1978). To date, however, the use of seismic signals for communicating biologically significant information has been demonstrated unequivocally in only three species of vertebrates other than the white-lipped frog of Puerto Rico: the blind mole-rat of Israel (*Spalax*), the bannertail kangaroo rat of the southwestern United States (*Dipodomys*) and the Cape mole-rat of South Africa (*Georychus*). Although subterranean rodents are known to produce low-frequency sounds by foot- and head-drumming, very little is known of the details of these behaviors. We examined the propagation characteristics of both seismic and auditory signals in the natural habitat of the Cape mole-rat (*Georychus capensis*), a subterranean rodent from the Cape Region of Southern Africa. This solitary animal lives its entire life underground and communicates with members of its own species by drumming its hind feet inside its burrow at ~20 Hz (Fig. 3). Using an electro-

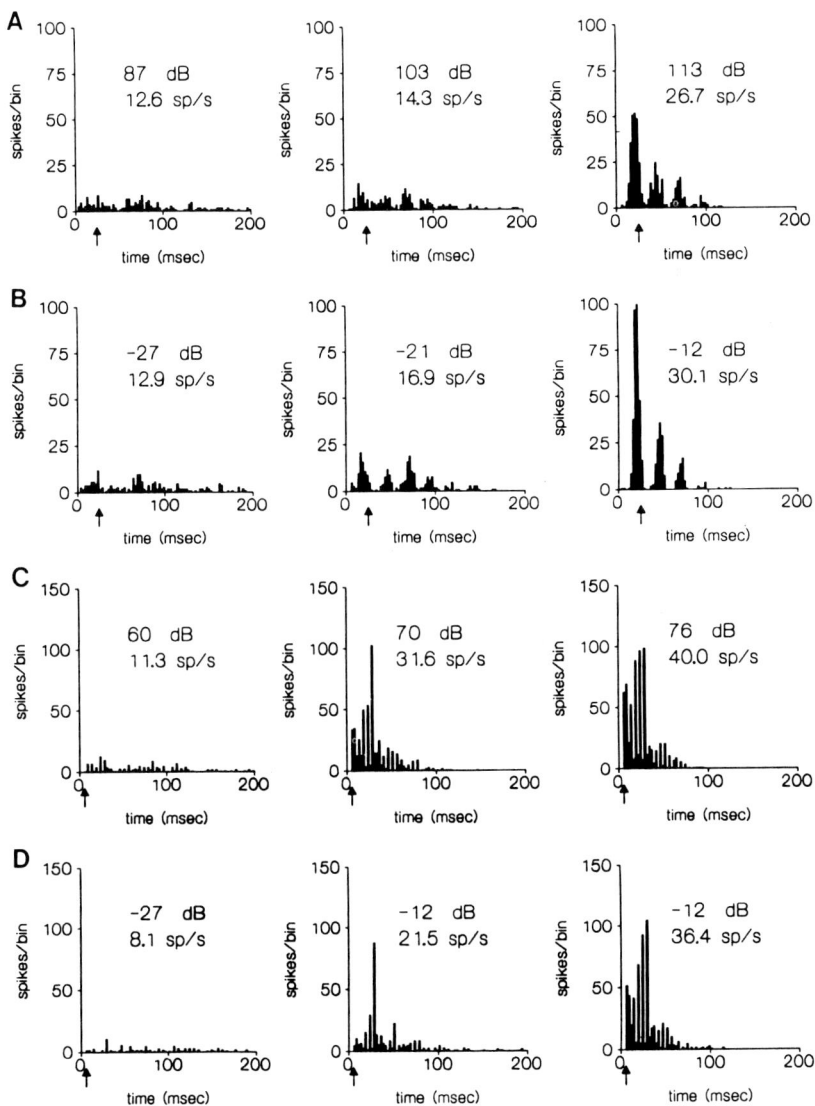

Fig. 2. Interval histograms of the response of a single saccular fiber from *Leptodactylus albilabris* to airborne sound (row A) and substrate vibration (row B) stimulation at its CF (40 Hz) at three sound and acceleration levels. Rows C and D illustrate the response of an amphibian papillar (AP) fiber (CF = 180 Hz) to both sound and vibration. In each histogram the upper number refers to the sound or vibration level (in dB SPL and re 1 cm/sec², respectively) and the lower number is the mean spike rate in spikes/sec. The stimulus period is indicated by an arrow (from Christensen-Dalsgaard and Narins, 1993).

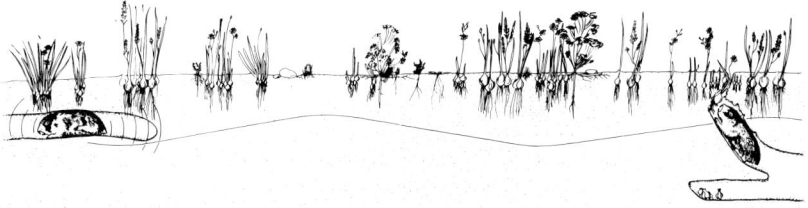

Fig. 3. Subterranean habitat of the Cape mole-rat, *Georychus capensis* near Capetown, South Africa. Footdrumming by the male (left) produces a signal with both acoustic and seismic components, each of which propagates outward from the source and attenuates at a substrate-specific rate. At a distance of 1 m, the amplitude of the acoustic component is comparable to the background noise, whereas at the position of the nearest neighbor (3–4 m, at right), the amplitude of the substrate particle velocity is at least 6 dB above the ambient noise. Original drawing by K. Bolles.

mechanical transducer, simulated mole-rat thumps were produced over an abandoned natural burrow and the maximum peak-to-peak substrate particle velocities at several points over a neighboring burrow were obtained (Narins *et al.*, 1992). Each waveform in Fig. 4 shows the average response of a vertically polarized geophone to 50 artificial thump stimuli measured at one position over the neighboring burrow. The measured waveforms at all distances tested in this experiment (up to 5.7 m) are uniformly above the ambient noise level. For comparison, octave-band sound levels measured within a freshly excavated artificial burrow located 1 m from the thump generator were indistiguishable from ambient levels. Thus Rayleigh waves, not airborne sound waves, appear to subserve intraspecific signaling in this solitary species. The neural mechanisms underlying the detection, transduction and encoding of surface waves in *Georychus* are unstudied. It is becoming clear that seismic communication and sensitivity to whole-body vibrations are more widespread among the vertebrates than had been previously imagined.

The common Malaysian treefrog, Polypedates leucomystax

Recently, Albert Feng, Jakob Christensen-Dalsgaard and I discovered that one species of Malaysian treefrog (*Polypedates*) exhibits a particularly unusual behavior. During courtship, females living in dense mats of floating vegetation perch conspicuously on a reed or blade of grass and tap their rear toes rhythmically. This activity, which

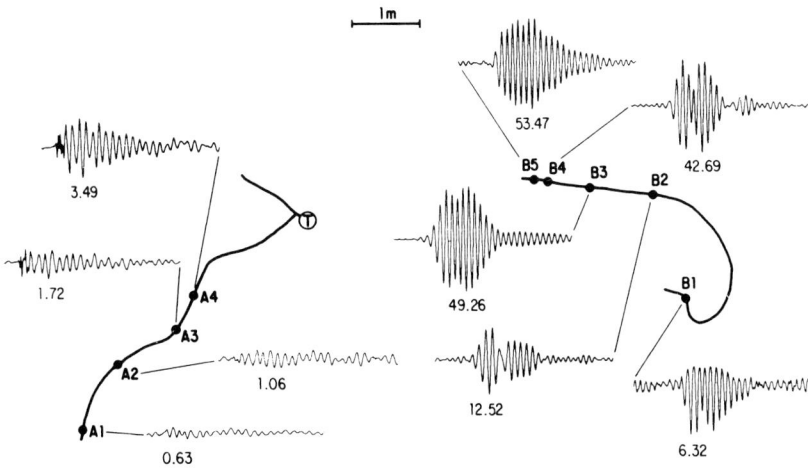

Fig. 4. Scale drawing of portions of two natural, vacant burrow systems of the Cape mole-rat. We measured the magnitudes of the vertical component of simulated thumps generated by the thumper at T measured by geophones placed at various locations above two mole-rat burrow systems, A and B. Each waveform represents the average velocity in response to 32 thumps. The value shown is the maximum peak-to-peak particle velocity of these averages and is a coefficient of 1×10^{-6} m/sec. Not all waveforms have the same vertical scale. Straight-line distances from the thumper to the recording sites are B1: 5.7 m; B2: 5.1 m; B3: 4.2 m; B4: 3.6m; B5: 3.4 m (from Narins *et al.*, 1992).

persisted for several minutes, occurred in the dark and was only occasionally accompanied by vocalizations. Males on neighboring reeds were observed to quickly locate and mate with the tapping female. Thus, it is likely that the toe tapping functions as a vibrational signal indicating the female's presence to neighboring males. This behavior is remarkable for two reasons: (i) females of this species are stationary and signal to males who then approach the female (in contrast to most other known treefrog species in which stationary males call and females localize them); (ii) this is the first known example of a vertebrate using seismic signaling via a substrate other than the earth. In this species, the attenuation characteristics of the seismic signals and the neural basis of their reception by conspecific males both merit further study.

Conclusions

Seismic signaling provides a means for animals to communicate with each other without the effort of overcoming the often formidable ambient acoustic noise levels produced by other calling individuals. It is also possible that various substrates in diverse habitats offer a relatively quiet channel waiting to be exploited by seismically active animals. It is becoming clear that seismic communication and sensitivity to whole-body vibrations are more ubiquitous among the vertebrates than had been previously imagined, and that this modality appears to be emerging as a neuroethological model for understanding the evolution of communication systems in noisy environments.

Acknowledgements

Supported by NIDCD Grant no. DC00222.

References

Bullock, T. H. (1973). Seeing the world through a new sense: Electroreception in fish. *Amer. Sci.* **61:** 316–325.

Capranica, R. R. (1965). *The Evoked Vocal Response of the Bullfrog.* Research Monograph 33, MIT Press, Cambridge, MA.

Christensen-Dalsgaard, J. and Jorgensen, M. B. (1988). The response characteristics of vibration-sensitive saccular fibers in the grassfrog. *J. Comp. Physiol.* **162:** 633–638.

Christensen-Dalsgaard, J. and Narins, P. M. (1992). Sound and vibration sensitivity of VIIIth nerve fibers in the frogs *Leptodactylus albilabris* and *Rana pipiens pipiens. J. Comp. Physiol.* **172:** 653–662.

Feng, A. S., Hall, J. C. and Gooler, D. M. (1990). Neural basis of sound pattern recognition in anurans. *Progr. Neurobiol.* **34:** 313–329.

Hartline, P. H. (1971). Physiological basis for detection of sound and vibration in snakes. *J. Exp. Biol.* **54:** 349–371.

Heiligenberg, W. and Rose, G. (1985). Neural correlates of the jamming avoidance response (JAR) in the weakly electric fish *Eigenmannia. Trends Neurosci.* **8:** 442–449.

Hopkins, C. D. (1988). Neuroethology of electric communication. *Ann. Rev. Neurosci.* **11:** 497–535.

Huber, F. and Thorsen, J. (1985). Cricket auditory communication. *Sci. Am.* **253:** 60–68.

Knudsen, E. I. (1983). Early auditory experience aligns the auditory map of space in the optic tectum of the barn owl. *Science* **222:** 939–942.

Konishi, M. (1983). Neuroethology of acoustic prey localization in the barn owl. In *Neuroethology and Behavioral Physiology* (F. Huber and H. Markl, eds), pp. 303–317. Springer-Verlag, Berlin.

Konishi, M. (1985). Birdsong: from behavior to neuron. *Ann. Rev. Neurosci.* **8:** 125–170.

Koyama, H., Lewis, E. R., Leverenz, E. L. and Baird, R. A. (1982). Acute seismic sensitivity in the bullfrog ear. *Brain Res.* **250:** 168–172.

Lewis, E. R. and Narins, P. M. (1985). Do frogs communicate with seismic signals? *Science* **227:** 187–189.

Lopez, P. T., Narins, P. M., Lewis, E. R. and Moore, S. W. (1988). Acoustically induced call modification in the white-lipped frog, *Leptodactylus albilabris*. *Anim. Behav.* **36:** 1295–1308.

Marler, P. and Sherman, V. (1983). Song structure without auditory feedback: emendations of the auditory template hypothesis. *J. Neurosci.* **3:** 517–531.

Moiseff, A. and Konishi, M. (1981). Neuronal and behavioral sensitivity to binaural time differences in the owl. *J. Neurosci.* **1:** 40–48.

Narins, P. M. (1990). Seismic communication in anuran amphibians. *Bio-Science* **40:** 268–274.

Narins, P. M. and Lewis, E. R. (1984). The vertebrate ear as an exquisite seismic sensor. *J. Acoust. Soc. Am.* **76:** 1384–1387.

Narins, P. M., Reichman, O. J., Jarvis, J. U. M. and Lewis, E. R. (1992). Seismic signal transmission between burrows of the Cape mole-rat, *Georychus capensis*. *J. Comp. Physiol.* **170:** 13–21.

Neuweiler, G. (1984). Foraging, echolocation and audition in bats. *Naturwissenschaften* **71:** 446–455.

Pires, A. and Hoy, R. R. (1992). Temperature coupling in cricket acoustic communication II. Localization of temperature effects on song production and recognition networks in *Gryllus firmus*. *Fed. Proc.* **37:** 2316–2323.

Pollak, G. D. and Casseday, J. H. (1989). *The Neural Basis of Echolocation in Bats.* Springer-Verlag, Berlin.

Ross, R. J. and Smith, J. J. B. (1978). Detection of substrate vibrations by salamanders: inner ear sense organ activity. *Can. J. Zool.* **56:** 1156–1162.

Schwartz, J. J. and Megela Simmons, A. (1990). Encoding of a spectrally-complex communications sound in the bullfrog's auditory nerve. *J. Comp. Physiol.* **166:** 489–499.

Simmons, J. A. and Dear, S. P. (1991). Computational representations of sonar images in bats. *Curr. Biol.* **1:** 174–176.

Suga, N. (1989). Principles of auditory information-processing derived from neuroethology. *J. Exp. Biol.* **146:** 277–286.

Walkowiak, W. (1992). Acoustic communication in the fire-bellied toad: an integrative neurobiological approach. *Ethol. Ecol. Evol.* **4:** 63–74.

Subject Index

373